BIOLOGY: BRAIN & BEHAVIOUR

The Senses and Communication

Springer
Berlin
Heidelberg
New York
Barcelona
Budapest
Hong Kong
London
Milan
Paris
Santa Clara
Singapore
Tokyo

Tim Halliday (Ed.)

The Senses and Communication

With 129 Figures

Springer in association with

Unless otherwise stated, all contributors are (or were at the time this book was written) members of The Open University

Academic Editor

Tim Halliday

Authors

Peter Bailey (University of York)
Tim Halliday
Mike Harris (University of Birmingham)
Heather McLannahan
Robin Orchardson (University of Glasgow)
David Robinson
Dick Stephen (University of Leicester)
Vicky Stirling

External Assessors

Richard Andrew, School of Biological Sciences, University of Sussex (Series Assessor)
Clive Catchpole, Zoology Department, Royal Holloway and Bedford New College, University of London (Book Assessor)

Biology: Brain & Behaviour series

1. Behaviour and Evolution
2. Neurobiology
3. **The Senses and Communication**
4. Development and Flexibility
5. Control of Behaviour
6. Brain: Degeneration, Damage and Disorder

CIP Data applied for
Die Deutsche Bibliothek - CIP - Einheitsaufnahme
The Senses and Communication/The Open University. Tim Halliday (ed.). -Berlin; Heidelberg; New York; Barcelona; Budapest; Hong Kong; London; Milan; Paris; Santa Clara; Singapore; Tokyo: Springer 1998

Published by Springer-Verlag, written and produced by The Open University

Cover design: *design & production* GmbH, Heidelberg

Printed and bound by Kyodo Printing Co (S'pore) Pte Ltd

ISBN 3-540-63775-3 Springer-Verlag Berlin Heidelberg New York

This text forms part of the Open University *Biology: Brain & Behaviour* series. The complete list of texts which make up this series can be found above. Details of Open University courses can be obtained from the Course Reservations and Sales Office, PO Box 724, The Open University, Milton Keynes MK7 6ZS, United Kingdom: tel. (00 44) 1908 653231. Alternatively, much useful course information can be obtained from the Open University's website: http://www.open.ac.uk

3.1

SPIN 10654330 #39/3137 – 5 4 3 2 1 0

CONTENTS

	Preface	**1**
1	**Introduction**	**4**
1.1	Introduction	4
1.2	Sense organs	4
1.3	Sensation and perception	10
1.4	Communication	13
	Summary of Chapter 1	15
	Objectives for Chapter 1	16
	Questions for Chapter 1	16
2	**Animal communication**	**17**
2.1	Introduction	17
2.2	Auditory communication in frogs	17
2.3	Communication and information	31
	Summary of Chapter 2	38
	Objectives for Chapter 2	38
	Questions for Chapter 2	39
	References for Chapter 2	39
	Further reading for Chapter 2	39
3	**Human hearing and human speech**	**41**
3.1	Introduction	41
3.2	Hearing	42
3.3	Auditory pathways in the brain	53
3.4	Production of speech	56
3.5	Perception of speech	66
3.6	Phonetic processing of speech	68
3.7	Language understanding	71
	Summary of Chapter 3	73
	Objectives for Chapter 3	74
	Questions for Chapter 3	74
	Further reading for Chapter 3	74

4	**Vision**	**77**
4.1	Introduction	77
4.2	Image formation	79
4.3	Image description: luminance	82
4.4	Image description: spectral composition	113
4.5	Image description: motion	120
4.6	Image description: binocular vision and stereopsis	127
4.7	Modularity in the early stages of vision	132
4.8	Later descriptive processes	136
4.9	Interpretative processes	140
4.10	Concluding comments and unanswered questions	143
	Objectives for Chapter 4	144
	Questions for Chapter 4	145
	Further reading for Chapter 4	146
5	**Touch and pain**	**147**
5.1	Introduction	147
5.2	The neurophysiology of sensory systems	148
5.3	The value of touch	165
5.4	Pain	169
5.5	Measurement of pain	176
5.6	The psychology of pain	183
5.7	Modulation of afferent signals by 'gate control'	186
5.8	Endogenous analgesic systems	188
5.9	The treatment of pain	195
	Summary of Chapter 5	206
	Objectives for Chapter 5	207
	Questions for Chapter 5	208
	References for Chapter 5	210
	Further reading for Chapter 5	210
Epilogue		**211**
General further reading		**214**
Answers to questions		**215**
Glossary		**221**
Acknowledgements		**227**
Index		**229**

PREFACE

The Senses and Communication, like any other textbook, is designed to be read on its own, but it is also the third in a series of six books that form part of *SD206 Biology: Brain & Behaviour,* a course for Open University students.

Each subject is introduced in a way that makes it readily accessible to readers without any previous knowledge of that area. Questions within the text, marked with a □, are designed to help readers understand and remember the topic under discussion. (Answers to in-text questions are marked with a ■.) The major learning objectives are listed at the end of each chapter, followed by questions (with answers given at the end of the book) which allow readers to assess how well they have achieved these objectives. Key terms are identified in bold type in the text; these are listed, with their definitions, in a glossary at the end of the book. Key references are given at the end of each chapter, where appropriate. A 'general further reading' list, of textbooks relevant to the whole book, is also included at the end.

The study of the brain and behaviour is an experimental science. This means that it involves the collection of observations, the formulation of specific hypotheses to explain those observations and the carrying out of experiments to test (confirm or falsify) those hypotheses. Throughout this book, these different aspects of the investigative process are emphasized, often through the use of in-text questions in which the reader is invited to engage in the process of deductive reasoning themselves. An understanding of the scientific method, as it applies to the behavioural and brain sciences, is an important aim of this book.

The focus of this book is the integration of brain and behaviour. It looks in detail at the senses of hearing, vision, touch and pain and how the sense organs function and the ways in which animals, including humans, integrate the information provided by their sense organs to direct their behaviour. Within the nervous system there is communication between sense organs, the brain and various parts of the body, such as muscles and limbs, that are involved in behaviour. There is also communication between individuals, a process in which sense organs are intimately involved.

Chapter 1 introduces the subject of communication and provides a brief description of sensory receptors and perception. Humans use a relatively small suite of the possible communication systems and, in Chapter 2, the wide range of such systems used in the natural world is illustrated by examples of animal communication.

Human hearing and human speech is a major research area in both physiology and psychology. Chapter 3 deals with the subject from both perspectives, starting with a description of the structure and function of the human ear and an explanation of how we produce the sounds that constitute speech. The perception of speech is a very complex process and the second part of Chapter 3 provides an introduction to this fascinating area of human behaviour.

Vision too is a complex process and a book of this size can only offer an overview of the subject. In Chapter 4, the process of image formation is described, followed by a substantial section on visual processing: image description. Luminance, spectral composition, motion detection and binocular vision are all described. The final section of the Chapter includes a number of questions which we would like answers to but for which, as yet, there is insufficient information. However, vision has given us some of the most detailed insights into brain function that we have to date.

The final Chapter deals with touch and pain. While touch is a recognized sense, with sight, hearing, taste and smell being the others, pain is not included. However, pain is a very important sensation, so discussion of pain and touch together is appropriate in a book on the senses and communication.

Before you begin to read this book, there are some important points that you should bear in mind.

1 Experiments on animals

The use of living animals in research is a highly emotive, contentious and political issue. You are no doubt aware of the strong views held by animal liberationists. There is also considerable debate among scientists concerning what kinds of experiments and procedures are acceptable and what are not. Most scientists working with animals seek to minimize any suffering that animals may experience during experiments and each researcher makes his or her own judgement as to whether the suffering caused by an experiment is justified by the scientific value of the results that the experiment yields. The ethics of animal experimentation is not simply a matter of individual judgement, however, but is a matter of concern for society as a whole. In Britain and many other countries, all researchers work within strict guidelines enforced by government; for example, the Home Office licenses all animal experimentation in the UK. Some academic societies, such as the Association for the Study of Animal Behaviour, and many institutions, such as medical schools, have Ethical Committees that oversee animal-based research. In this book, a number of experiments are described; this in itself raises ethical issues because reporting the results of an experiment may be thought to be giving tacit approval to that experiment. This is not necessarily true and it should be pointed out that some of the experiments described were carried out several years ago and a number of them would not be carried out today, such has been the shift in opinion on these issues within the biological community. Paradoxically, certain experiments carried out many years ago, such as those on the effects of maternal deprivation on young monkeys, produced such strong and distressing effects on their subjects—results that were not generally anticipated—that they have had a substantial impact on the kind of experiments that are permitted today.

2 Latin names for species

A particular individual animal belongs to various categories. If you own a pet, it may, for example, be categorized as a bitch, a spaniel, a dog, a mammal, or an animal. Each category is defined by particular features that differentiate it from other, comparable categories. The most important level of categorization in biology is at the level of the species. When a particular species of animal is referred to in this book, its Latin name is also given, e.g. earthworm (*Lumbricus terrestris*).

CHAPTER 1
INTRODUCTION

1.1 Introduction

The biological study of communication and of the senses involved in communication is a vast subject, worth a course all to itself. This book presents a number of specific topics in this general area, selected to cover a variety of senses, and to illustrate a number of the more important general principles involved in sensory physiology and communication. Chapter 2 looks at communication between animals, with particular emphasis on the role of sense organs in communication. Its emphasis is largely ethological, but it includes a case study of vocal communication in frogs, the objective of which is to show how studying a specific phenomenon from several different perspectives (neuroanatomy, neurophysiology and ethology) yields a very complete picture of how a communication system works in nature. Chapter 3 is concerned with hearing, both in humans and other animals, and with human speech, and covers both neurophysiological and psychological aspects. Chapter 4 is concerned with psychological as well as neurobiological aspects of a sensory system; it deals particularly with vision, a sense that is especially well developed in humans, with particular emphasis on colour vision. In Chapter 5, attention is on communication within the body rather than communication between animals; it discusses the sense of touch and goes on to look at the perception of pain. In this chapter, the material is primarily concerned with neurophysiology, but a number of other aspects of pain are dealt with. Finally, a brief 'epilogue' summarizes the book and draws together a number of points raised in the various chapters.

This introductory chapter discusses a number of essential concepts necessary for understanding the rest of the book. It looks, first, at some general properties of sense organs, then discusses the nature of perception, and finally discusses the concept of communication, both within the nervous system and between animals, and at the way that it is studied by brain and behaviour scientists.

1.2 Sense organs

An animal's sense organs have been described as its 'windows onto the world'. This phrase encapsulates two important features of sensory systems. First, sense organs 'look out' at the environment and gather in sensory information and, secondly, this process is highly selective. It is important to remember that, in this context, the 'environment' includes both an animal's external world and its internal state, though it is with the former that this book is mostly concerned. As emphasized in Book 1, Chapter 2, sense organs do not provide an animal with information about all aspects of the environment, but only with those aspects to which its sense organs are specifically receptive. Typically, they are specially adapted to gather information that is of particular biological importance to the animal, for example that relating to its food, predators, potential mates (Book 1, Chapters 2 and 9) and, especially in the context of motivation, aspects of its internal state such as the level of body fluids (Book 1, Chapter 7).

1.2.1 Receptors

The term 'sense organ' covers a wide range of sensory structures, from those consisting of single receptors, like the touch-sensitive mechanoreceptors in the skin, to those containing millions of receptors, like the human eye. Receptors are specialized according to the kind of energy to which they respond (Book 2, Section 2.4.1). Photoreceptors, in the eye for example, contain pigments that are chemically altered by light. Mechanoreceptors, such as those involved in the senses of touch and hearing, respond to deformation of the cell membrane. Chemoreceptors, such as those involved in smell and taste, respond to particular chemical compounds that come into contact with the cell membrane.

In Book 2, Chapter 2, receptors were described as 'excitable cells' that transduce a variety of forms of energy into electrical signals that are passed on to other cells in the nervous system. Before going further, it is important to recall a number of general properties of receptors.

□ What is meant by the term 'transducer'?

■ A transducer is a structure that converts one type of energy into another.

A light bulb converts electrical energy into light; a touch receptor converts physical forces into electrical signals.

□ What are the initial electrical signals produced within receptors called?

■ Receptor potentials.

□ In what important respect does a receptor potential differ from an action potential?

■ It is *graded*. The magnitude of a receptor potential is proportional to the strength of the stimulation received by the receptor, whereas an action potential is an all-or-none phenomenon.

□ To what other phenomena in the nervous system are receptor potentials comparable to in being graded rather than all-or-none?

■ Synaptic potentials.

If you have not been able to answer these questions, you should re-read Book 2, Section 2.4.1.

At some point in the nervous system, receptor potentials produced by receptors are translated into action potentials in the neurons that make up sensory pathways. There is variation among different kinds of receptors in terms of the point at which this translation occurs. In a touch receptor in the skin, output is in the form of action potentials, which are generated if the amount of stimulation received by the receptor exceeds a certain threshold level (Figure 1.1a). (Skin receptors are not separate cells, but are the specialized ends of neurons. Within these receptors the conversion of a receptor potential into an action potential is brought about by electrical changes within the cell.) By contrast, light and vibration receptors produce graded receptor potentials and their output is also graded, in that the amount of transmitter released at the synapse is proportional to the strength of the receptor potential. It is at the next cell in the pathway that this stimulation may or may not generate an action potential (Figure 1.1b).

Many receptors are active all the time, not just when they are stimulated, producing weak receptor potentials or action potentials at a low rate; they are described as being 'spontaneously active'. When they are stimulated, the frequency of electrical signals that they produce increases or decreases (e.g. in the eye). As you will see in later chapters in this book, sensory systems are particularly sensitive to *changes* in incoming information, rather than to levels of stimulation that are constant.

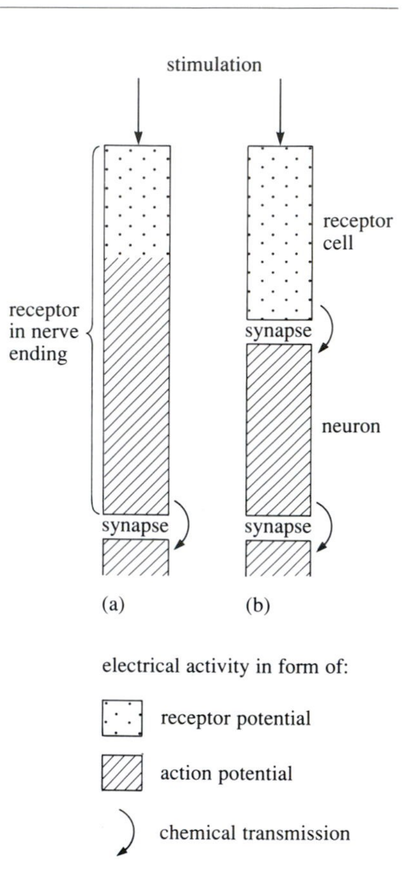

Figure 1.1 Two kinds of sensory pathway showing the relationship between receptor and action potentials. (a) Translation of a receptor potential into an action potential occurs within a nerve ending. (b) Translation occurs at the synapse between the receptor and the neuron with which it makes contact.

1.2.2 Sensory coding

All forms of sensory stimuli show a great deal of variation. One important kind of variation is in terms of the *intensity* of stimulation; sounds may be loud or faint, light may be bright or dim, for example. It may be very important for an animal to be able to distinguish between stimuli according to their intensity, as, for example, in the moth evading a bat, described in Book 1, Section 1.1.2. Another important kind of variation relates to the fact that some stimuli are arrayed along a spectrum: sounds vary in frequency (pitch) and light varies in wavelength (colour). It is generally important that animals be able to differentiate between stimuli according to their position along such a spectrum. The way that sense organs work and are arranged in such a way that they can detect these kinds of variation is called **sensory coding**.

Figure 1.2 illustrates sensory coding in a single neuron that conducts the output from a single receptor in the ear of a frog. The V-shaped plot in the Figure is called the *tuning curve* of that receptor, and is determined in the following way. A frog is anaesthetized and placed in front of a loudspeaker and a very fine electrode is inserted into a single neuron in its auditory system. A pure tone of a particular sound frequency, such as 200 Hz (see Box 2.1) is played through the speaker at low intensity (volume); no electrical activity is detected in the neuron with the electrode. The intensity of the sound is gradually increased until the neuron begins to fire; this defines the threshold intensity, for that particular sound frequency, at which that cell will fire. Then a different frequency is selected, the process is repeated, and another threshold is determined. Repeated measurements of this kind eventually produce a tuning curve like that in the Figure.

□ For the neuron in Figure 1.2, what is the threshold when the frequency of the sound is 200 Hz?

■ 90 dB. (dB is the abbreviation for decibels, the unit of sound intensity; see Box 2.1.)

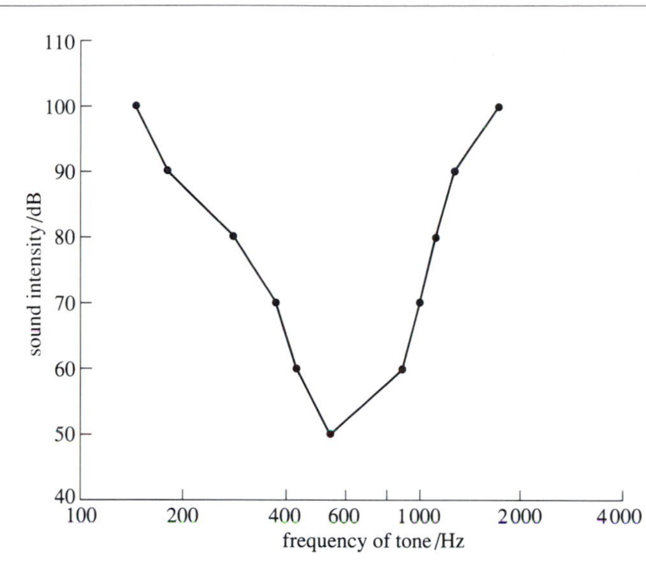

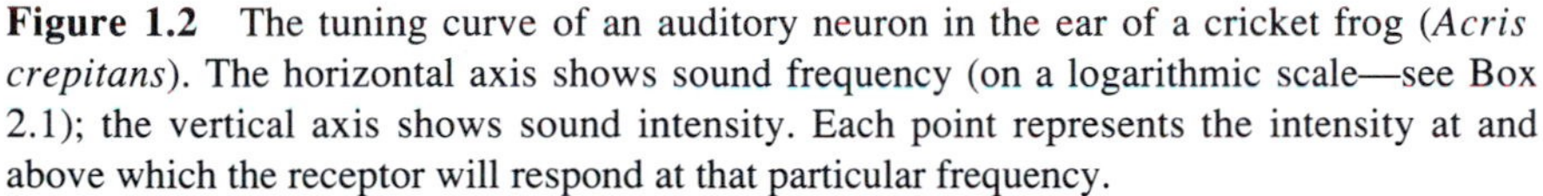

Figure 1.2 The tuning curve of an auditory neuron in the ear of a cricket frog (*Acris crepitans*). The horizontal axis shows sound frequency (on a logarithmic scale—see Box 2.1); the vertical axis shows sound intensity. Each point represents the intensity at and above which the receptor will respond at that particular frequency.

Box 2.1 Units of sound measurement

Hertz (Hz) The unit used for measuring the frequency of any physical process that takes the form of repeated cycles or waves (e.g. sound). A frequency of 1 Hz is equivalent to 1 cycle per second. Middle C on a piano has a frequency of 256 Hz. For healthy, young humans, the lowest frequency (deepest pitch) that can be heard is about 20 Hz, the highest frequency is about 16 000 Hz (usually expressed as 16 kilohertz or 16 kHz).

Decibel (dB) The unit used for measuring the intensity (or volume) of a sound. The decibel scale is *logarithmic* (i.e. every time the decibels increase by a fixed amount (6 dB), the intensity of the sound doubles, thus, 13 dB is twice as loud as 7 dB, which in turn is twice as loud as 1 dB, and 13 dB is four times as loud as 1 dB). A road drill generates a sound intensity of about 120 dB, a level that causes both discomfort and physical damage in the auditory system; people talking produce about 60 dB.

The lowest point of the curve (the bottom of the V) is called the 'best frequency' of that receptor, because it is the frequency to which the receptor is most sensitive. Put another way, the neuron is said to be 'tuned to' that frequency.

□ What is the best frequency of the receptor in Figure 1.2?

■ About 550 Hz.

Note, however, that this particular neuron will respond to other frequencies, provided that they are at a relatively high intensity. This receptor will in fact respond to any sound with a combination of frequency and intensity that falls within the area enclosed by the V-shaped curve.

This example illustrates two general properties of receptors; first, that they are tuned to a specific category of stimulus and, secondly, that they respond only when the intensity of that stimulus exceeds a threshold. These properties form the basis of sensory coding, especially when, as in ears and eyes, receptors are not isolated but form part of an extensive array of similar detectors, each with slightly different tuning curves.

There are two basic ways in which animal sense organs code for the intensity of a stimulus. The first is based on individual receptors, each of which generates receptor potentials in proportion to stimulus intensity. Thus, a strong stimulus causes single receptors to produce stronger receptor potentials than a weaker stimulus. The second kind of intensity coding occurs where receptors form arrays. If each receptor has a different threshold then the stronger the stimulus, the more receptors will be stimulated to respond. Once again, the moth's ear (described in Book 1, Section 1.1.2) provides a simple example of this kind of coding; a weak stimulus causes only one receptor to respond, a strong stimulus causes both receptors to respond.

Sensory coding across a spectrum of stimuli is achieved by an array of receptors each of which is tuned to a different point in that spectrum. For example, humans see a range of colours because they possess light detectors that respond best to different wavelengths of light (Chapter 4), and frogs can hear a range of sounds because they have a large number of receptors tuned to different frequencies of sound (Chapter 2).

Sensory coding across a particular spectrum is achieved within limits that vary from one species to another. For example, humans can see light between red and violet on the electromagnetic spectrum; some snakes can 'see' in the infrared and many insects and birds can see in the ultraviolet (Book 1, Section 2.3.1). Sensory coding for intensity involves a particular aspect of sensory systems, their capacity to differentiate between one stimulus and another, as discussed in the next section.

Another important aspect of the processing of sensory information, in addition to coding for intensity and frequency, is locating the source of stimuli. It will clearly be adaptive for an animal to know *where* a predator is, rather than simply that it is close by, for example. Many sensory systems contain features that make it possible for the animal to have, within its nervous system, a topographic map of its world. As discussed in later chapters of this book, the way that this mapping works varies from one sensory modality to another. ('Sensory modality' is an alternative term for 'sense'; see Book 1, Section 2.3.2.) Some widespread and obvious features of sense organs contribute, however, to the topographic processing of sensory information. For example, having two ears, separated by the width of the head and pointing in different directions, makes it possible to locate a sound source because the information from that source will be slightly more intense, and will arrive slightly earlier, at one ear than the other.

1.2.3 The signal-to-noise ratio

In many natural contexts, it is very important that animals are able to detect particular aspects of their environment and differentiate between them and other incoming information. This is important, for example, if an animal is looking for food items that are camouflaged against their background. It is especially important in communication between animals, a context in which it may be very important that signals such as mating calls, alarm calls, and so on are reliably and accurately detected. Differentiating between a particular signal and other information in the same sensory modality requires that an animal's sensory system be able to distinguish between the signal itself and what is generally called **noise**. Noise is a term used for all sensory modalities, not just sound, and refers to any energy that tends to obscure or interfere with sensory input of biological significance.

The importance of the signal-to-noise ratio can be illustrated by a familiar example. Cars are equipped with sidelights that act as signals in reduced light, indicating the position and width of a car. If a car approaches you with its sidelights on during daylight, you will generally see the car itself long before you are aware that the sidelights are on. In good light, the signal-to-noise ratio is low. In darkness, however, the sidelights are visible before you can see any features of the car; the signal-to-noise ratio is high. The signal-to-noise ratio is an important factor in communication systems between animals, where certain features, of both behaviour and sense organs, can be regarded as adaptations that serve to maximize the ratio.

□ In the example of car lights, what constitutes noise?

■ The light present in the environment.

□ Continuing with this example, how can the signal-to-noise ratio be improved when the level of noise is high?

■ By turning on headlights, as well as sidelights.

1.2.4 Chemoreception

Some of the general features of sensory systems can be seen in chemoreception, which includes the senses of taste and smell. Chemoreceptors are cells that are specialized to detect and identify particular chemical compounds and to measure the concentration of those compounds. In being responsive to chemical stimuli, they represent a modification of a property that is common to all cells (Book 2, Chapter 2). Some chemoreceptors are located within the body (enteroceptors) and respond to compounds such as nutrients, hormones, oxygen and carbon dioxide. Others are exteroceptors and respond to chemicals in the external world.

The senses of taste and smell (olfaction), involve very similar chemoreceptors. In most animals, both respond to chemicals in solution; consequently, the tissue containing chemoreceptors (e.g. the olfactory epithelium which is the single layer of cells lining the nose or covering the tongue) is always kept moist. Taste and olfaction differ in that olfactory receptors typically have a much lower threshold than taste receptors and so are responsive to much lower concentrations of chemical stimulus. They are generally anatomically distinct; humans for example, have separate noses and mouths; the housefly (*Musca domestica*) has olfactory receptors in the antennae on its head and taste receptors on its feet. They are also neurologically distinct in mammals, in that they are connected to the brain by different cranial nerves (Book 2, Section 2.6.3). Olfactory information reaches the brain via the olfactory nerve and taste information via the facial and glossopharyngeal nerves.

Some chemoreceptors are highly specific, others are responsive to broad categories of compounds. The olfactory receptors of male silkmoths (*Bombyx mori*) are responsive only to bombykol, the attractant secreted by the female (Book 1, Section 2.7.2). The very large, feathery antennae are covered in millions of tiny *sensilla*. Each sensillum consists of a fluid-filled projection, with numerous pores opening to the outside world. Bombykol molecules enter through the pores and stimulate the chemoreceptors inside (Figure 1.3). So sensitive are these receptors, that each one will fire in response to a single molecule of bombykol.

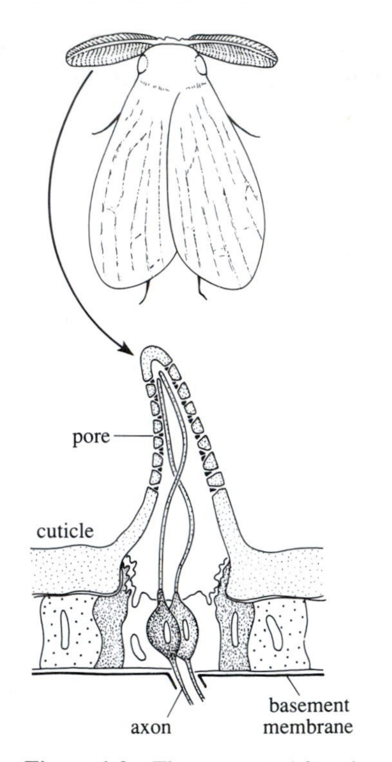

Figure 1.3 The antennae (above) and a single sensillum of a moth (below).

In humans, the tongue is covered with numerous chemoreceptors called taste buds. Different regions of the tongue are sensitive to one of four general categories of taste: sweet, sour, salty and bitter (Figure 1.4a). The wide variety of compounds that will stimulate the taste buds results in humans perceiving sugar and saccharin as being very similar, despite large chemical differences. Human taste buds consist of small clusters of receptor cells, called papillae (Figure 1.4b), within which the actual receptor cells are replaced every few days.

1.3 Sensation and perception

Imagine a pond surrounded by small children who are throwing things into the water. Each object creates a characteristic pattern of expanding circles on the surface; large objects create large waves with considerable distances between them while small objects create small, closely-packed ripples. Sometimes several objects land at once, so the surface of the pond is covered by a complex, ever-changing pattern.

Now, imagine that you have to identify all the objects landing in the pond. To make the task interesting, you are not allowed to see or hear the children, or the objects, or even the surface of the pond. In fact, all you can do is dip the index finger of each hand into the pond and feel the ripples drifting past.

You would be quite justified in believing this task to be impossible but, in fact, it is a fair analogy with what you do when you hear things. Sound sources are just mechanical disturbances that set up waves of pressure variation in the air, rather like the pattern of ripples on the pond. Instead of your fingers, your only contact with these pressure variations is through two small membranes, one in each ear, which vibrate as the waves drift past. Yet from this impoverished stimulus, your auditory system can recognize and locate all the complex and different sounds you hear.

The first point of this analogy is obvious: perception is very complex—it just seems simple because we are very, very good at it.

The second point is a little less obvious and has to do with *why* perception is so complex. Distance senses, like sight and hearing, have no direct contact with the world. Information is brought to them indirectly by light or pressure variation and our perceptual systems make sense of the resulting complex and often fragmentary pattern of stimulation.

Technically, objects out there in the world are called *distal* (far) objects. The pattern of information, carried by light, pressure variation and so forth, which actually reaches the sense organs is called the *proximal* (near) stimulus and *it is not necessarily anything like* the distal object (compare the pattern of ripples with the objects thrown into the pond). The task of the perceptual scientist is to understand how, given only proximal stimuli, the brain reconstructs the rich world of distal objects in which animals, including humans, live.

In recent years scientists have taken two very different approaches to this task, each stemming from a different view of the complex relationship between proximal stimuli and distal objects. The first approach acknowledges that distal and proximal phenomena are very different from each other but maintains that, nonetheless, proximal stimuli contain all the information needed for perception. In terms of the pond analogy, every perceivable feature of an object has *some* effect upon the pattern of ripples and so, in principle, a complete description of the pattern of ripples contains within it a complete description of the perceivable features of the object. This approach, pioneered by J. J. Gibson, is called *direct perception* and it views perception as essentially a process of picking up information from the environment. The second approach is much more *indirect*. It maintains that, because the relationship between proximal stimulus and distal object is so complex, simple picking up of information is really not enough. Even if they can derive a complete description of the proximal stimulus, perceptual systems still face the more subtle and difficult task of working out what produced it. This latter approach, which is the one adopted throughout this book, views perception as an active process, endlessly trying to make sense of the fleeting glimpses of reality provided by the sense organs.

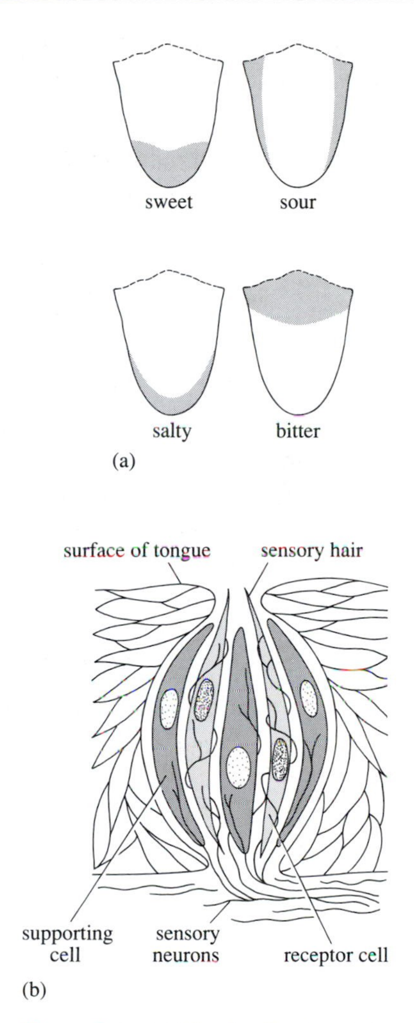

Figure 1.4 (a) The distribution of the regions of maximum sensitivity to different tastes on the human tongue. (b) The structure of a single taste papilla on the human tongue, shown in vertical section.

Broadly speaking, according to the indirect approach, it is often useful to think of perception as a two-stage process: the first stage is concerned with deriving a useful description of the proximal stimulus, the second with interpreting this description in terms of the distal object which might have caused it. Furthermore, the distinction between the descriptive and interpretive stages is the same as the distinction between *sensation* and *perception*. Sensations may be complex but they are essentially descriptive and they deal only with the proximal stimulus. Perceptions are interpretive and they deal with the objects in the external world that produce the proximal stimulus.

To make this distinction clearer, look at Figure 1.5 (*overleaf*). If you have not seen this before it will probably appear at first as a meaningless collection of black and white blobs. But if you persist, it will suddenly leap out at you as a Dalmatian dog snuffling among leaves. That 'leaping out' marks the switch between sensation and perception. Before you saw the Dalmatian you had only a description of the image in terms of black and white blobs. Seeing the Dalmatian is just the visual system working out what produced it; now you can describe the dog, rather than just its image.

Figure 1.5 A fragmentary figure. If you look at the black blobs for a while you will suddenly see a Dalmatian dog.

The distinction between descriptive processes (sensation) and interpretive processes (perception) is also often couched in terms of 'bottom-up' and 'top-down', or 'data-driven' and 'concept-driven' processing. All these terms refer to roughly the same distinction and are frequently used interchangeably. Bottom-up processing refers to the idea that the description of the proximal stimulus must be completed before any interpretive processes are brought to bear. Top-down processing suggests that sometimes knowledge about the distal object may make the descriptive process easier or even unnecessary. Although the distinction and its associated terminology can help you think about perception, you should not take them too far. Even though a particular approach may be primarily concerned with one aspect of the distinction, it does not really make much sense to claim that perception is generally a bottom-up, or conversely, generally a top-down process. Both types of processing are clearly involved. Moreover, the extent to which they are involved is not fixed but will depend upon the precise context and stimulus. When faced with a new and unexpected stimulus, for example, you may need to derive a rich and sophisticated description before you can make much sense of things. One might reasonably claim that this particular situation is dominated by a bottom-up approach. But when entering a familiar room, you have detailed expectations of its contents and may need only a cursory description to confirm them. In this situation, then, perception would be predominantly top-down.

1.4 Communication

Neurons interact within the nervous system to produce coordinated movement and responses on the part of an individual animal, and two or more animals interact to produce various kinds of social behaviour, such as courtship, fighting, cooperative hunting and parental behaviour. Both processes involve *communication*, between neurons at one level, between individual animals at another. This section looks briefly at what is meant by the word communication and at what communication between neurons and between individuals has in common.

At its simplest, communication means that two entities, A and B, share something, as in the phrase 'communicating rooms' to describe rooms which share a door. In biology, however, it means something more specific than this, namely the passing of something, called *information*, from A to B. Information is an abstract concept and is not something that can be directly observed, although it can be coded in a form that can, such as action potentials, a visual display, a sound or a smell. The passing of information from A to B can only be inferred from the fact that the activity or behaviour of A influences the activity or behaviour of B (Book 1, Section 2.4.2). Use of the word 'information' creates potential confusion because it has both an everyday and a scientific meaning. As defined in *information theory*, information is measured in terms of items that reduce the uncertainty of the receiver of the information. If you arrive at a station hoping to take a train to somewhere, you need two items of information, the time that the train will leave and the platform it will leave from. The person at the information desk will, hopefully, provide you with both items of information and so reduce your uncertainty about the departure of your train.

In fact, information is defined more precisely than is implied by the use of the word 'item'. In some contexts information is measured in the form of 'bits'. One bit of information is the information which answers a 'yes' or 'no' question. Suppose that there are eight platforms at the station, but that the porter can only provide information in the form of 'yes' and 'no' answers to your questions. You could ask a series of questions of the form 'does my train leave from platform x' until you get the answer 'yes'. You would have to ask anything up to seven questions before you knew your platform, depending on when you ask the appropriate question. In fact this problem can be solved by asking only three questions, as follows:

1 Is my platform number between 1 and 4? Answer: 'no'.

2 Is my platform number between 5 and 6? Answer: 'yes'.

3 Is my platform number 5? Answer: 'yes'.

If there are eight possibilities of which only one is correct, the answer to the problem thus requires three bits of information. The relevance of this example to communication systems is that it shows that the efficiency with which information can be gained in a particular context depends on how that information is extracted. Just as a platform number can be determined more efficiently by asking one kind of question rather than another, so the capacity of sense organs to extract information from the external world depends on how receptors are arranged and linked together.

Studying communication, either in the nervous system or between animals, involves an attempt to analyse how much of the information possessed by entity A is passed on to B. Because neurons and individual animals are, to varying degrees, independent entities with properties of their own, the information that they pass on may differ in many ways from that which they receive. Information may be lost, amplified or distorted at any stage, i.e. it is 'processed', in its passage from one entity to another.

In the nervous system, the aspect of information processing that is of primary concern is how the huge amount of information that typically enters sense organs, such as the eyes, is analysed within the various parts of the system so that animals can respond appropriately to specific objects and events. How do animals extract the information that is most relevant to them from all the mass of information that is gathered by their sense organs?

Many of the aspects of communication that are of interest at the behavioural level are illustrated schematically for acoustic communication in Figure 1.6. This has two parts, which are conceptually very different. The black arrows represent what human observers can detect directly. One animal produces a signal, in this instance a call, which passes through the air to another animal that hears the call. The medium through which the signal passes, in this case air, is called the *channel*. Because the call has travelled through a noisy environment it will typically be weaker and distorted in some way by the time it reaches the receiver. If the receiver responds it may be possible to deduce something about the other, abstract, part of the system, which concerns the nature of the information contained in the signal, referred to as the *message*. If the receiver rushes towards the caller and attacks it, it could be inferred that the call was 'aggressive'; if the receiver approaches the caller and squats down in front of it, the call could be interpreted as a mating signal. By studying the various kinds of signal shown by members of a given species, and the various responses shown by different categories of receiver (such as males and females) it is possible to describe the communication 'repertoire' or 'language' of that species.

There is an important element that must be added to Figure 1.6, the 'eavesdropper'. In the context of neurophysiology, the eavesdropper is the researcher with his or her microelectrode, intercepting and attempting to analyse the action potentials that pass from one neuron to another. In ethology, the eavesdropper is the observer attempting to understand the language of animals. There is another, natural, kind of eavesdropper, however. Neuronal activity and animal signals create energy that can be detected and responded to by other animals. Some predatory fish find crabs by detecting the electrical activity created by their nervous systems and muscles, and crabs have the ability to switch off all neuromuscular activity, including their heart beat, if they detect a fish before it detects them. In the jungle of central America, male tungara frogs (*Physalaemus pustulosus*) are eaten by a bat that finds them in the dark by listening for their mating calls.

The presence of eavesdroppers of many kinds raises important questions about what communication is and highlights the fact that it is not as easy to define as it might at first appear. An owl flying over a field at night finds mice by listening for the rustling sounds they make as they move through the grass. Is the mouse communicating with the owl? To the information theorist, it is; the sound of the mouse reduces the owl's uncertainty about the location of prey. On the other hand, no-one would suggest that mice have evolved rustling sounds in order to attract the attention of owls.

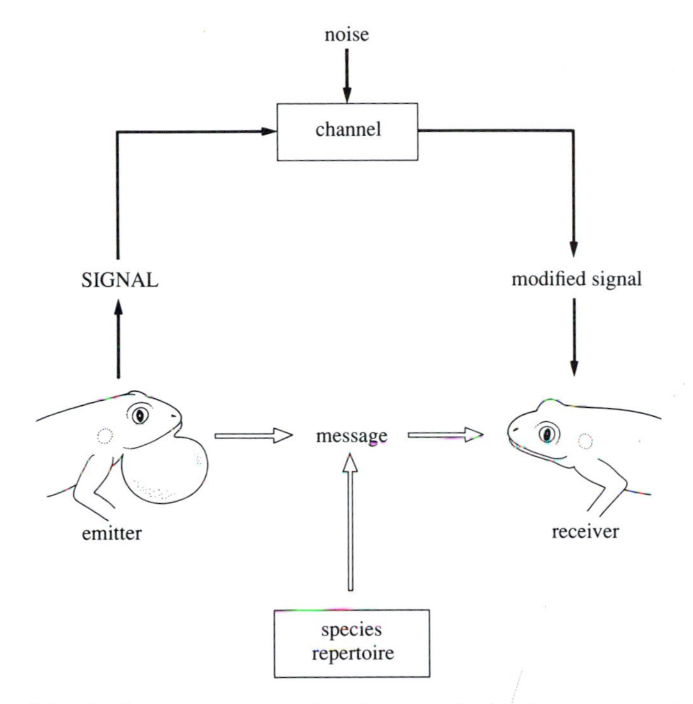

Figure 1.6 A schematic representation of communication between two animals.

Biologists do not agree on this issue and the question of whether the perception of information by an eavesdropper is classified as communication or not depends largely on the question being addressed. It is of interest, for example, to study how mice move in such a way as to keep the sounds they make to a minimum, and how tungára frogs modify their calls so as to minimize the chance of being eaten while not missing out on the chance to mate. (This will be discussed in Chapter 2.) When such questions are being addressed, the reduction of information transfer becomes as important as features that enhance the reliability with which it is transferred from one individual to another.

Summary of Chapter 1

Animals possess a variety of sense organs that respond to particular features of their external and internal environment. Receptors transduce sensory stimuli into receptor potentials that are converted within sensory pathways into action potentials. Some of the properties of receptors are illustrated by a neuron in the frog's ear which has a distinctive tuning curve; this defines the range of stimuli to which that neuron will respond and illustrates two important properties of receptors, specificity and threshold effects. To be maximally effective, the sensory systems of animals must be adapted to maximize their capacity to discriminate between biologically important information (signals) and other information (noise). A distinction is made between sensation, the nature of the stimulation received by sense organs, and perception, the interpretation that the animal's nervous system puts on that stimulation. Communication involves the passing of information from one entity to another. In nervous systems the entities are neurons; at the level of the whole animal, they are different individuals.

Objectives for Chapter 1

After reading this chapter, you should be able to:

1.1 Define and use, or recognize definitions and applications of each of the terms printed in **bold** in the text.

1.2 Distinguish between receptor potentials and action potentials and describe different places where they are generated in sensory pathways.

1.3 Explain what is meant by a tuning curve and use it to illustrate how a single receptor can code for the frequency and intensity of a sound. (*Question 1.1*)

1.4 Describe two ways in which a sense organ may code for the intensity of a stimulus. *(Question 1.2)*

1.5 Distinguish between sensation and perception.

Questions for Chapter 1

Question 1.1 (*Objective 1.3*)
What value would you give for the intensity threshold of the frog auditory receptor, the tuning curve of which is shown in Figure 1.2?

Question 1.2 (*Objective 1.5*)
Describe, briefly, how the intensity of a stimulus may be coded in the nervous system by (a) a single receptor, and (b) an array of receptors.

CHAPTER 2
ANIMAL COMMUNICATION

2.1 Introduction

The first part of this chapter consists of a 'case study', of vocal communication in frogs. This has been selected for two reasons. First, it illustrates how sense organs, ears in this instance, have become specially adapted during evolution for a particular communicative role. Secondly, it demonstrates a major theme of the course, namely that a complete understanding of behaviour requires analysis at several different levels. This particular case study shows how an understanding of the neurophysiology and neuroanatomy of the frog's auditory system enhances our understanding of phenomena observed at the behavioural level. Thus, it seeks to show how some of the questions about the causation and function of behaviour raised in Book 1 can be answered by understanding some of the physiological mechanisms and anatomical arrangements covered in Book 2.

The second part of the chapter considers the amount of information that may be contained within animal signals and discusses evidence concerning how much of that information is actually extracted when signals are used in communication between animals.

2.2 Auditory communication in frogs

Frogs and their calls are particularly suitable material for studying both the behavioural and neurophysiological basis of communication. Frogs produce very simple sounds, compared with birds or humans, for example. The sounds are associated with simple and readily observed patterns of behaviour and frogs have a nervous system that is readily accessible to neurophysiological investigation, including recording from single neurons.

In tropical and semitropical parts of the world, frogs will congregate in their thousands at certain times of year in and around ponds and other bodies of water. Frogs require water to mate and lay their eggs and, because ponds are few and far between, many frogs, belonging to more than twenty different species, may congregate in a very confined area. In temperate regions, such as Europe and North America, there are not so many species of frogs, but their breeding aggregations can nonetheless be very spectacular.

At these breeding aggregations (called *choruses*), males call very loudly, producing a sound that is quite deafening to humans nearby and which may be audible to the human ear over 1 km away. The calls produced by males are very stereotyped and species-specific (Book 1, Section 2.8). A major function of these calls is to attract and stimulate females of the same species. A female faces a formidable problem in a large, multi-species chorus. She must identify and locate a male of her own species by extracting and identifying the sound appropriate to her species from the intense and complex background noise. Research on auditory communication in frogs has focused on three major questions:

1 How do females identify their species-specific call?

2 How do females locate a particular male?

3 Do females discriminate among males of their own species on the basis of their calls, i.e. do females choose their mates?

Frogs have excellent eyes and a sense of smell, so it is possible that they could use vision and olfaction as well as hearing. However, frogs typically call, move and mate at night, when vision will be of less use than in daylight, and experiments using recorded calls have shown that the call alone is sufficient for a female to be able to identify, locate and select a male of her own species.

Look back at Figure 1.6 (Section 1.4), which shows a generalized scheme for a communication system. It has two parts. The upper part concerns what human observers, or 'eavesdroppers', can observe. A male frog produces a call that travels through the air to a female that locates and approaches him. As it passes from male to female, the call is subject to noise. One kind of noise, that from other males of the same and different species, has already been mentioned; another form of noise, distortion due to environmental effects, will be discussed later. The lower part of Figure 1.6 concerns the more abstract aspect of communication: what information is being passed from male to female? Of importance here is the repertoire of calls that frogs use. In many species, males produce only one kind of highly stereotyped call. Because this attracts females, it used to be called the mating call but, because other males may also respond to it in a variety of ways, for example by attacking a caller or by moving away from him, it is now called the **advertisement call**. Some species, however, have a more extensive repertoire that includes other calls. Some frogs have one or more *aggressive calls* that males direct towards one another during competition for calling sites. While females may be attracted to these in some species, in others they will move away from them. Many species have a *release call* that males give when they are clasped by another male, which responds by releasing his grip. Release calls may also be given by females that are not sexually receptive, for example after they have laid all their eggs.

2.2.1 The physiology of calling

Frog calls are generated by vibrations of the vocal cords, located in the larynx. Most frogs call with their mouths firmly closed; they inhale a large quantity of air that is forced back and forth over the vocal cords. As it is pushed outwards from the lungs, the air enters one or two vocal sacs which become inflated (Figure 2.1). The vocal sac increases the effectiveness of the call, not by amplifying (adding energy to) it, but by 'coupling' the sound to the air. It has a similar function to the soundboard of a piano or the box of a violin. Effective coupling means that a large proportion of the sound energy generated within a frog's vocal apparatus is transferred to the air. Frog calls range in duration from brief clicks of 5 to 10 ms to continuous trills that last several minutes. In terms of sound quality, they range from pure-tone whistles to a weird variety of croaks, grunts and trills that variously resemble the quacks of ducks, belches and computer-games machines. Although of small size—some frogs are only 2 cm long—males produce calls that are so loud that, at close range, they are painful to the human ear. This is evidence that they are very efficient at producing sound energy, but it also reflects the fact that calling males maintain a very high rate of energy expenditure. Not surprisingly, calling males can become exhausted after lengthy periods of sustained calling. This effect may not be important in species that call only at night and over just a few days, but in species with an extended breeding season, exhaustion becomes an important factor in determining a male's mating success.

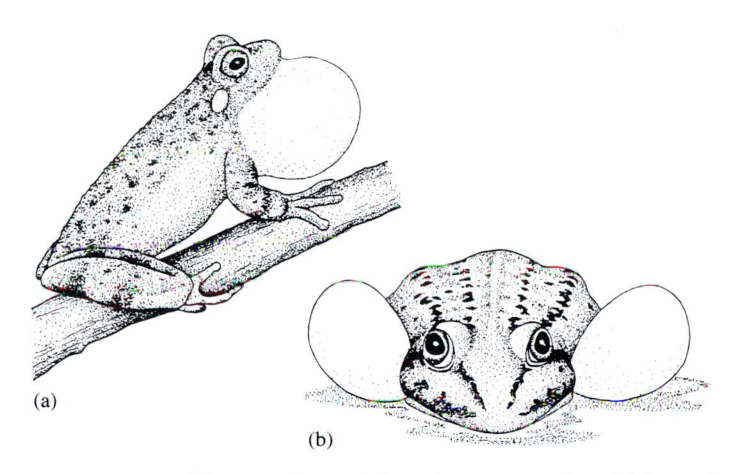

Figure 2.1 Male frogs calling. (a) The north American gray treefrog (*Hyla versicolor*) has a single vocal sac and calls from an elevated position in vegetation. (b) The European edible frog (*Rana esculenta*) has paired vocal sacs and calls while half-submerged in water.

Within the call of a given species, there is more sound energy present at one particular frequency than at any other. This frequency is called the **dominant frequency** of the call. For a particular species, dominant frequency shows rather little variation among individuals and so is characteristic for that species. In general, larger-bodied species have a lower dominant frequency than smaller-bodied species. The reasons for this are complex and beyond the scope of this discussion. Suffice it to say that larger objects characteristically produce sounds of lower frequency; for example, a cello is larger and produces lower frequencies than a violin and to get really good low frequencies (bass) from a hi-fi system, you need very large speakers. This relationship between body size and dominant frequency also holds *within* many species; larger males produce calls with a lower dominant frequency, a relationship which, as discussed later, is relevant to the question of whether female frogs choose their mates.

2.2.2 The properties of frog calls

The advertisement calls of male frogs represent the conversion of muscular work into sound energy. The way that sound energy is contained within a call can be analysed with respect to two parameters, the structure of the call over time, its *temporal patterning*, and the distribution of energy across sounds of different pitch or frequency, its *frequency spectrum*. Figure 2.2 shows these two ways of looking at two hypothetical calls. Figure 2.2a consists of graphs called sonagrams which show sound energy plotted against time. Both calls consist of a rapid series of discrete pulses with a consistent or stereotyped period of time, called the *inter-pulse interval*, between them; these calls sound like rapid trills. Species A produces one long trill, with a relatively long inter-pulse interval, species B produces three short trills, with a relatively short inter-pulse interval. This form of analysis takes no account of the frequency of the sounds, unlike Figure 2.2b which does, but which ignores their temporal structure. The sound produced by species A breaks down into three frequency peaks, f_1 to f_3, so that, in effect, it is a three-note chord. The call of species B, however, contains only two frequency peaks, one of which, at the lower end of the scale (f_1) is common to species A, while the other, f_4, is higher than either of the higher-frequency peaks in the call of species A.

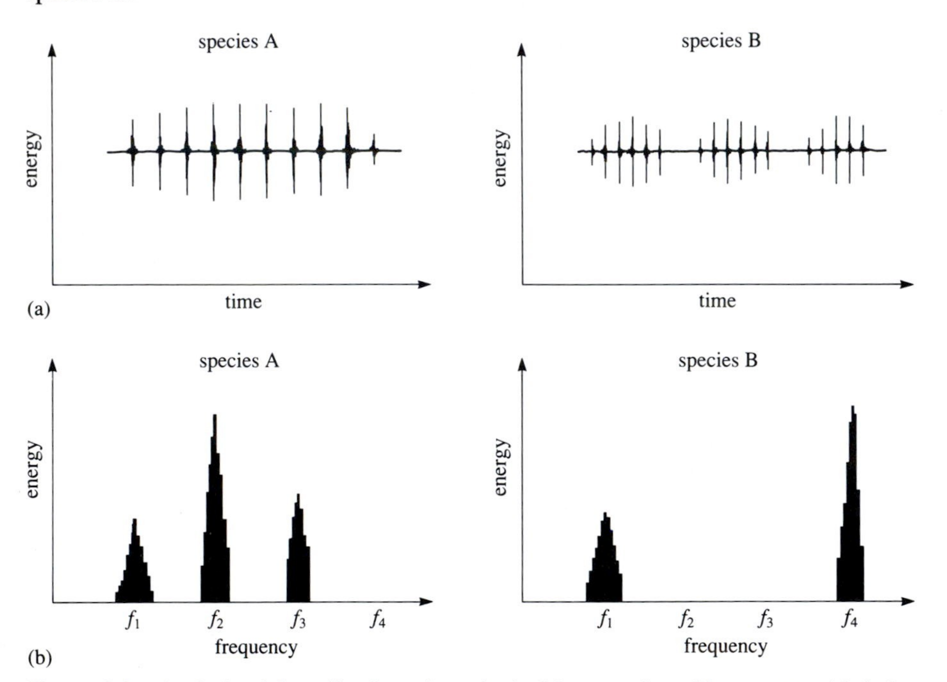

Figure 2.2 Analysis of the calls of two hypothetical frog species with respect to (a) their temporal patterning and (b) their frequency spectra. See text for explanation.

This kind of analysis of the properties of frog calls reveals features that could, *potentially*, be used by females to discriminate between calls of one species and another. A female of species A could, for example, discriminate between a conspecific call and one produced by a male of species B in four different ways.

□ List four features of the calls shown in Figure 2.2 that could be used to discriminate species A from species B.

■ 1 Species A has a call of longer duration than species B.

2 Species A has a longer inter-pulse interval than species B.

3 The call of species A contains frequency peaks f_2 and f_3 which are absent from the call of species B.

4 The call of species A lacks f_4 which is present in the call of species B.

Any of the features 1 to 4, provided they are stereotyped within each species, could provide reliable cues to correct species identification. Whether or not, in a given species, females use one or several of such features, is a matter for detailed experimental investigation. A very powerful way of doing this, pioneered by H. Carl Gerhardt in the USA, is to use synthetic calls, generated by a computer, and to offer female frogs a choice of two such calls that differ only in one feature, with all other features kept constant. By presenting a series of choices, with different call features contrasted in each choice, it is possible to determine which feature or features are used by females to identify their species' call.

In general, the results of experiments of this kind indicate that several call features have to be 'correct' if females are to respond to them by approaching them. The frequency spectrum of a call is very important for species recognition, for reasons that will become apparent below, and so are such structural features as the inter-pulse interval. Other features of calls may be important, not so much in species recognition as in preference between males within a species. For example, females of some species have been found to prefer loud calls to quieter ones, in other species they prefer longer calls to short ones, and in others, calls that are produced at a high rate are preferred to those produced at a low rate.

2.2.3 The frog's auditory system

The first stage in the physiology of hearing in frogs involves the tympanum, a large circular membrane that can be clearly seen behind the eye of many frogs (Figure 2.3). Sound causes the tympanic membrane to vibrate. The exact nature of this vibration depends on the intensity of the sound and the direction of the sound source. The ears of frogs are linked internally by a passageway that connects the inner ears so that each tympanum receives vibrational stimuli from the outside and from the inside. Differences in the relative pressure of the vibrations from a given sound source received by the inside and outside of the two tympanic membranes enable a female to determine the direction of a calling male, so that she can move towards him. This mechanism, called a pressure-gradient system, is very different from the mechanism that humans use to locate sound sources, to be described in Chapter 3. When moving towards a sound source, a female hops, listens for a while, turns directly towards the sound source and then hops again, repeating this sequence until she reaches the male frog (Figure 2.4).

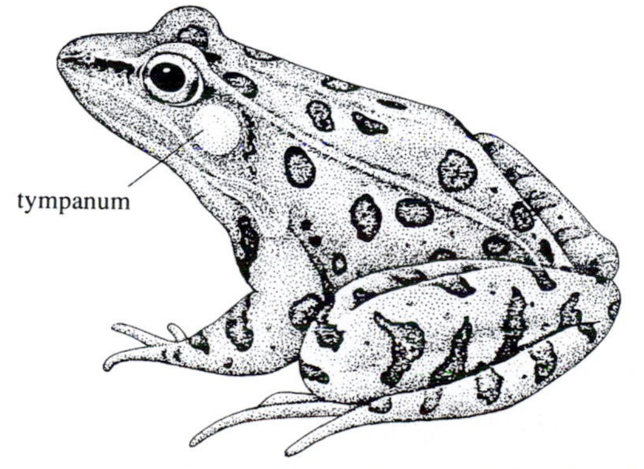

Figure 2.3 A southern leopard frog (*Rana sphenocephala*), from the USA, showing the large tympanic membrane just behind the eye.

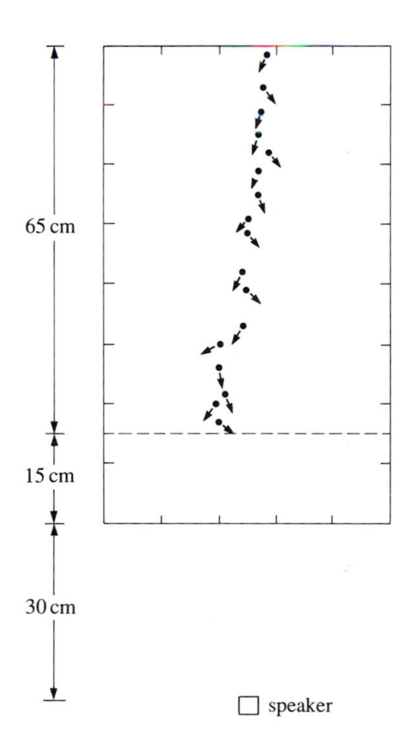

Figure 2.4 The path of a female frog (*Colostethus nubicola*) towards a loudspeaker playing the advertisement call of her species. The path consists of a series of hops, denoted by arrows. Note that, as the frog approaches the speaker, she tends to turn slightly, alternately to right and left.

2.2.4 Neuroanatomy and neurophysiology of the frog's ear

Figure 2.5 shows a schematic diagram of a frog's right ear. Vibrations of the tympanic membrane are transmitted by the bones of the middle ear (the plectrum, the columella and the operculum) to a fluid-filled capsule, the inner ear or otic capsule. As a result, the fluid in this capsule vibrates. The auditory nerve (part of the 8th cranial nerve, see Book 2, Section 8.4.1) has a number of branches within the otic capsule. Of particular importance are two branches that innervate two hearing organs, each consisting of many auditory receptors. These organs are called the amphibian papilla and the basilar papilla. The significance of this division of the auditory receptors into two, anatomically distinct organs will become clear shortly.

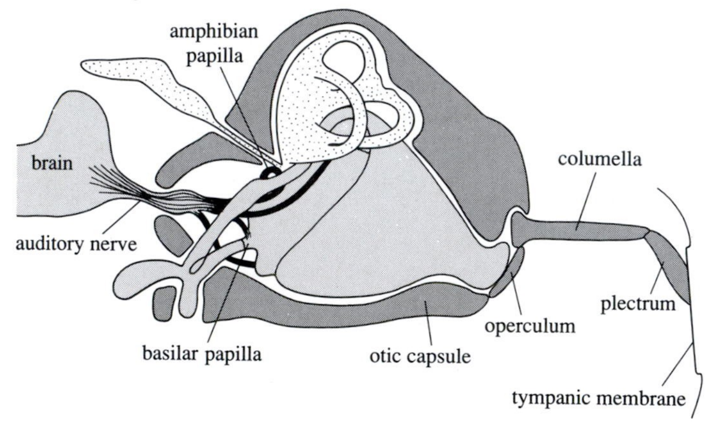

Figure 2.5 Diagram of a frog's right ear, viewed from behind, showing a vertical section through the brain (left), the inner ear (middle) and the middle ear and tympanic membrane (right).

Chapter 1 of this book described how it is possible to record the activity, and the selective response, of a single neuron in a frog's auditory system and to derive a tuning curve for an individual auditory receptor. Figure 1.2 shows the tuning curve for a neuron in the ear of a cricket frog (*Acris crepitans*); it has a best frequency of about 550 Hz. When the same procedure is repeated with a large number of receptors, it emerges that not all are tuned to the same frequency. Figure 2.6a shows a histogram of the number of receptors in the cricket frog's ear that are tuned to different frequencies. There are two distinct groups of receptors, one tuned to around 550 Hz, the other to a much higher frequency, about 2 750 Hz. What this means is that the cricket frog's ear is sensitive to only a few sound frequencies and that it is in effect deaf to sounds intermediate between those frequencies. Figure 2.6b shows a similar picture for another species, Couch's spadefoot toad (*Scaphiopus couchi*).

☐ What difference in the auditory system of cricket frogs and Couch's spadefoot toads is revealed by a comparison of Figure 2.6a and b?

■ Many more neurons were sampled in (b) but two distinct peaks in the distribution are apparent. Like the cricket frog, the spadefoot toad has a number of receptors tuned to low frequencies around 550 Hz, but its high-frequency receptors are tuned to around 1 300 Hz, as opposed to 2 750 Hz in the cricket frog.

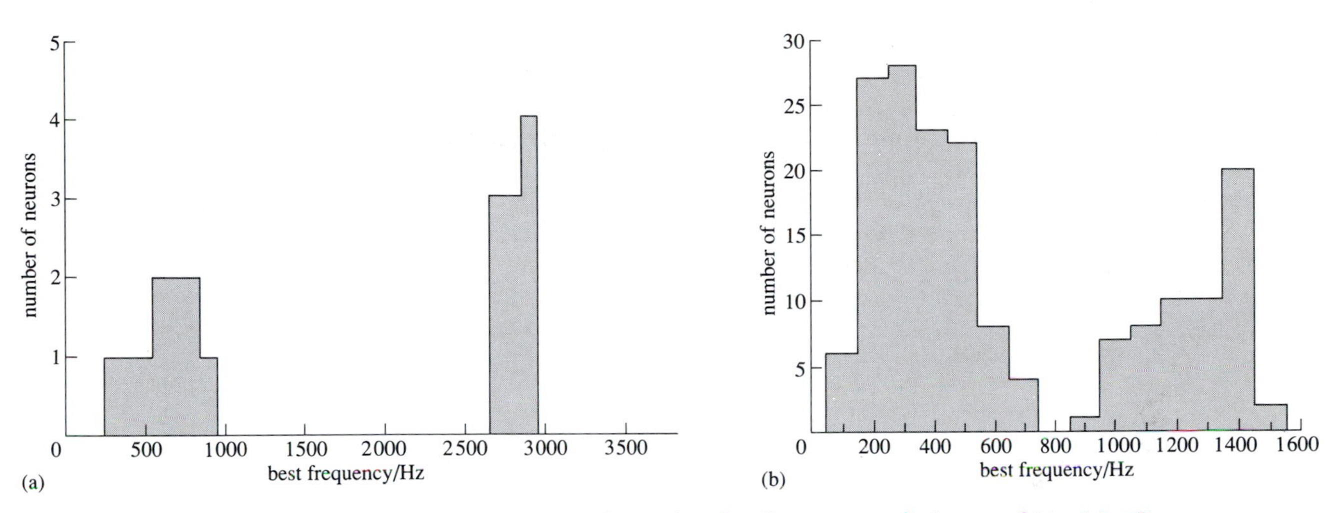

Figure 2.6 Frequency histograms of the best frequencies of a number of auditory neurons in the ears of (a) cricket frog (sample size = 20 neurons) and (b) Couch's spadefoot toad (sample size = 176 neurons).

A low-frequency peak in the sensitivity of frogs ears, tuned to around 500 to 1 000 Hz, is found in a wide variety of frog species. Where species generally differ is in the tuning of the high-frequency peak. The biological significance of this becomes clear when the frequency sensitivity of the female's ear is compared with the frequency spectrum of the male's call. There is typically an exact match; the female's ear is 'tuned' to the frequencies present in the conspecific male call.

You will recall from Book 1, Section 2.7.2, that a distinction was made between central and peripheral filtering of sensory information in the nervous system.

□ What kind of filtering is taking place, with respect to sound frequency, in a frog's auditory system?

■ Peripheral filtering. The frequencies that a female hears are determined largely by the tuning properties of the receptors in her ears.

The matching between the frequency spectrum of male calls and the tuning of females' ears can be remarkably precise. For example, across North America, some frogs show geographical variation in their calls, such that males in one locality have slightly deeper calls than those in others. A common reason for this is that frogs of a given species are, on average, slightly larger in one locality than another and, as described above, body size and call frequency tend to be positively correlated. Females show a parallel shift in frequency sensitivity such that their ears are tuned specifically to the calls of males in their immediate locality.

The coqui frog (*Eleutherodactylus coqui*) of central America gets its name from the fact that the male produces a two-note call, consisting of a low-frequency (1 200 Hz) 'co' followed by a high frequency (2 000 Hz) 'qui'. Figure 2.7 (*overleaf*) shows the frequency sensitivity of male and female coqui frogs.

□ How do the frequency sensitivities of the sexes compare, especially at 1 200 and 2 000 Hz?

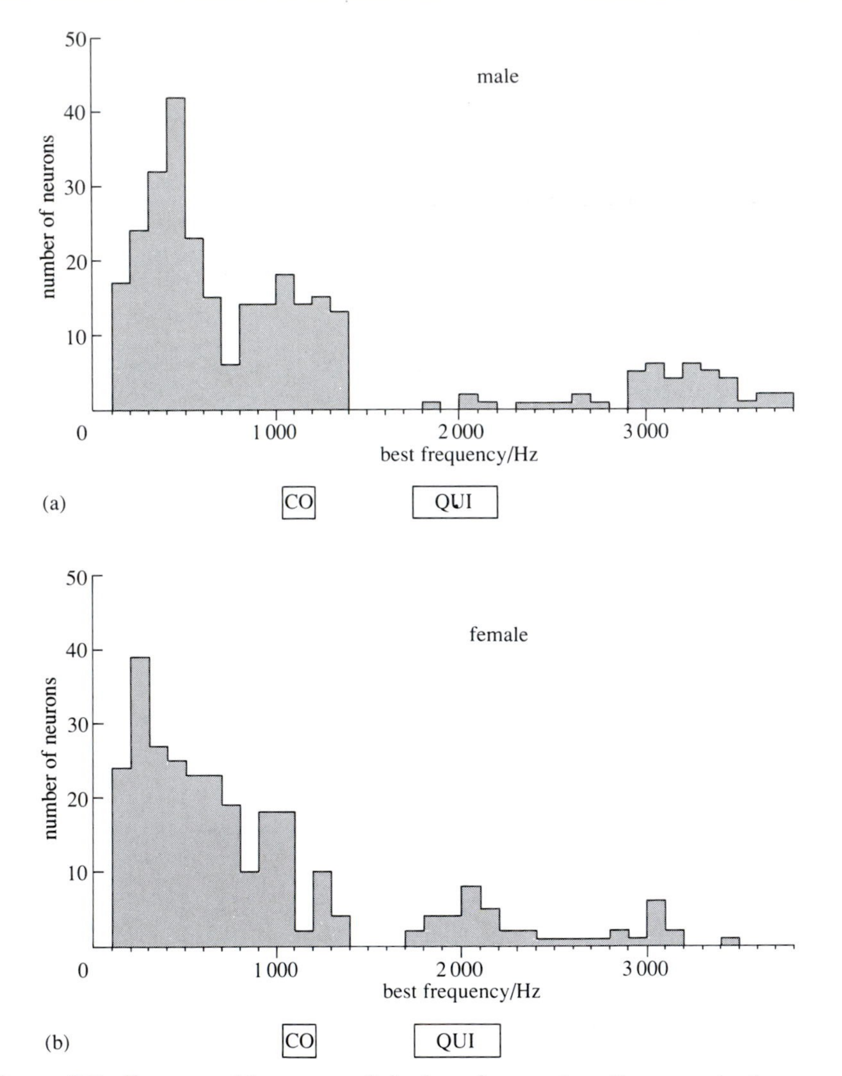

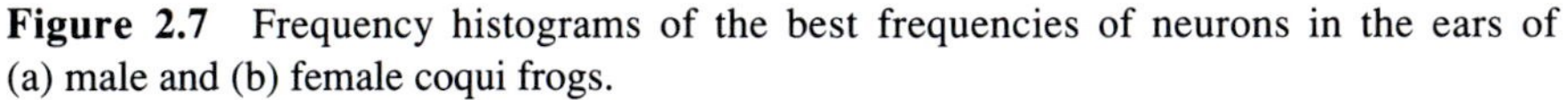

Figure 2.7 Frequency histograms of the best frequencies of neurons in the ears of (a) male and (b) female coqui frogs.

■ Males have a large number of receptors tuned to around 1 200 Hz, but females have few. Females have several receptors tuned to around 2 000 Hz but males have few.

In this species, then, there is a sex difference in the frequency sensitivity of the ear, with the ears of males being tuned to the first part of the call, the female's ears being tuned to the second part. Males thus selectively hear the 'co' part of the call; their response is to move away from another male so that calling males are widely spaced within the forest. Females selectively hear the 'qui' note and they respond by approaching the sound. Thus, the two major functions of the advertisement call, attraction of females and repulsion of males, are separated into two parts of a single call, and are differentially and specifically coded for in the auditory systems of the two sexes.

The discontinuity in the frequency sensitivity of frogs' ears, with the existence of distinct high- and low-frequency peaks, is detected by neurophysiological recordings from individual neurons in the auditory nerve and reflects the specific neurophysiological properties of individual receptors. The distinction is also apparent at the neuroanatomical level. The population of high-frequency neurons is located in the basilar papilla, the low-frequency neurons in the amphibian papilla (Figure 2.5). Thus, there is a correspondence between neuroanatomy and the selectivity of frogs' responses to sound frequencies.

2.2.5 Integrating neurobiology and behaviour

What relevance do the features of the frog's auditory system, in terms of neuroanatomy and neurophysiology, have to a female's mating behaviour? Figure 2.8 summarizes what happens in the auditory system of a female American green treefrog as she approaches a frog chorus. Green treefrogs commonly breed at the same time and in the same places as other species, some of them with quite similar calls. A female therefore has to identify, as well as locate, males of her own species and avoid males of other species. The male green treefrog's call is a trill that has a frequency profile emphasizing two frequencies, 1 000 and 3 000 Hz (top left in Figure 2.8). The 3 000 Hz peak and the inter-pulse interval of the trill are species-specific for green treefrogs; the 1 000 Hz peak is present in the calls of other species.

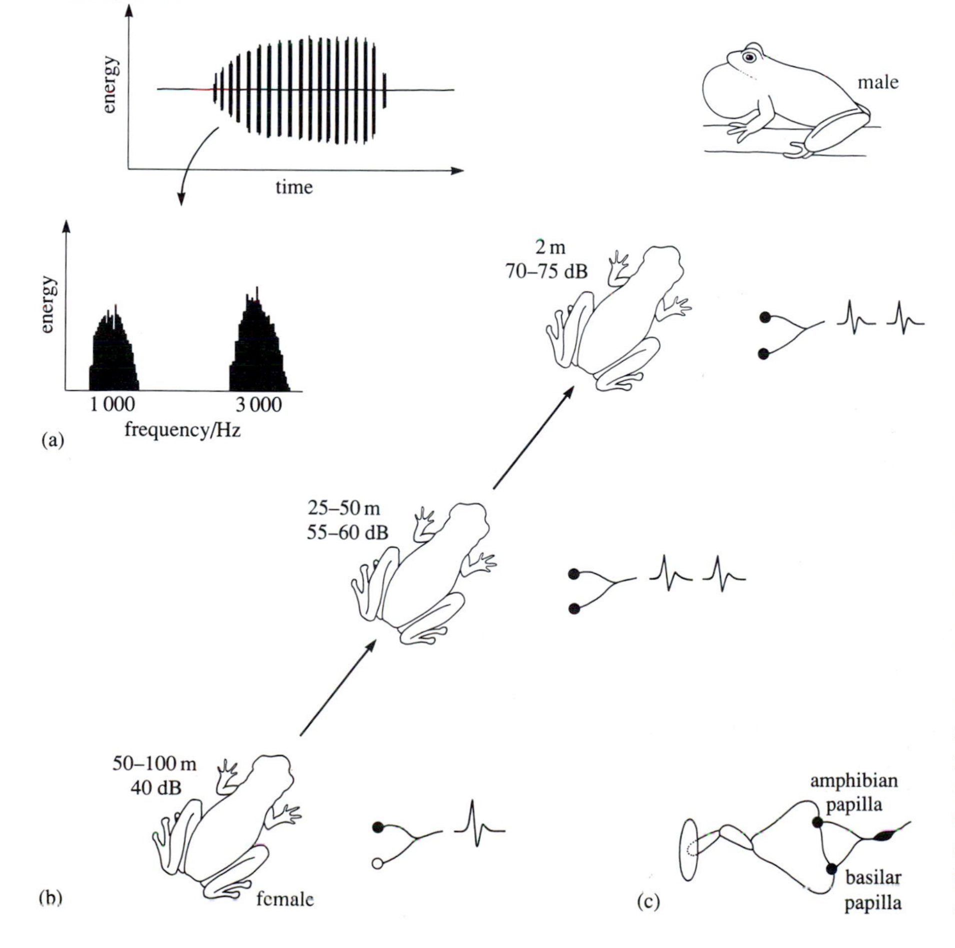

Figure 2.8 Diagram summarizing the information detected by the ear of a female green treefrog as she approaches a calling male from a distance of 100 m to 2 m. (a) The male's call consists of a trill containing sound at two frequencies, 1 000 and 3 000 Hz. (b) Activity in amphibian papilla and basilar papilla receptors as the female approaches the male. (c) Diagram of the female's ear (showing tympanum, ear bones and otic capsule) and its connections, through the amphibian papilla and the basilar papilla to the auditory nerve. See text for full explanation.

Females approaching a chorus from a distance receive only a weak stimulus that contains just a part of the frequency profile of the call. An important feature of sound is that it is attenuated (its intensity is reduced) with distance, and high frequencies are attenuated more than low frequencies. A female listening at a distance of 50 to 100 m from the chorus (lower left in Figure 2.8) thus receives a sound with an intensity of about 40 dB that contains a reasonable low-frequency signal but a very weak high-frequency signal. As a result, only the female's low-frequency receptors, located in the amphibian papilla of her ear, will detect sound and send action potentials to the brain. Because several species have calls containing this low-frequency peak, the only information that her ears provide her with at this distance is that there are male frogs nearby. As she hops closer, reaching a distance of about 25 to 50 m from the chorus, she will start to detect the high-frequency (3 000 Hz) signal that is characteristic of her own species; her high-frequency detectors, located in the basilar papilla, will also become active. Now she has the information that conspecific males are nearby and she can move preferentially to parts of the pond where male green treefrogs, rather than other species, are calling. As she enters a group of conspecific males, the temporal structure of the call becomes important and the species-specific inter-pulse interval of green treefrogs becomes an important recognition cue. Identification of this probably involves central rather than peripheral filtering.

Thus, the female's ear provides her with all the information that she needs to identify and locate a conspecific male. This information is made up of several components, each of which is encoded by neurophysiologically and anatomically distinct parts of her auditory system.

2.2.6 Noise and the frog communication system

Effective auditory communication between a particular male and a particular female frog is subject to three major sources of environmental noise, apart from sounds produced by other animals, such as insects and birds: the sound of frogs of other species, the sound of other frogs of the same species, and distortion of the male's call as it passes through the air.

The chance that a female might confuse the call of another species with that of her own is minimized by the peripheral filtering described above; females are largely deaf to sound frequencies that fall outside the range of frequencies in conspecific calls. Sound is not the only cue, however, that females use in locating a conspecific male. Figure 2.9 shows where calling males of seven species of Australian frogs are typically found during a chorus; each species has a characteristic calling site, ranging from the centre of the pond to places on dry land some metres from the pond's edge.

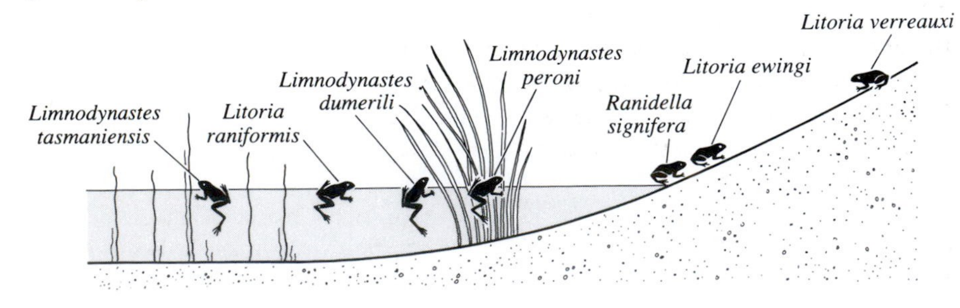

Figure 2.9 Characteristic calling sites of seven species of frogs in a pond in Australia.

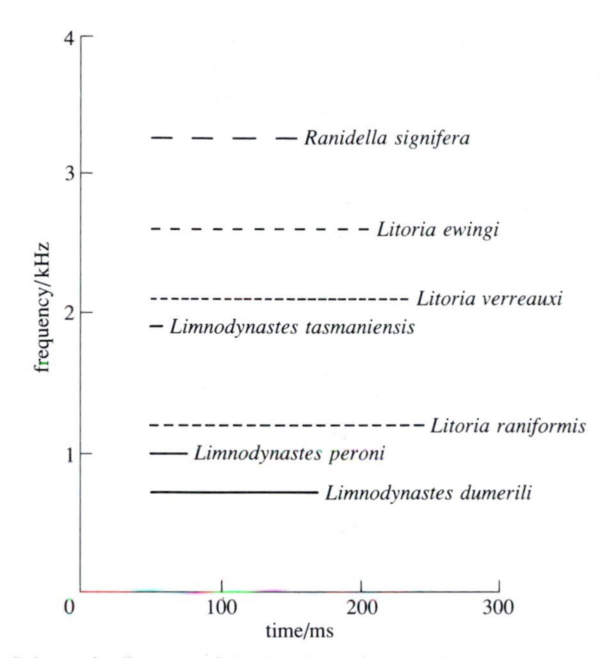

Figure 2.10 Schematic diagram of the dominant frequencies and temporal structure of the calls of the seven Australian frogs shown in Figure 2.9.

Comparison of Figures 2.9 and 2.10 reveals that frogs that have similar calls typically call from very different parts of the pond, and that species that call close to each other have very different calls. For example, *Ranidella signifera* and *Litoria ewingi* both call from the edge of the water (Figure 2.9), but their calls differ both in terms of dominant frequency and temporal structure (Figure 2.10). Thus, to locate conspecific males, females do not have to rely solely on their ability to locate males acoustically; by going to the appropriate part of the pond they can increase their chances of finding a conspecific male.

2.2.7 The effect of environmental temperature

Whereas mammals and birds are endothermic (body heat is generated from within and a fairly constant body temperature is maintained), amphibians are ectothermic (body heat mainly comes from external sources such as sunlight, and body temperature may vary considerably). (The terms 'warm-blooded' and 'cold-blooded' are no longer generally used by biologists to label these alternatives because many ectotherms can, in practice, have a similar body temperature to endotherms for much of the time.) In an ectothermic animal like a frog, many physiological processes speed up as body temperature increases. Temperature does not significantly alter the frequency spectrum of frogs' calls but it does alter their temporal structure. For example, the trill rates of two North American treefrogs, *Hyla versicolor* and *Hyla chrysoscelis*, are strongly positively correlated with environmental temperature (Figure 2.11). Because trill rate is an important species-recognition cue for females of these species, this effect is a source of potential confusion. Consider, for example, a female *Hyla versicolor* listening to a call with a trill rate of 30 pulses per second (look carefully at Figure 2.11). This could be being produced, either by a very cold *Hyla chrysoscelis* male, or by a very warm

Hyla versicolor male. It is, in fact, very unlikely that two such males would be present at the same time, since this would require there to be a range in environmental temperature within the pond of 15 to 28 °C. Females can, potentially, resolve the ambiguity, by taking account of their own temperature, since this will be similar to the temperature of the pond and therefore of any male calling in the pond. Gerhardt (1978) tested this idea by comparing the preferences of *Hyla versicolor* females for calls of different trill rates at different temperatures. Females were placed between two loudspeakers, one playing a trill with 15 pulses s^{-1}, the other 24 pulses s^{-1}. Identical tests were carried out at two temperatures, 16 °C and 24 °C. The results are shown in Table 2.1.

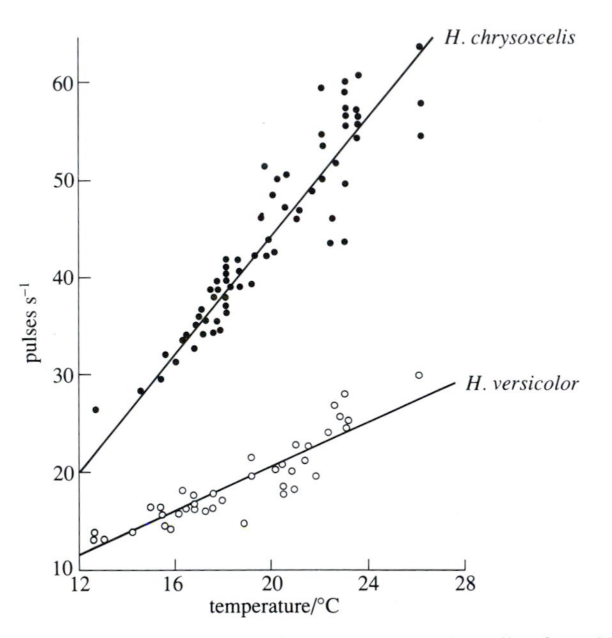

Figure 2.11 The effect of environmental temperature on the calls of two North American treefrogs, *Hyla versicolor* and *Hyla chrysoscelis*.

Table 2.1 Responses of female *Hyla versicolor* in discrimination tests.

Test temperature	Number of females responding to calls with pulse rate of	
	$15\ s^{-1}$	$24\ s^{-1}$
16 °C	12	0
24°C	1	10

☐ Do the data presented in Table 2.1 indicate that females alter their call preference according to the environmental temperature?

■ Yes, they do. At the lower temperature, they show a clear preference for the slower call. At the higher temperature they prefer the faster call.

□ Given that females show this effect, study Figure 2.11 and deduce whether it is likely that a female *Hyla versicolor* with a body temperature of 16 °C will be attracted to males of the wrong species.

■ It is very unlikely. A female at 16 °C prefers a slow call and will not be attracted by the faster rate of *Hyla chrysoscelis*.

This effect, whereby the female's call preference changes with temperature in a way that parallels the way the male's call is affected by temperature, is called temperature-coupling. It provides just one example of how the frog communication system is adapted to counter the potentially confusing effect of environmental noise.

2.2.8 Eavesdroppers

For male frogs, calling involves intense competition between males. A general reason for this is that, as discussed in Book 1, Section 9.2.2, sexual reproduction typically involves competition among males because, unlike females, they can greatly enhance their fitness by mating more than once. A specific reason is that male frogs typically spend more time at a chorus than females; females visit briefly to mate and lay their eggs, whereas males call over an extended period. Consequently, on any one night, there are typically many more calling males than there are receptive females. Thus, to attract females, an individual male must call more effectively than his rivals. He must call more often, more loudly and with more attractive calls. All these factors mean that he must allocate more of his energy to calling.

In several species, including the green treefrog and the British natterjack toad (*Bufo calamita*), individual males call on some nights and not on others. On their 'nights off' they are often present in the chorus but remain silent. Instead of calling they move towards calling males and take up a 'satellite' position from which they can intercept and mate with females that approach the calling male. As a result of the behaviour of these satellite males, calling males lose opportunities to mate that they would otherwise have.

In some species, there is another category of eavesdropper: predators. In making himself conspicuous to females, a calling male frog also makes himself conspicuous to a variety of predators, including water bugs, fish, larger frogs and turtles. Choruses of the central American tungara frog (*Physalaemus pustulosus*) attract a predatory bat (*Trachops cirrhosus*) that swoops down open-mouthed and plucks a calling male from his perch (Figure 2.12, *overleaf*). The calls of tungara frogs show a great deal of variation (Figure 2.13, p. 30). Some calls consist of one element, a 'whine'; others include one or more additional 'chucks'. A number of factors determine how complex the call a male will produce at any given time will be.

□ From what you have read earlier in this chapter, what cost do you suppose that male tungara frogs incur by producing more complex calls?

■ Complex calls, with additional 'chucks', probably require more energy to produce than simple calls.

Figure 2.12 A male tungara frog about to be devoured by a bat.

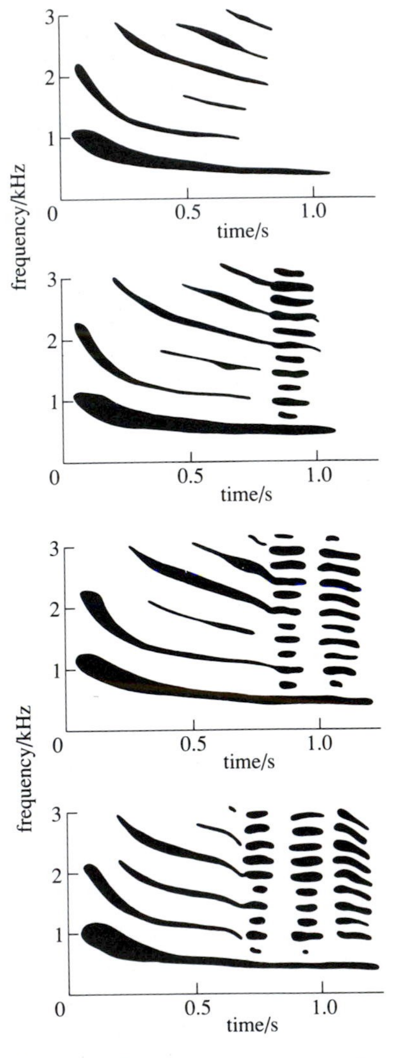

Figure 2.13 Sonagrams of calls of male tungara frogs, ranging from a simple call (top) consisting of a 'whine' to more complex calls that include one, two or three additional 'chucks'.

Saving energy is one reason why male tungara frogs produce simple rather than complex calls. A detailed study of this species by M. J. Ryan (1985) has revealed two additional factors. Ryan found that there is a relationship between call complexity and the number of males in a chorus; the more males are calling, the more chucks, on average, are included in their calls.

□ How do you account for this effect?

■ Complex calls may be more attractive to females than simple calls and, when more competitors are present, males must increase the attractiveness of their calls if they are to mate.

Ryan tested this hypothesis by conducting two-way choice tests in which individual females were placed between two loudspeakers, one playing a simple call, the other a complex call. The hypothesis was confirmed; females approached the complex call in preference to the simple call. There are two possible reasons for this preference. One is that a complex call is more 'attractive', meaning more sexually stimulating, to females than a simple call. The other is that, because the complex call contains more sound frequencies, especially low frequencies, it is easier for a female to locate. Indirect evidence supporting this second hypothesis was provided by another observation made by Ryan. He observed that bats are more likely to take males producing complex calls than those making simple calls.

Since the sexual attractiveness of male frogs is not a factor of relevance to bats, it suggests that they find more complex calls easier to locate.

Male tungara frogs are, in some way, able to detect the presence of bats. On nights when bats are numerous and active, males call less often, sometimes shutting up altogether, and they produce less complex calls. Thus the complexity of a male tungara frog's call represents a trade-off between three factors: conserving energy, being more attractive than rival males, and avoiding sudden death. It's a difficult life for a frog!

Summary of Section 2.2

Male frogs produce advertisement calls that are species-specific, and which elicit specific and different responses from male and female receivers. Calling is energetically expensive and, in general, larger males produce calls of lower frequency than smaller males. In terms of both their frequency content and their temporal structure, it is found that advertisement calls show a number of features that provide a number of cues by which receivers can discriminate one species from another and, within a species, one individual male from another. Female frogs locate a male on the basis of variations in the intensity of the sound that they hear. The ears of females contain receptors that are specifically tuned to the frequencies present in the calls of conspecific males. Low and high frequency receptors are located in different parts of the inner ear. As a female approaches a male, different parts of her auditory system are stimulated so that, as she gets closer to him, she has more and more information about his identity. The species-specificity of male calls, combined with species-specificity in terms of the exact locations where males call, reduces the risk of females approaching and mating with males of the wrong species. Male calls are affected by temperature and the hearing characteristics of females change with temperature in such a way that correct species identification is maintained over a wide temperature range. Predators may use male calls to locate their prey and male tungara frogs alter their calls in the presence of predatory bats so as to reduce the risk of predation.

2.3 Communication and information

Chapter 2 of Book 1 and the first part of this chapter have tended to emphasize the stereotyped nature of animal signals, in relation to their species-specificity. In the context of species-recognition, signals that give unambiguous information are clearly very important if animals are to avoid hybrid matings. Some signals produced in the context of mating are indeed highly stereotyped, like those produced by crickets and other insects, but those of many animals are not. A completely stereotyped, species-specific signal can only give a receiver the single item of information that the signaller is or is not a member of the same species. The question that is addressed in this section is whether the variation that occurs in those signals that are not totally stereotyped contains more subtle information about the signaller that can be extracted and responded to by a receiver.

This is an issue of currently burgeoning interest in ethology and reflects a shift in emphasis since the days of Tinbergen and Lorenz. They tended to emphasize the *species* as the unit of behavioural description, whereas contemporary ethologists are much more concerned with behavioural variation between *individuals*. There is now abundant evidence, some of it described below, that many signals do contain subtle information and that animals that receive such signals respond to that information. For example, you read in Book 1, Section 9.3.3 that peahens respond preferentially to peacocks whose trains contain large numbers of eye-spots. The peacock's sexual display is a species-specific signal that is stereotyped to a degree, but the male's train contains subtle information that females use as a basis for mate choice.

2.3.1 Plumage variation in birds

In many species of birds, species-specific plumage patterns show considerable variation, and there is evidence that, in some species at least, differences in plumage can signal differences in dominance status. In the house sparrow (*Passer domesticus*), for example, the size of the black throat patch characteristic of males varies substantially from one male to another (Figure 2.14) and it has been shown by Anders Pape Møller (1987), working in Sweden, that the larger the throat patch, the higher the male's dominance status. Similarly, dominance status in winter flocks of great tits (*Parus major*) is related to some extent to the size of the black 'bib' on the chest. Siewert Rohwer, who was the first researcher to study plumage variation and dominance (in an American species, the Harris sparrow, *Zonotrichia querula*), coined the term 'badge' of status for such plumage signals.

Figure 2.14 Variation in dark plumage on the throat of male house sparrows.

Status signalling is not the whole story of badges, however. Møller also showed that males with larger throat patches enjoyed greater reproductive success, because they were more attractive to females (Møller, 1988). Evidence for this came from two observations:

1 Males with larger patches obtained mates earlier in the spring compared with those with smaller badges.

2 Males that remained unmated throughout the breeding season had smaller patches than those that obtained mates.

These observations are consistent with the hypothesis that female house sparrows, first, detect variation between males in the size of their throat patches and, second, that they prefer as mates those with larger patches. They do not prove conclusively, however, that the relationship between patch size and female preference is quite that simple.

□ What other factor might be involved in mate choice by female house sparrows?

■ The male's behaviour. It is possible that males with large patches behave differently from those with small patches and that it is this variation that differentially attracts females.

Further studies by Møller (1988, 1990) do indeed indicate that patch size is not the only feature on which females base their choice of mates. First, high-status males with larger patches obtained territories containing better nest-sites (i.e. deeper holes in trees), and so females may have been attracted by the quality of males' territories rather than their patch size. Second, males with larger patches engage in more sexual behaviour than those with small patches and so may have been more attractive because of their behaviour rather than their appearance.

The observation that females prefer males with larger badges raises a question about the *function* of females' behaviour. Why is it adaptive for females to mate selectively with such males? One possibility is that badge size is related to some other factor, like territory quality, as described above, that has a direct effect on a female's fitness.

Studies like those of Møller have been carried out on several bird species, generally with similar conclusions. Plumage variation is often correlated with behavioural variation, making it difficult to determine whether or not plumage variation in itself is the cue used in competitive interactions and/or mate choice. However, where appropriate experiments have been carried out to tease the effects of appearance and behaviour apart, it is often the case that appearance does play a role.

The role of variation in signals in aggressive interactions is taken up again in Book 5, Chapter 4.

2.3.2 Dancing bees

Section 4.3.7 of Book 1 gave examples of animals giving calls that warn others about the presence of predators. This is a very simple form of communication in which the information passed from signaller to receiver relates to the outside world, that is, it is not directly related to the signaller. The classic example of communication about the outside world is the complex language used by honey-bees (*Apis mellifera*) to convey to fellow hive-members the precise location of food, first described by the Austrian ethologist Karl von Frisch in the 1930s. Some worker bees go out each fine day in search of new sources of nectar and pollen. On returning to the hive they perform one of two dances, either the *round dance* or the *waggle dance*. In the round dance, the worker that has found food simply rushes round in circles; this dance is performed if the food she has found is less than 50 m from the hive. Other workers pick up the odour of the food source from food that she has brought back with her, fly out of the hive and search for the food source. This dance thus just gives the message 'this food is available nearby'.

If the new food source is at a greater distance from the hive, the food-scout performs the more elaborate waggle dance (Figure 2.15b, *overleaf*). This is described as a *symbolic language* because the dancer encodes specific kinds of information in particular aspects of her dance. The dance is performed on the vertical surface of the comb in the dark interior of the hive. The angle between the central axis of the dancer's figure-of-eight movement and the vertical plane (angle *A* in Figure 2.15a) is the same as that between her path back from the food source to the hive and a line between the sun and the hive. During the straight part of the dance, the dancer waggles her abdomen and produces bursts of sound; the number of waggles and of sound bursts is proportional to the distance (*D* in Figure 2.15a) between the hive and the food source. Other workers in the hive follow the scout's movements with their antennae and then, equipped with the information contained in the dance, they leave the hive and find their way directly to the food source.

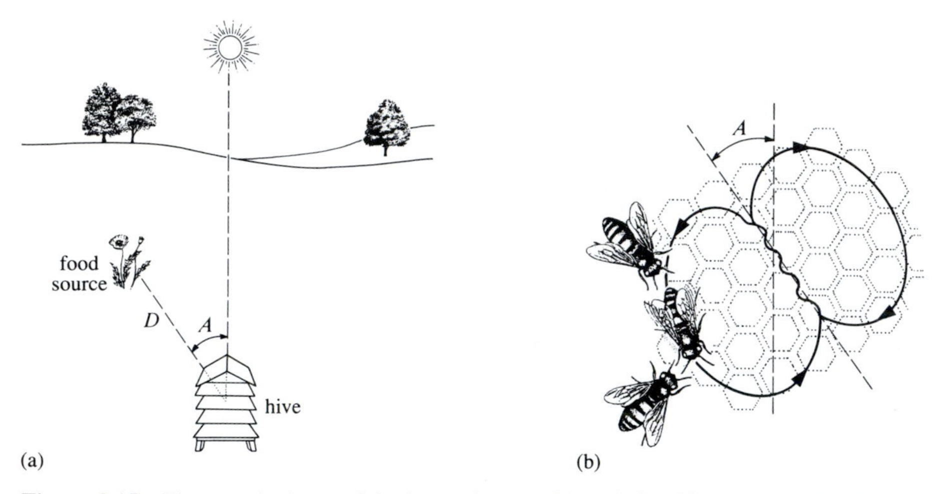

Figure 2.15 The waggle dance of the honey bee. (a) The relationship between hive, food source and the sun. (b) A worker performing the dance on the vertical comb, attended by two recruits.

The dance language of bees is in fact more complex than this brief account suggests. For example, the scout, if she has to contend with a strong cross-wind on the way back to the hive, dances in such a way that the direction she indicates is not the actual direction to the food source but is the course that the recruits must take to counter the effect of the cross-wind. You might also feel that a communication system that is dependent on the sun will have severe limitations, particularly in the UK. In fact, bees do not require a direct view of the sun. They are sensitive to polarized light and open sky contains a distinctive pattern of light polarization such that, given only a view of a small patch of blue sky, a bee can deduce the position of the sun hidden behind cloud.

A further complication is that bees from different localities code distance information slightly differently; bees from one area are said to have a different *dialect* from those in another. This effect emerged from a series of experiments carried out by von Frisch's student, Martin Lindauer, who showed that Swiss bees recruited by an Austrian scout flew in the right direction but did not fly for the correct distance! The reason for this is not clear but a possible reason is that certain details of bee language are learned by each generation from the last. Birds are known to show song dialects that result from individuals copying features of their songs from close neighbours. This provides an example of cultural, as opposed to genetic, transmission of behaviour from one generation to another (Book 1, Section 10.5).

2.3.3 Echolocation by bats

The final example in this section continues the theme of how animals use the information encoded in very fine variation in signals, but in a very different context. Here, instead of one animal communicating with another, animals (in this case, bats) signal to themselves. As described briefly in Book 1, Section 1.1.2, bats find insect prey at night by sending out ultrasonic (very high frequency) sounds and listening for the echoes off the insects' bodies. They use the same system, called echolocation, to avoid trees and other objects in their environment. This is a rather special form of communication in that a feeding bat is both the signaller and the receiver; it is communicating with itself. It is also a very sophisticated and complex system, to which the account that follows does scant justice.

The basic mechanism involved in echolocation is that the bat produces a cry, this bounces off an insect, and the bat hears the echo through its ears. To locate an insect, a bat requires two kinds of information, its direction and its distance. Directional information is a relatively simple matter; because bats emit sound in a forward-directed beam and because their ears are positioned so as to pick up sounds in front of them, any insect that the bat detects will typically be in front of it.

□ Bearing in mind that sound takes time to travel through air, and that its intensity attenuates with distance, what two features of an echo could a bat use to determine the distance between it and an insect?

■ (i) The time delay between making a noise and hearing the echo; the greater the distance, the greater the delay. (ii) The degree of attenuation; the further away the insect, the weaker will be the echo returning from it.

The echolocation calls of bats vary, depending on whether they are just 'cruising' about or whether they are actively hunting for insect food. While cruising, a bat is moving through an environment containing large objects, such as buildings and trees, that are relatively immobile, and a low rate of call production is sufficient to provide it with an 'acoustic map' of its environment. Catching a small, moving insect, however, requires a lot more information than avoiding a house and, when a bat is hunting, echolocation calls are emitted at a much higher rate. These are known as 'buzzes' by biologists who 'observe' the behaviour of bats using high-frequency listening devices called bat detectors.

Some species of bat produce brief calls and, by using the time delay before the echo arrives, can locate an insect. Sound intensity information is less reliable than the time delay for determining distance because sound is attenuated by a number of other factors, such as the humidity and temperature of the air.

Because sound attenuates with distance, a bat must produce a very loud call in order to be certain of hearing a detectable echo. This creates a problem. If the bat hears its own very loud calls it will be temporarily 'deafened' (if its ears are not actually damaged) at the very time when it needs its ears to be maximally sensitive in order to hear the faint echo. Some species of bat have evolved a muscle called the stapedius muscle which loosens the connections between the ear bones as a call is being produced and then tightens them up immediately the call has ceased. This effectively 'switches off' the bat's ears while it is producing its own call and turns them on again so that it is maximally sensitive to the echoes of its own call. (A stapedius muscle fulfilling the same function is found in the human ear as described in Chapter 3.)

Other bats produce relatively long, less loud calls, with the result that the echo begins to arrive at the ear before the call has finished. As a result, the ear hears an overlap between the call and the echo. This might appear to be a source of confusion but in fact is not, because the call takes the form of a downward frequency sweep. The call starts at a relatively high frequency that falls during the call. Consequently, what the bat hears during the later part of its call is the frequency it is currently emitting plus a slightly higher frequency in the echo (Figure 2.16). The further away the insect is, the greater the discrepancy between the outgoing and the incoming frequencies (compare (a) and (b) in Figure 2.16).

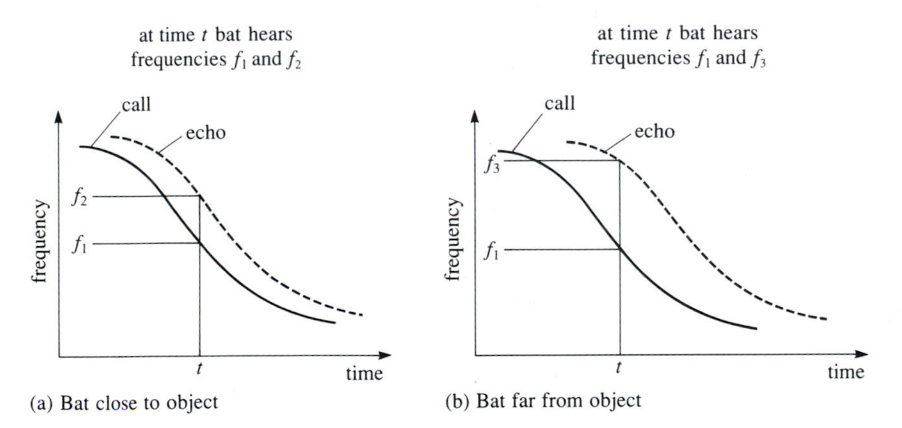

Figure 2.16 How having a frequency sweep in its call enables a bat to determine the distance between itself and an object. When a bat is close to an object (a), at a given time t, it hears frequencies f_1 and f_2. Further from the object (b), it hears frequencies f_1 and f_3.

An alternative form of echolocation has evolved in certain species, such as the British greater horseshoe bat (*Rhinolophus ferrumequinum*). Instead of producing very short pulses of sound with a frequency sweep, it produces long pulses of a constant frequency. The apparent frequency of the echo that it receives back varies with the speed of the bat. If the bat is approaching a brick wall at speed, it will receive an echo that is very different in frequency from its outgoing call. As it slows down to avoid the wall, however, the frequency of the echo falls, getting closer to that of the outgoing sound. This apparent change in frequency produced by movement is called the Doppler effect, after a scientist named Doppler who described the phenomenon after doing experiments with a brass band carried past him on a railway truck! You will be familiar with the Doppler effect if you have ever stood beside a railway line as a train passes or beside a road as a car passes. As the train goes past, its sound appears to drop markedly in frequency and, the faster the train is moving, the more marked is this change in frequency.

Now consider a horseshoe bat approaching a flying insect. Both animals are moving, so the echo that the bat hears will be shifted in frequency as a consequence of the bat's movement and will also be shifted as a consequence of the insect's movement. Thus, to locate the insect, the bat must integrate both reafferent and exafferent information (Book 1, Section 2.3). The task that the bat has in evaluating this information in creating a perceptual map of its environment seems formidable. There is not much information available about how a bat might achieve this feat. However, the greater horseshoe bat is known to be able to perceive the fluttering change in frequency that is produced in echoes by the movement of an insect's wings. This fluttering signal stands out amongst the general acoustic 'clutter' of the environment; there is a higher signal-to-noise ratio than there would be from simple echolocation pulses.

This account gives only an outline of the complexity of the echolocation systems of bats. That they have become so complex and sophisticated during evolution is due, in part, to the fact that detecting small, moving objects by means of sound cues alone is no simple task. An additional factor is that insects are not passive participants in this process; they have evolved a number of adaptations that reduce the risk of their being detected by a bat. You should recall from Book 1 (Section 1.1.2) that, because of the attenuation of sound with distance, a moth has a good chance of hearing a bat before the bat detects it. Many moths are very hairy, with the result that they do not reflect clear echoes back to the bat. Others, in response to hearing a bat, can vibrate part of their hard external skeleton, to produce a very loud sound that masks the fainter echoes of the bat's calls. Some moths are distasteful and signal the fact by giving out distinctive ultrasonic sounds; such moths are generally avoided by hunting bats. Finally, other, palatable, moth species gain protection by mimicking the sounds produced by distasteful species. This complex interplay between the adaptations shown by bats to catch insects, and those shown by insects to avoid capture, provides an example of coevolution (Book 1, Section 2.8).

2.3.4 Conclusion to Section 2.3

The examples described in this section have been selected to provide a sense of the many and diverse ways in which animals can emit, receive and process information. They show that the relationship between external stimuli, sensory systems and behaviour is much more complex and sophisticated than that suggested by the simple stimulus–response model of behaviour introduced in Chapter 2 of Book 1. Many other examples could be given to illustrate the ability of animals to make very fine discriminations between stimuli. It has been shown, for example, that, during behavioural interactions, animals can determine one another's motivational state very precisely; they can differentiate between members of their immediate social group and those of another group; they can discriminate their kin (genetic relatives) from non-kin; they can recognize one another as individuals. In non-social contexts they can discriminate between what is good to eat and what is not and between what is an enemy and what is not, despite all kinds of adaptations on the part of their food and their predators to prevent them from so doing.

These abilities are all based on the fact that the sense organs of many animals are capable of extracting very detailed information from signals and from the environment and thus of making very fine discriminations. The way that this is achieved in the context of hearing, vision and touch is discussed in the next three chapters of this book.

Summary of Chapter 2

The relatively simple ears of frogs and the very simple sounds produced by male frogs provide a good system for looking at communication at both the behavioural and the neurobiological level. In particular, this system reveals how sense organs are 'tuned' to signals of particular biological relevance to a given species. It also illustrates how communication between animals can be constrained by environmental factors, in this instance temperature effects and predation.

The second part of the chapter described a number of examples to show how the sensory systems of animals can extract very precise information from signals, based on the fact that most signals are inherently variable in form. Many birds use variations in plumage patterns in aggressive and sexual interactions such that they vary their behaviour towards different individuals; bees use variation in symbolic dances to convey information about their environment; bats use variations in the echoes they receive from their calls to determine the exact location of insects and other objects.

Objectives for Chapter 2

After reading this chapter, you should be able to:

2.1 Define and use, or recognize definitions and applications of each of the terms printed in **bold** in the text.

2.2 Recall the neurobiological features of the ears of frogs that enable them specifically to respond to calls of their own species. (*Question 2.1*)

2.3 Describe how, in the green treefrog, different parts of the auditory system are involved in species recognition when a female is at different distances from a calling male. (*Question 2.2*)

2.4 Describe what is meant by temperature coupling in the context of frog calling and hearing.

2.5 Describe how male tungara frogs alter their calling behaviour in relation to the number of males that are calling and to the presence or absence of bats. (*Question 2.3*)

2.6 Describe evidence that birds use variation in plumage patterns as 'badges' of social status and in mate choice, and discuss problems in interpreting this evidence.

2.7 Describe how bees encode distance and direction information in their tail-waggle dance.

2.8 Compare and contrast loud, short calls incorporating a frequency sweep with softer, longer, fixed-frequency calls, as a means for echolocating insects by bats. (*Question 2.4*)

Questions for Chapter 2

Question 2.1 (*Objective 2.2*)
What two neurobiological features of frogs mean that they are selectively responsive to conspecific calls?

Question 2.2 (*Objective 2.3*)
Why, as a female green treefrog approaches a calling male, is her amphibian papilla stimulated before her basilar papilla?

Question 2.3 (*Objective 2.5*)
How will the calling behaviour of male tungara frogs differ when (a) there are few calling males and bats are active, and (b) there are many calling males and bats are absent?

Question 2.4 (*Objective 2.8*)
How can a greater horseshoe bat that is approaching an insect differentiate between one that is approaching it and one that is flying away from it?

References

Gerhardt, H. C. (1978) Temperature coupling in the vocal communication system of the gray tree frog, *Hyla versicolor, Science,* **199**, pp. 992–994.

Møller, A. P. (1987) Variation in badge size in male house sparrows *Passer domesticus*: evidence for status signalling, *Animal Behaviour,* **35**, pp. 1637–1644.

Møller, A. P. (1988) Badge size in the house sparrow *Passer domesticus*: effects of intra- and intersexual selection, *Behavioural Ecology and Sociobiology,* **22**, pp. 373–378.

Møller, A. P. (1990) Sexual behaviour is related to badge size in the house sparrow *Passer domesticus*: *Behavioural Ecology and Sociobiology,* **27**, pp. 23–29.

Ryan, M. J. (1985) *The Tungara Frog,* University of Chicago Press.

Further reading

Halliday, T. R. and Slater, P. J. B. (eds.) (1983) *Animal Behaviour*, *Vol 2*, *Communication*, Blackwell Scientific Publications.

CHAPTER 3
HUMAN HEARING AND HUMAN SPEECH

3.1 Introduction

Sound communication between animals is often very conspicuous to the human observer and so there has been a substantial amount of work carried out on, for example, bird song and frog calls, as you saw in Chapter 2. However, advanced technology has made it possible to investigate sounds animals produce that are beyond human hearing and it is clear that these sounds have great significance. Bats use high-frequency sound for navigation and for detecting prey; whales use very low-frequency sound to communicate over distances of hundreds of miles underwater. Humans rely on the sense of hearing for a lot of communication and in this chapter we shall consider human hearing and speech. In doing so you will draw on your knowledge of neurophysiology and brain structure derived from Book 2.

The chapter is divided into four sections. Section 3.2 describes the structure of the ear in humans and the process by which sound is detected, converted to neural information and passed to the brain. We then turn to the production of speech, in particular the relationships between the shapes of the vocal tract (the structures that generate speech) and the type of speech sounds they produce, and the control of the vocal tract movements required for fluent speech. Section 3.3 is concerned with speech perception, beginning with the peripheral analysis of speech sounds carried out by the auditory system, through phonetic processing, to the recognition of words. The final short section focuses on understanding language.

Much more information about our surroundings comes from the visual system than the auditory one. Light waves are much better at providing detailed information because, as you will learn in Chapter 4, light has a very small wavelength and therefore very great resolution (the ability to distinguish small objects). By comparison, using echoes, most of us can detect cliffs and large buildings but, without training, very little else. The area of the brain devoted to processing visual information is substantially greater than that devoted to processing auditory information. This comparison is not intended to diminish the importance of the auditory sense. For example, deafness has profound effects upon people who suffer from it. The chief loss in deafness is the ability to communicate in real time—writing and reading take much longer than speech. So, this description of human hearing and speech concentrates on the communication aspects of hearing and sets aside other uses to which humans put the sense of hearing.

3.2 Hearing

The human vocal apparatus and ear together form a matched sound communication system. Most of the acoustic power generated by the voice lies in a frequency band extending from approximately 200 Hz to 5 kHz (Hz and kHz were explained in Section 1.2.3). The ear has its greatest sensitivity over the same frequency interval with a maximum sensitivity in the range 1.5–3 kHz, although the range of hearing is greater, reaching 18 kHz or more, in some young people. In this section the anatomical structures and the physiological mechanisms responsible for the acoustic properties of the ear will be examined. The basic properties of sound are described in Box 3.1.

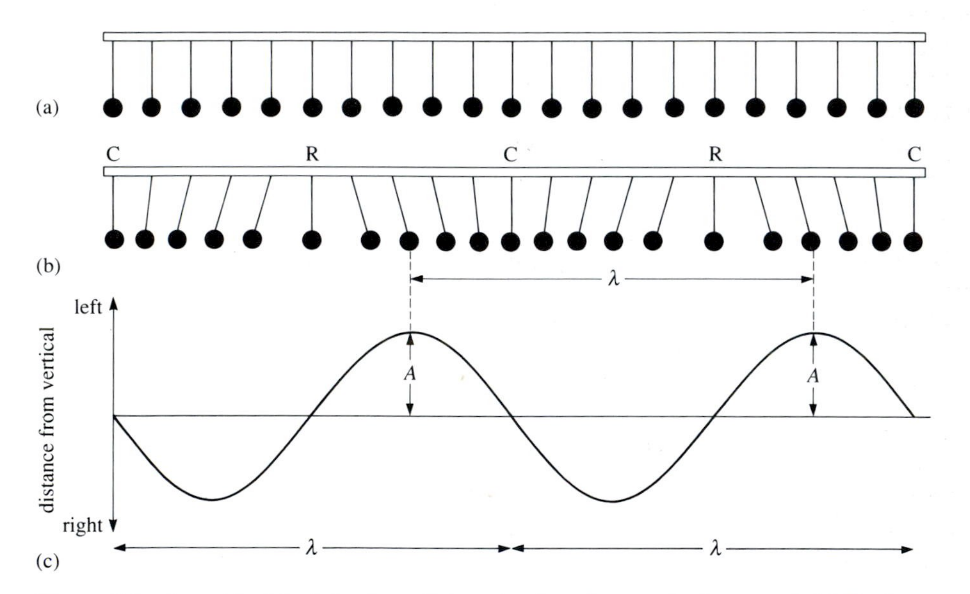

Figure 3.1 A model of the behaviour of air molecules when conducting sound. (a) Steady state. (b) A wave travelling along the line of balls. C marks an area of compression and R marks an area of rarefaction. (c) A plot of the distance moved by a ball against time. A shows the maximum displacement of the ball (the amplitude) and λ shows the distance between one peak of amplitude and the next (the wavelength).

Box 3.1 Physical characteristics of sound

Sound energy is transmitted through air by vibration of air molecules. Imagine a row of balls representing air molecules, suspended on a rod (Figures 3.1a and b). Swing the first ball and release it and it will contact the next one, setting it in motion. This in turn will set the next in motion and after a short time the last ball in the row will be moving backwards and forwards. This movement creates areas where the balls are very close (C in Figure 3.1) and areas where the balls are widely separated (R). These are equivalent to areas of *compression* of air molecules and areas of *rarefaction* (decompression) of air molecules.

The graph in Figure 3.1c shows a plot of the distance each ball has moved from its resting point, against time. This gives a wave-form that is similar to a sound wave-form. A represents the maximum displacement of the ball (called the amplitude) and λ represents the wavelength (the distance between one peak of amplitude and the next). The number of peaks that are generated each second is the frequency—for example 5 peaks per second (5 Hz).

If the rod with the balls on was very long indeed, the amplitude of the movement of the balls would gradually decrease along the rod due to friction between the balls, and hence loss of energy (attenuation). This happens as sound travels through air, which is one reason why distant sounds are faint.

☐ Can you think of another reason why distant sounds are faint?

■ Sound does not travel in straight lines. It radiates away from a source like ripples on a pond radiating away from a stone splash. So the sound energy generated by the loudspeaker in a television spreads out to fill the room with sound but the sound energy per unit area of wall is much less than the sound energy per unit area of loudspeaker.

There are three characteristics of sound that influence what humans perceive when they hear it, each of which matches one of the physical characteristics described in the previous paragraphs—the loudness or amplitude, the pitch or frequency and the tone quality or wave-form. In addition, however, the performance of the ear affects what is heard. Humans do not hear equally well at all frequencies so, if two frequencies of the same amplitude are played, one may sound louder than the other due to the characteristics of the ear.

A sound may be made up of a range of frequencies—like pressing all the notes on the piano at once! The range of frequencies is referred to as the frequency *spectrum* of the sound, similar to the light spectrum in sunlight that is revealed when you see a rainbow.

3.2.1 The ear

Figure 3.2 shows a section of the human ear illustrating the main structures which make up the organ. The ear consists of two discrete parts. The first part comprises the outer and middle ear whose function is to ensure effective transfer of sound wave energy in the external air into sound wave energy in the fluids contained in the inner ear. The inner ear, or **cochlea**, makes up the second part of the ear. The cochlea contains the transduction apparatus which converts the sound waves in the cochlear fluids into action potentials passing up the auditory nerve (part of the 8th cranial nerve) to the brain.

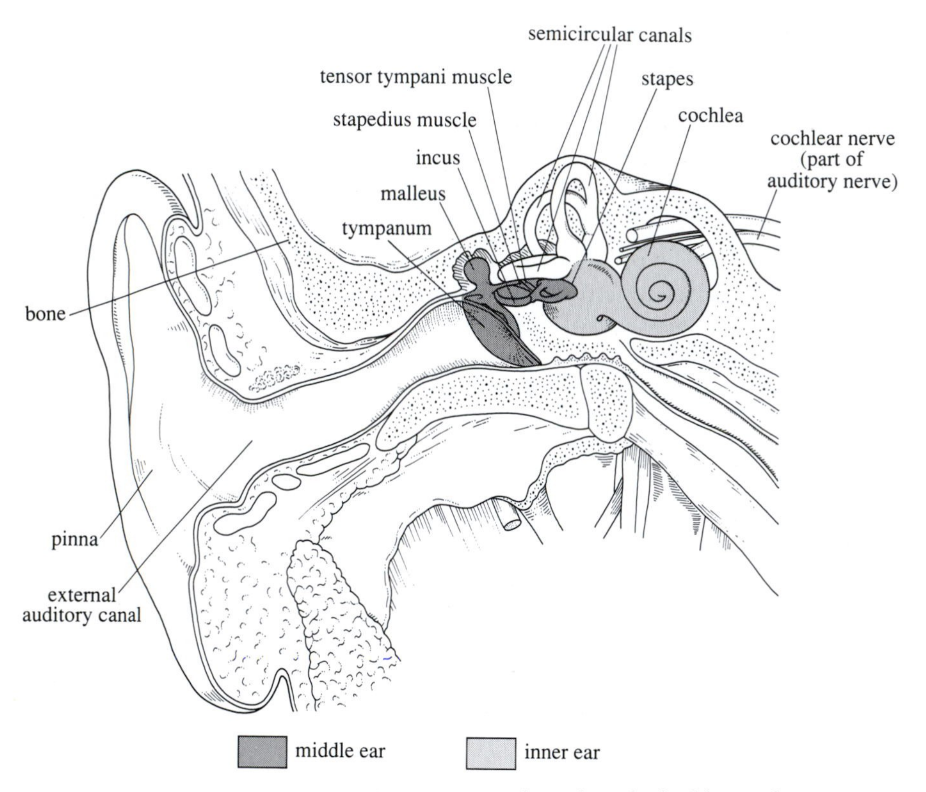

Figure 3.2 A section through the human ear to show the principal internal structures. (Source: *Tissues and Organs: A Text-Atlas of Scanning Electron Microscopy*, by Richard G. Kessel and Randy H. Kardon. Copyright © 1979 by W. H. Freeman and Company. Reprinted by permission.)

The ear is not simply a mechanical device that changes acoustic information into neural information. The ear is the first stage in a complex information transduction and filtering system. Recent research has clearly shown that the ears are much more than a pair of sensitive microphones attached to the sides of the head. The cochlea is able, like the retina of the eye (Chapter 4), to pre-process the incoming auditory information and convert it to a form for further processing by the auditory cortex of the brain. The details of this pre-processing are not yet fully understood.

3.2.2 The outer ear

The outer ear consists of a flange of skin-covered cartilage, the *pinna*, and the external auditory canal leading to the *tympanic membrane* or **tympanum**, known in everyday language as the eardrum (Figure 3.2). The convolutions of the pinna are important for coupling sound waves in the external air to the sound waves that are generated in the external auditory canal. This coupling depends on the frequency and direction of the incident sound and provides each ear on its own with a certain amount of directional sensitivity. At high frequencies the wavelength of the sound is small, compared with the size of the pinna, and considerable directionality is present. In cats, the pinna can amplify the sound if a high-frequency sound source is in line with the axis of the pinna. Humans use low frequency predominantly (and hence long wavelength), so the pinna has only a limited value in directionality. However, experiments have shown quite good localization of sound by human subjects using only one ear and the importance of monaural sound localization may be greater than laboratory experiments have shown.

□ How do you suppose that humans get directional information about sound sources?

■ Directional information can be derived from the difference in signals from the two ears. A sound coming from one side of the head will reach one ear before the other, for example. This is called binaural processing.

□ Where in the nervous system might binaural processing take place?

■ One obvious site is the auditory cortex. However, the earliest stage at which binaural processing can be carried out is in the superior olivary nucleus, which is situated in the lower part of the pons, and many other sub-cortical nuclei have binaurally-driven cells.

The brain areas and pathways associated with hearing will be described in Section 3.3. Before describing these it is necessary to understand the basic operating principles of the ear and the way in which sound is converted to action potentials which code information that the brain can process.

3.2.3 The external auditory canal

Look again at Figure 3.2. The external auditory canal runs from the pinna into the bones of the skull. The canal is relatively straight, about an inch long and a quarter of an inch wide and at the end is an elastic membrane, the tympanum. Along the canal are fine, outwardly pointing hairs which, together with a wax secretion, keep dust and insects at bay.

The tympanum and canal act together as an amplifier that generates an increase, or *gain*, in sound intensity (see Box 3.2, *overleaf*) at the tympanum (approximately 20 dB) over the important frequencies of speech (500 Hz–5 kHz). This frequency-dependent gain only operates for sound in free space, that is, a sound source at a distance from the ear with nothing but air intervening.

Box 3.2 Measuring sound intensity

Sound intensity is a measure of the amount of sound energy striking a particular area over a particular length of time. Intensity is measured in watts per square metre ($W\,m^{-2}$). The human ear can detect sounds of intensity levels as little as one millionth of a millionth of a watt per square metre, whereas the loudest tolerable sound has an intensity of $1\,W\,m^{-2}$. This is a huge range and so a logarithmic scale is used for intensity—the decibel scale (dB), named after Alexander Graham Bell, the inventor of the telephone. On this scale 0 dB is taken as the threshold of human hearing, which gives a loudest tolerable level of 120 dB. Although 120 dB is *tolerable*, damage to the hair cells in the ear will occur. Lower intensities than this can cause damage—temporary hearing impairment has been measured for up to three months after a single evening at a rock concert.

Humans do not actually perceive all sounds that are at the same intensity to be equally loud. A sound at a frequency of 1 000 Hz and an intensity of 40 dB is audible but a sound at 60 Hz and 40 dB is not. This relationship between frequency and intensity for the human ear is illustrated in Figure 3.3, which shows the auditory characteristics of the human ear. The Figure also relates these characteristics to sounds that you might encounter. The curved lines on the graph show lines of equal *perceived* loudness.

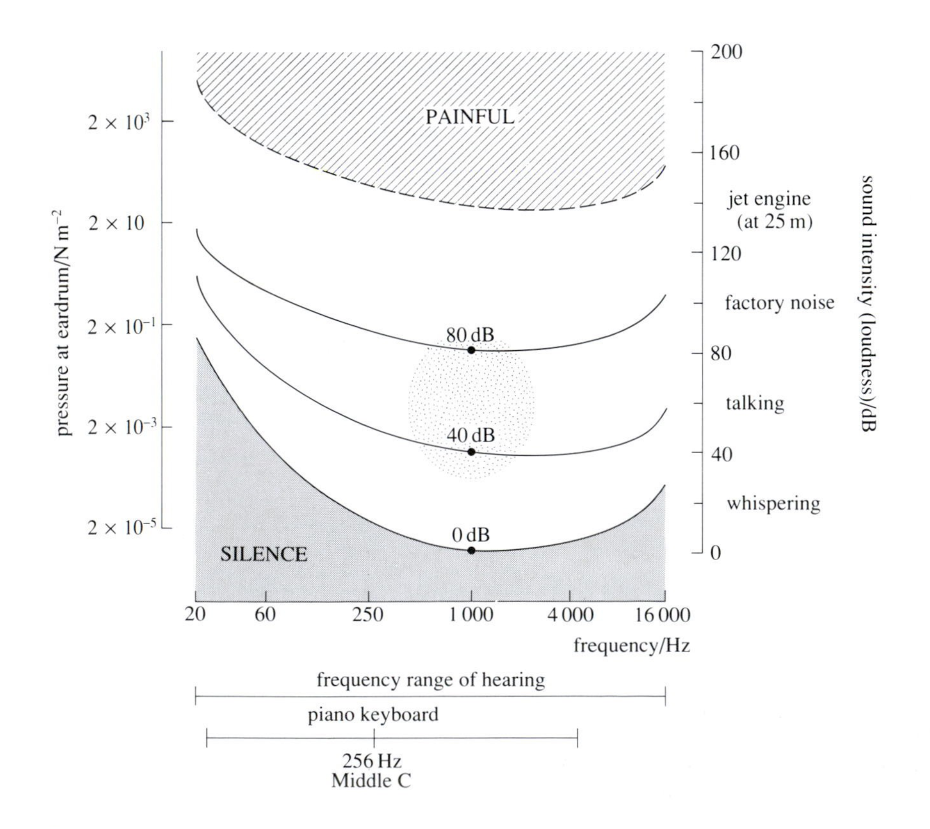

Figure 3.3 Graph to show some characteristics of human hearing. The stippled area shows the approximate range of frequencies and intensities of sound occurring in speech. The pressure at the eardrum is measured in newtons per square metre ($N\,m^{-2}$). The newton is a measure of force.

□ Can you think of any conditions under which this frequency-dependent gain would be low or absent?

■ When wearing earphones which project into the ear canal. The frequency-dependent gain is much reduced for sound heard through these earphones, which is why music heard through them can sound so different from the same music played through a conventional loud speaker.

3.2.4 Frequency response of the tympanum

The tympanum is a cone-shaped elastic membrane attached at its edge to the internal end of the external auditory canal. The tympanum bows in and out as sound strikes it so that, within limits, it moves at the same frequency as the sound waves. Attached to the inner surface of the tympanum is an elongated part of a small bone called the malleus (meaning hammer). This is the first of three bones, the **ear ossicles**. The attachment of the remaining two ossicles, the incus and the stapes, is shown in Figure 3.4. Together, these three bones couple the tympanum to a second, smaller membrane called the *oval window*. The air contained in the external auditory canal and the mass of these three bones act as a load attached to the elastic tympanum. As the tympanum moves in and out, so the stapes will move back and forth, driving the oval window in and out.

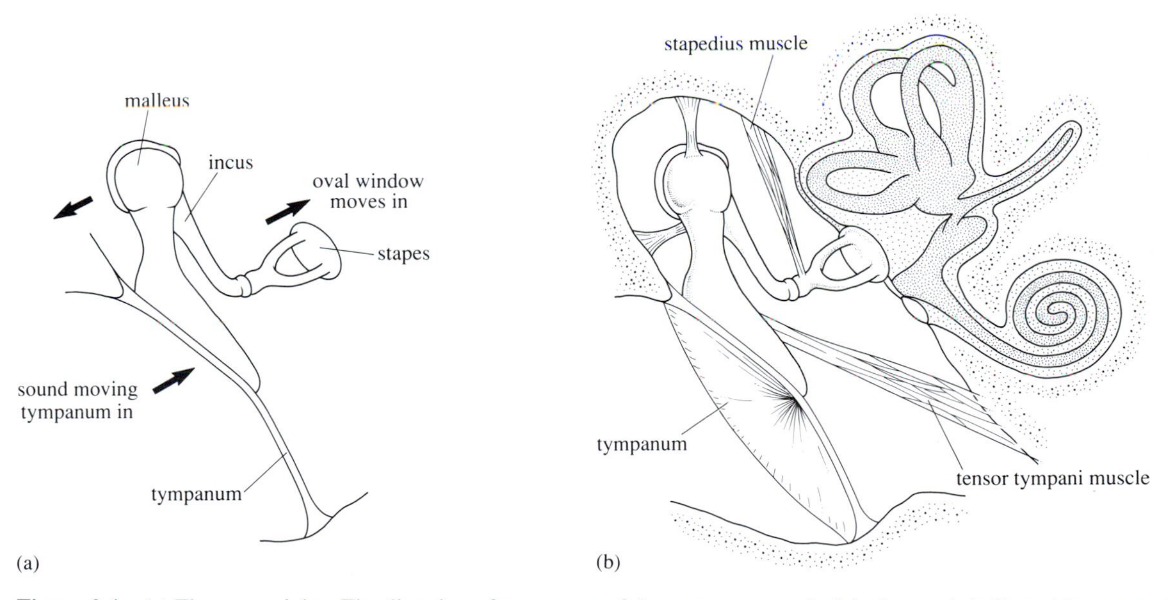

Figure 3.4 (a) The ear ossicles. The direction of movement of the tympanum, and of the bones, is indicated by arrows for the arrival of a sound pressure wave at the tympanum. (b) The position of the stapedius muscle and tensor tympani muscle. The semi-circular canals and cochlea are also shown.

The principal physiological function of the tympanum and ossicles is to match precisely the load on the tympanum, which results from the air mass in the external canal, to the load on the end of the stapes created by the fluid that lies behind the oval window. This matching ensures that energy transfer from the tympanum to the ear ossicles is achieved with minimum loss. The mechanical arrangement of the ear ossicles provides amplification. The system operates predominantly by virtue of the difference in area of the tympanum and the end of the stapes that contacts the oval window—the *stapes footplate*. The total pressure gain across the middle ear is approximately 22 : 1.

Acoustic power transmission through the middle ear can be modified by the activity of the middle ear muscles (see Figure 3.4). This helps to counteract the effect of very loud sounds. A muscle called the tensor tympani (Figure 3.4b) pulls on the malleus and increases the tension on the tympanum, increasing its resistance to movement. The stapedius muscle (Figure 3.4b) pulls on the neck of the stapes, immobilizing the movement of the stapes footplate. The stapedius muscle contracts reflexly in response to sound with a pressure exceeding 80 dB received by either ear. This reduces the sound energy reaching the cochlea. These reflexes are also activated prior to speaking, effectively preventing self-stimulation. (The stapedius muscle fulfils a similar function in bats; Section 2.3.3.) Conversely, the faint sounds of somebody whispering can be enhanced by mental concentration on listening to the speaker. The effect of the concentration is that the stapedius muscle is relaxed and sound energy transfer to the cochlea is increased.

3.2.5 Cochlea mechanics

The piston-like movements of the stapes footplate produce pressure waves in the fluid of the cochlea, behind the oval window. In Figure 3.4b there are three semicircular canals, the function of which is to provide information about position and movement, and the coiled cochlea, whose function is auditory.

The cochlea is a coiled tube with two and three quarter turns. If you were to 'uncoil' it, it would look something like Figure 3.5a. The fluid (perilymph) in two of the canals, the *scala vestibuli* and the *scala tympani*, is continuous. Figures 3.5b, c and d show that the internal structure is highly complicated. There are three fluid-filled canals. As the oval window moves in and out in response to the movement of the stapes bone, pressure waves are set up in the fluid. These are transmitted to the *basilar membrane,* causing it to ripple. Between the basilar membrane and a second membrane, the *tectorial membrane* (Figure 3.5d) are sensory cells called the auditory **hair cells**. These are mechanoreceptors (Book 2, Section 2.4.1). As the basilar membrane ripples, the hair cells are stressed and the tips of the hairs attached to the tectorial membrane are distorted. The hair cells connect with neurons of the cochlear nerve which convey the information about incoming sounds to the central nervous system (CNS).

Low frequency sound produces maximum displacement of the basilar membrane at the apical end of the cochlea (i.e. the end closest to the helicotrema) and high-frequency sounds produce maximum displacement at the other end. The size of the basilar membrane movement is extremely small. High intensity sound may produce displacements of 10–100 nm (1 nm is 1 thousand millionth of a metre). At sound pressures very close to the threshold of hearing the movements are of atomic dimensions. In a completely silent and echoless environment you could, *theoretically*, hear nothing but the movement of molecules in the basilar membrane. However, the very slight noise that can sometimes be heard under such conditions is the noise of blood flowing through blood vessels near the ear.

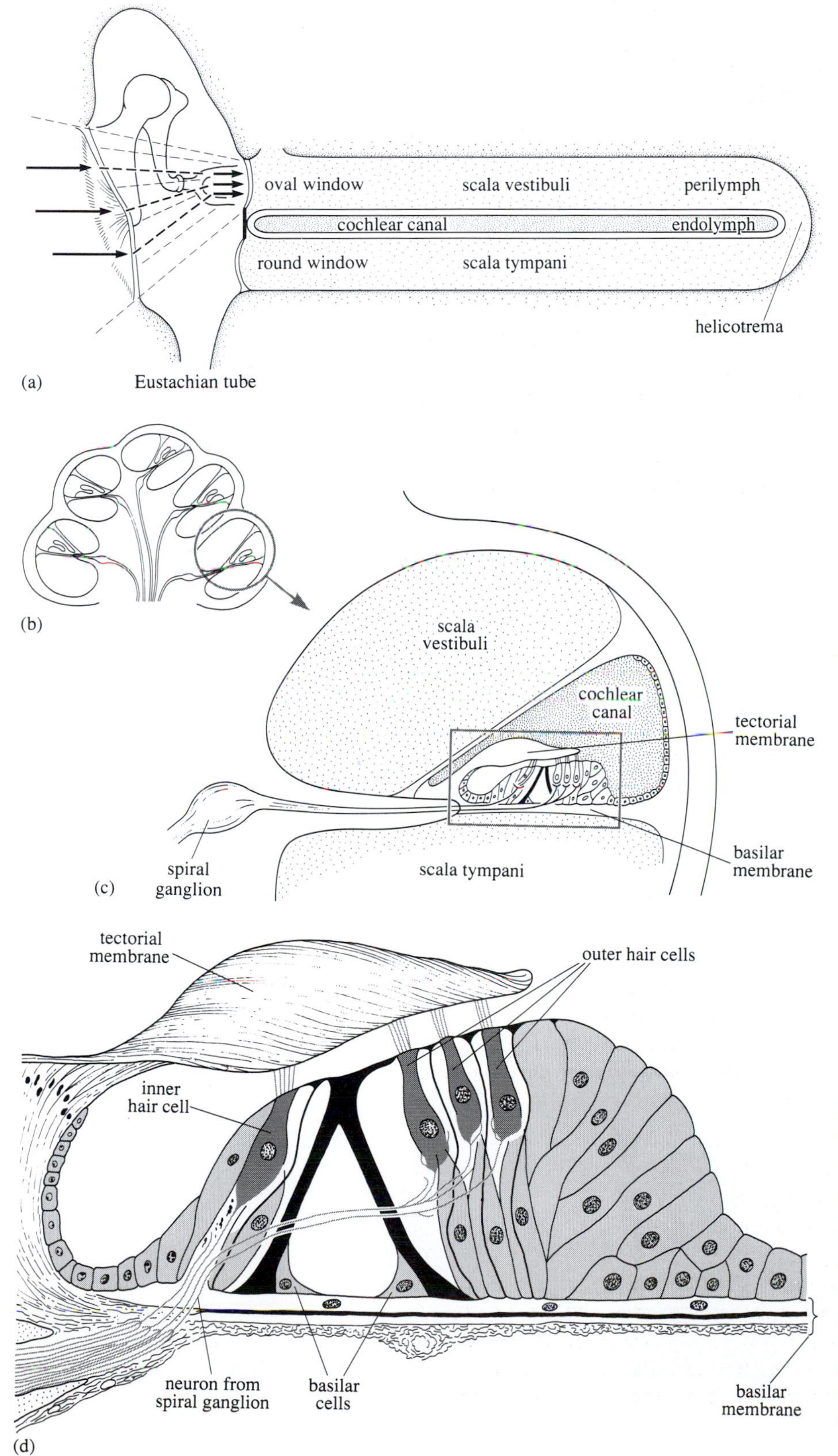

Figure 3.5 (a) Diagrammatic representation of an uncoiled cochlea showing how the fluid is continuous from the oval window to the round window. (b) Transverse section through the cochlea to show the coiled structure. (c) Enlargement of a single coil to show the hair cells and the two membranes. (d) A further enlargement to show the attachment of the hair cells to the membranes. The auditory hair cells are positioned between the basilar membrane and the tectorial membrane.

3.2.6 Responses of auditory hair cells

The auditory hair cells are subjected to forces as the basilar membrane vibrates and the hairs bend. The bending depolarizes the hair cells. This depolarization is a receptor potential and it causes a release of transmitter that generates an action potential in a neuron in the cochlear nerve (Section 1.2.1).

The auditory hair cells in the cochlea are shown diagrammatically in Figure 3.5 and in scanning electron micrographs in Figure 3.6. The response characteristics

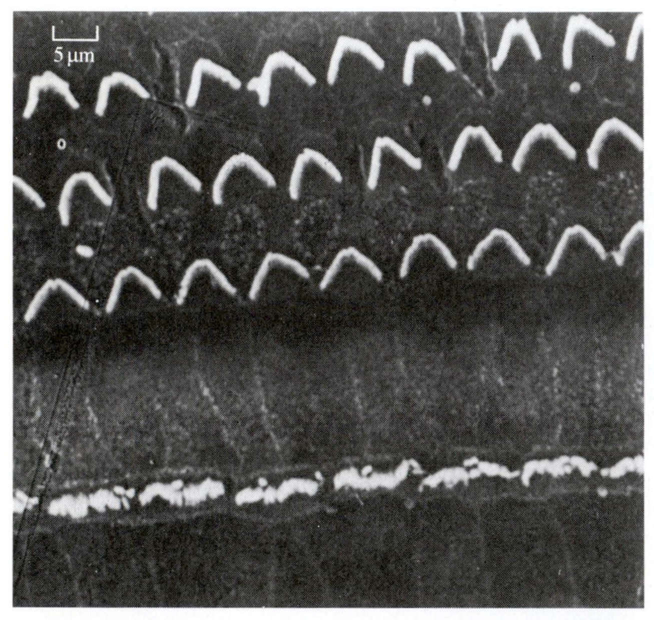

(a)

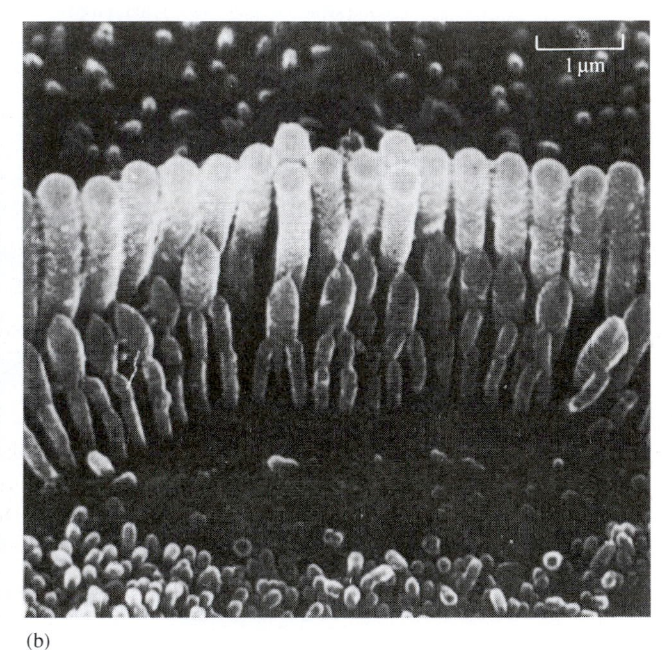

(b)

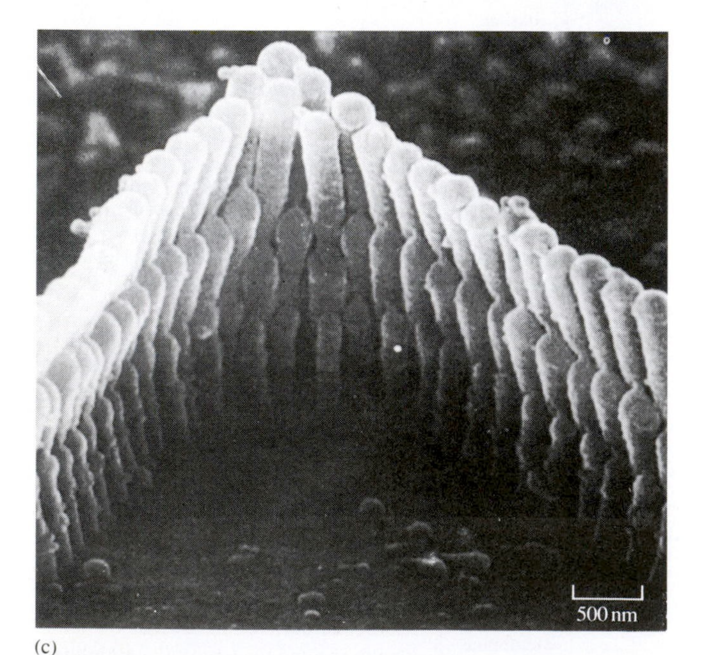

(c)

Figure 3.6 Scanning electron micrographs of auditory hair cells. (a) Three rows of outer hair cells and one row of inner hair cells in the cochlea of the guinea-pig. (b) Higher powered view of inner hair cells. (c) High powered view of outer hair cells.

of these hair cells can be measured by recording potentials from the cochlear nerve. Figure 3.7 illustrates the frequency response characteristics of a population of single cochlear neurons originating from different points along the length of the cochlea of the ear of the guinea-pig (*Cavia porcellus*). All the units recorded show a maximum sensitivity (black dot on each curve) at a specific frequency (the best frequency; Section 1.2.2). This means that they act as part of an acoustic filter, responding only to a limited range of the stimulating sound, like a radio that is tuned to a particular frequency. Cochlear neurons tuned to a low frequency arise from hair cells at the apical end of the basilar membrane and those tuned to a high frequency arise from hair cells at the basal end (close to the tympanum) (Figure 3.5a). The shapes of the cochlear nerve tuning curves match closely the frequency response characteristic of *inner* hair cells. Only about 10% of the neurons in the cochlear nerve make synaptic contact with the outer hair cells. Similar response curves, characteristic for the ear of the particular species, have been found in other animals such as frogs (Section 1.2.3). The response curves for human hair cells are assumed to be similar. From these observations the following conclusions can therefore be drawn.

(i) Inner cochlea hair cells are part of a remarkably sharply tuned acoustic filter.

(ii) The hair cells are organized along the length of the cochlea such that high-frequency units are at the basal end and low-frequency units are at the apical end.

(iii) The cochlea appears to behave like a vast organized array of parallel, sharply tuned acoustic filters. Complex sounds are analysed in terms of the individual frequency components which make up the sound. The frequency is then coded in terms of position along the cochlea and the time pattern of the neural impulses.

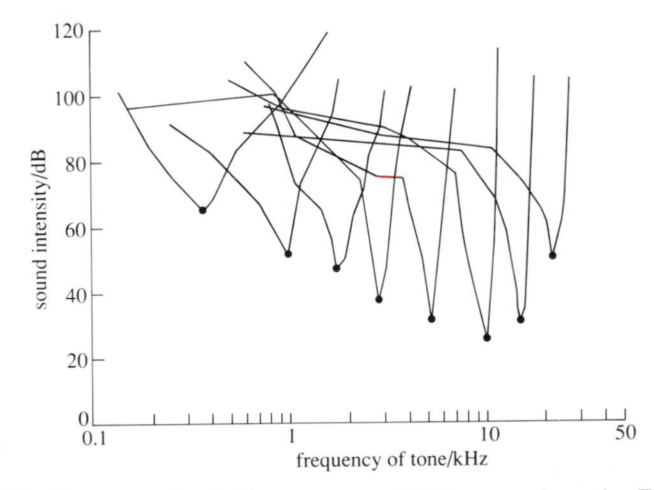

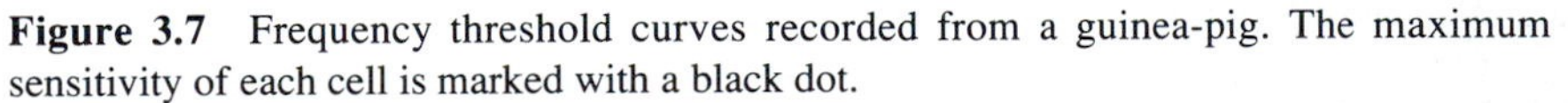

Figure 3.7 Frequency threshold curves recorded from a guinea-pig. The maximum sensitivity of each cell is marked with a black dot.

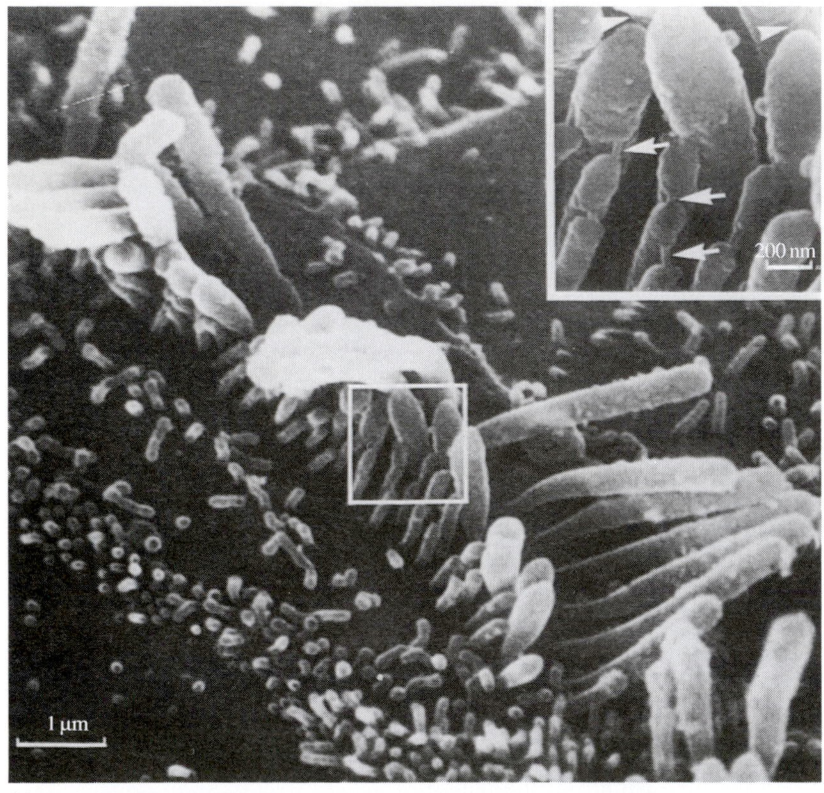

(a)

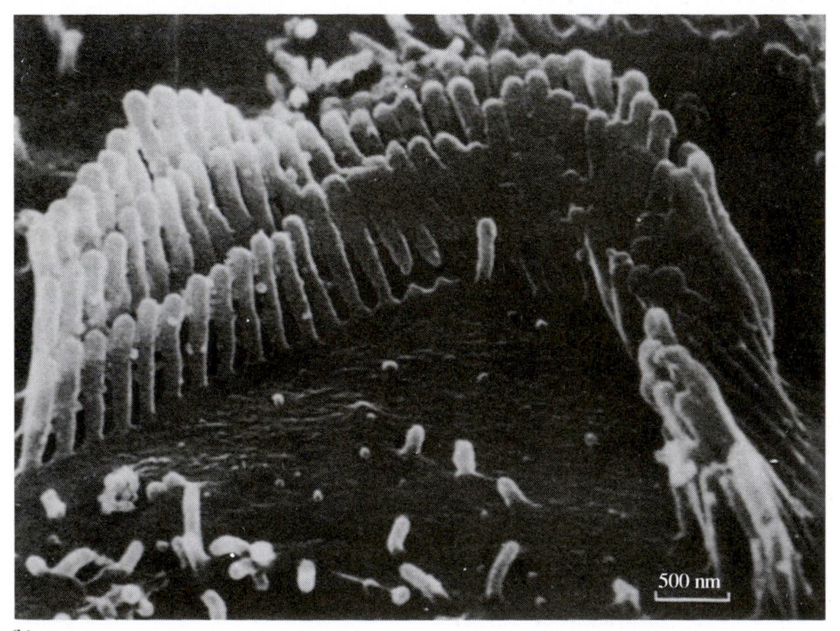

(b)

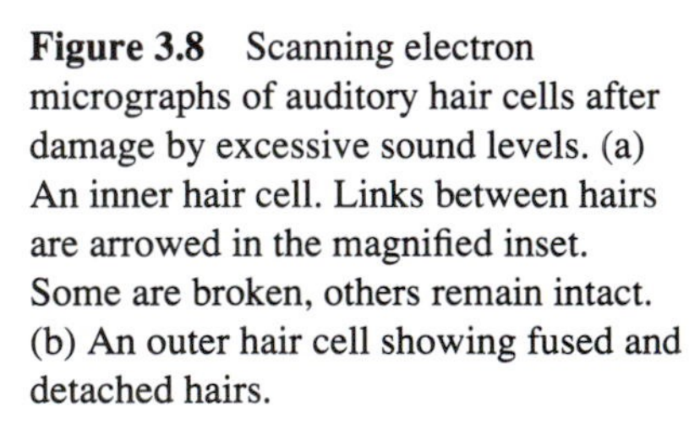

Figure 3.8 Scanning electron micrographs of auditory hair cells after damage by excessive sound levels. (a) An inner hair cell. Links between hairs are arrowed in the magnified inset. Some are broken, others remain intact. (b) An outer hair cell showing fused and detached hairs.

3.2.7 Hearing damage

The very fine hair cells are susceptible to damage. Figure 3.8 shows the results of exposure to high intensity sound. The hair cells become detached or fused—compare this with the undamaged ones in Figure 3.6. Some drugs, such as certain antibiotics, can also damage hair cells. Hair cell damage is generally irreversible.

Summary of Section 3.2

The human range of hearing is around 20 Hz to 20 kHz, wider than the range of linguistically-relevant frequencies in the human voice. The ear pre-processes sound information before it reaches the auditory cortex of the brain. The pinna of the outer ear provides limited directionality. The ear ossicles and tympanum couple the external air to the fluid in the cochlea, providing a pressure gain of 22 : 1. As the pressure moves through the fluid in the cochlea it produces movement in the basilar membrane. This vibrates the hair cells, which produces receptor potentials in them. These generate action potentials in axons of the cochlear nerve.

3.3 Auditory pathways in the brain

In Section 3.2.6 you read how the movement of hair cells produced action potentials in axons in the cochlear nerve and how individual neurons had different frequency characteristics. This section provides a brief description of the auditory pathways in the brain.

The cell bodies of the sensory neurons associated with the hair cells are located in the spiral ganglion (Figure 3.5c). Their peripheral processes connect to hair cells. Their central processes form the cochlear nerve, which is part of the 8th cranial nerve (Figure 3.9, *overleaf*). Each axon in the cochlear nerve divides into two at the ventral cochlear nucleus (Figure 3.10, *overleaf*). One branch synapses with cells in this nucleus while the other branch passes on to the dorsal cochlear nucleus. From the ventral cochlear nucleus axons pass across to the superior olivary nucleus on the opposite side (Figure 3.10) and then up to the inferior colliculi. There are also fibres from the cochlear nuclei which go direct to the opposite inferior colliculus. Finally, there are fibres which pass from the ventral cochlear nucleus up to the inferior colliculus on the same side via the superior olivary nucleus. Fibres interconnect the two colliculi. These interconnections provide the basis of directional processing for sound localization. Axons from each inferior colliculus pass via the medial geniculate nucleus of the thalamus to the auditory cortex.

Most of the research on the functioning of the auditory cortex has been carried out using cats since this area is exposed on the surface of their brain. In primates the auditory cortex is hidden and much less accessible. The research has shown that there is a relationship between the frequency of sound and activity in particular areas of the cortex. The auditory cortex is said to show **tonotopic organization**. Figure 3.11a (*overleaf*) shows a side view of the cortex of the cat with some auditory areas marked upon it. Figure 3.11b shows the primary auditory cortex, area AI, and the relationship between frequency and position of the cells. The cells in area AI are very sharply tuned to particular frequencies and this has enabled a map to be produced showing lines of cells that all respond to a particular frequency.

The mapping of sound frequency is not the same at all levels of the auditory pathway. At the cochlea a single frequency is represented by a small population of hair cells. In the cochlear nuclei it is represented by a two-dimensional sheet of cells. In the cortex, there is a one-dimensional strip of cells to represent the frequency, and there are several areas of the cortex where such strips of cells are found.

Behavioural studies of the auditory cortex have shown that it is involved in many functions. Usually, experiments have been of the lesioning type, comparing behaviour in normal animals with that of animals with a lesion (tissue removed or destroyed) in known parts of the cortex. The results from such experiments show that, contrary to expectation, functions such as frequency discrimination are *not* totally abolished by lesions in the cortex. In fact it appears that cortical lesioning upsets performance of auditory tasks that were initially difficult to learn but does not necessarily abolish the ability to process sound, at least in a rudimentary way. There remain whole areas of ignorance about the way in which the auditory cortex functions but the tonotopic organization of the auditory areas makes it a very attractive area of study.

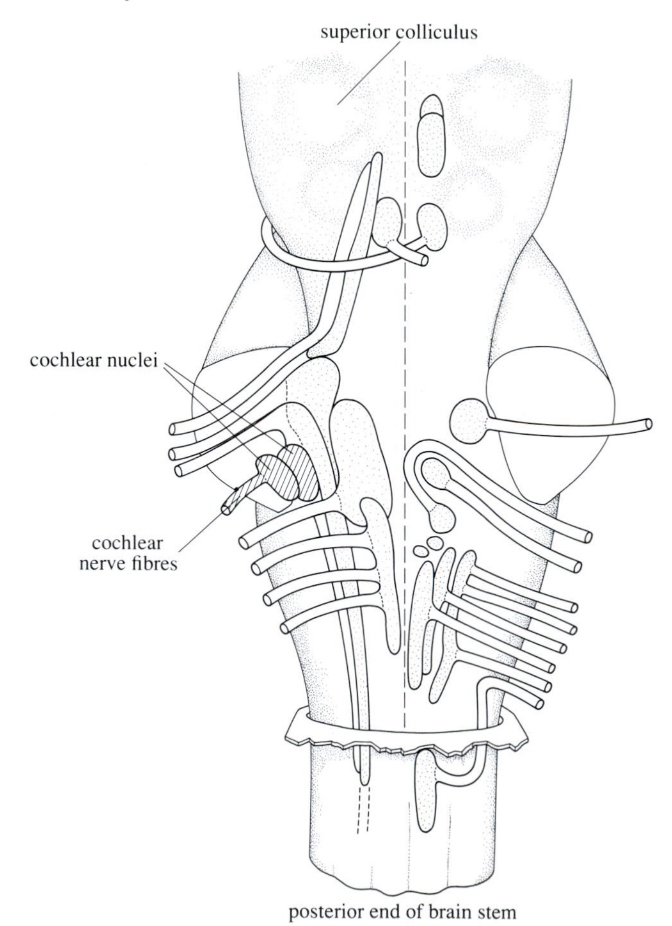

Figure 3.9 Cranial nerves and their nuclei (dotted) with the cochlear nerve fibres and cochlear nuclei cross-hatched. Only the sensory nerves and nuclei are shown on the left side; only the motor nerves and nuclei are shown on the right.

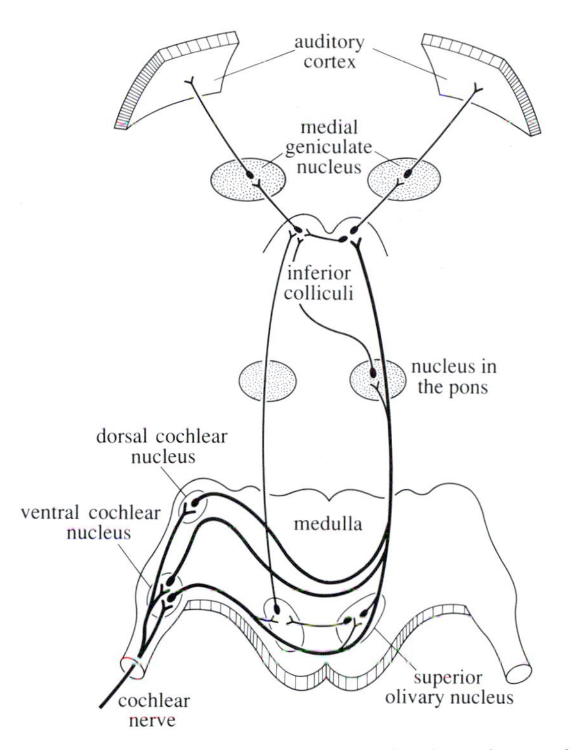

Figure 3.10 The auditory system in the brain. Note that the pathways shown are those of nerves, not of individual axons; where 'synapses' are shown, all the axons in the nerve terminate at that point.

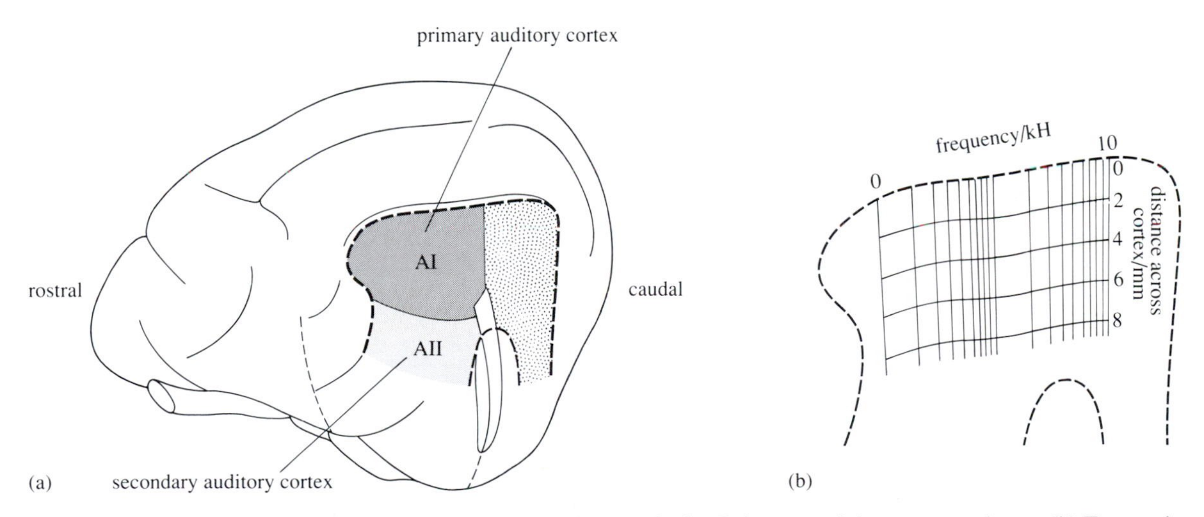

Figure 3.11 (a) Auditory areas on the cortical surface of the cat brain. Only some of the areas are shown. (b) Tonotopic organization of auditory area AI showing lines of cells that respond to the same frequency.

Summary of Section 3.3

Individual hair cells are in contact with the neurons that form the cochlear nerve. The axons of the cochlear nerve enter the brain stem and send branches to both the dorsal and ventral cochlear nuclei.

Axons from the ventral cochlear nucleus cross to the superior olivary nucleus on the other side of the brain stem and then pass up to the inferior colliculus.

Axons from the inferior colliculus connect it to the auditory cortex via the medial geniculate nucleus.

The role of the auditory cortex has been investigated both behaviourally and neurophysiologically. Cells in the cortex are arranged in auditory areas and, within some of these areas, single cells are tuned to specific frequencies. The cells are organized in a tonotopic fashion. Lesioning experiments in the cortex do not totally abolish the ability to process sound.

3.4 Production of speech

Humans are capable of producing a wide range of sounds, differing, for example, in loudness, pitch, timbre and rhythm. A particular spoken language uses a relatively small subset of these sounds to carry linguistic information. The objective of this section is to focus on how these sounds are created, how the different speech sounds are distinguished and how the sequences of sounds that characterize fluent speech are produced.

3.4.1 Vocal tracts and simple speech sounds

The peripheral structures that generate human speech (vocal cords, pharynx (throat), soft and hard palate, tongue, teeth, jaw, lips, nasal passages, etc.) are referred to as the **vocal tract**, and the larger moving parts (lips, tongue and jaw) as the major **articulators** in the vocal tract. All these structures are visible in the section through the vocal tract shown in Figure 3.12a. Broadly speaking, the talker has control over two inter-related aspects of the state of the vocal tract—its overall shape (which will include the disposition of the major articulators within the tract), and the type and amount of sound energy that is created and fed into the tract.

The sounds which comprise normal speech can be roughly divided into two types. The first type is the **voiceless sound**, produced with no vibration of the vocal cords. An example is 'ss', produced by forcing air through a narrow space between the tongue and the palate of the mouth. The second type is the **voiced sound** such as a vowel. The generation of voiced sounds requires the vocal cords and structures of the upper airways. It is on the generation of this second type of sounds that this chapter concentrates.

The generation of voiced sounds requires three basic components:

1 a steady pressurized air supply from the lungs;

2 an acoustic vibrator, the larynx, the function of which is to excite the air molecules contained in the upper airways. The larynx is situated at the upper end

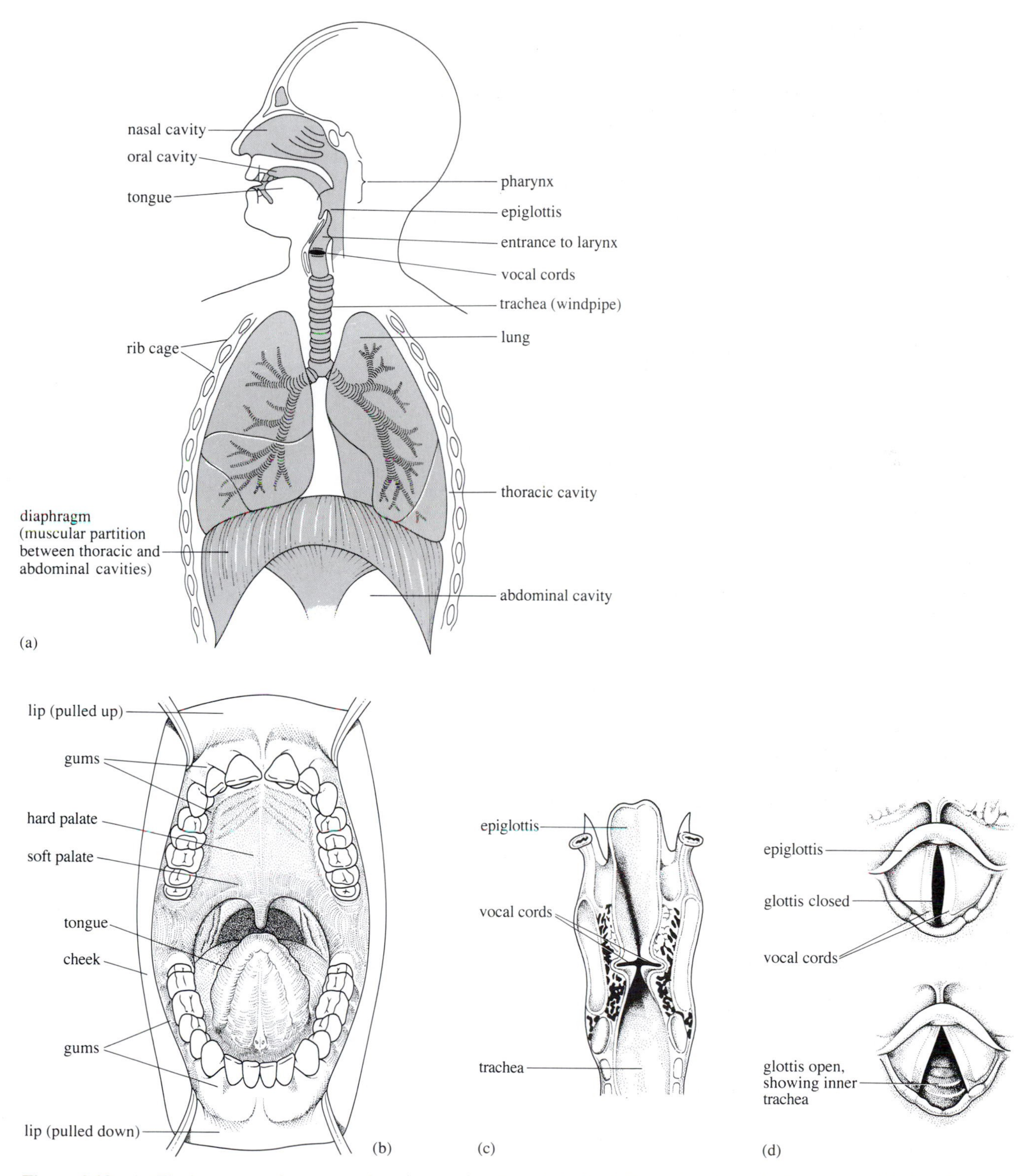

Figure 3.12 (a) The human vocal tract, seen here in a section through the head. Power for the production of sound derives from the positive air pressure provided by the lungs. (b) The mouth showing the position of the hard and soft palates. (c) Transverse section through the larynx, showing the position of the vocal cords. (d) The vocal cords viewed from above with the space between them (the glottis) open or closed.

of the respiratory tract. Figure 3.12b shows a view down the upper airway of the larynx. Prior to vocalization the vocal cords come into contact, the tension in the cords being controlled by the intrinsic and extrinsic laryngeal muscles. During vocalization the pressure of air from the lungs forces the vocal cords apart. Air flows through the space between the cords (the glottis) and the air pressure between the cords falls. This allows the vocal cords to close, shutting off the air flow, and the cycle repeats. The rate and duration of the cycle of opening and closing of the vocal cords is determined by the tension in the vocal cords and the pressure of the air supplied by the lungs. The pulses of air through the vocal cords so generated excite the air in the upper airways, causing it to vibrate;

3 the upper airways act as acoustic resonators, that is the air they contain vibrates maximally at a particular frequency—their resonant frequency. As a result, the resonant frequency is amplified. Of the upper airway resonators, the most important is made up of the upper vocal tract including the mouth. The resonant frequencies can be changed by altering the volumes of the cavities, e.g. by changing the position of the tongue. The resonance can also be varied by changing the area of the opening (the mouth) connecting the resonator to the outside air into which the sound is being projected. The resonant frequencies are roughly, and inversely, dependent upon the cavity size. Probably you are familiar with this generalization—large organ pipes produce low notes, and large animals tend to have low-pitched calls.

Usually when people produce a vowel, the sound source (or source of excitation) is provided by fluctuation in the air flow through the narrow aperture between the vocal cords. The rate of this fluctuation determines the perceived pitch of the vowel. The filtering action of the vocal tract has the effect of enhancing energy in the source at some frequencies—those that coincide with the resonant frequencies of vocal tract cavities—and of attenuating energy at other frequencies.

Figure 3.13 shows the extent to which the sound source is modified by the vocal tract for three vowels, [ee] (as in 'heed'), [ah] (as in 'hard') and [oo] (as in 'food'). Different vowels are distinguished in articulatory terms by differences in the gross configuration of the vocal tract. So, to produce a vowel like [ah] for example, one drops the lower jaw, opens the mouth fairly wide and places the tongue body fairly low down in the back of the mouth. By contrast, try saying the vowel [ee]. You will find that to say it requires the lower jaw to be high, the lips spread, and the tongue high and forward in the mouth. Figure 3.13a shows the frequency spectrum of sound energy generated by vocal cord vibration and Figure 3.13c the frequency spectrum of output sound waves for the three vowels.

The vowels are distinguished acoustically by the overall shape of the frequency spectrum (Box 3.1), and particularly by the frequencies of the spectral peaks. The spectral peaks arise from vocal tract resonances and are referred to as **formants**, identified by number (F1, F2, F3, etc.) with the first formant having the lowest frequency.

A formant is not associated specifically with the resonance of a particular vocal tract cavity; there is no simple unique relationship between the size of a specific cavity and the frequency of a particular peak in the output spectrum.

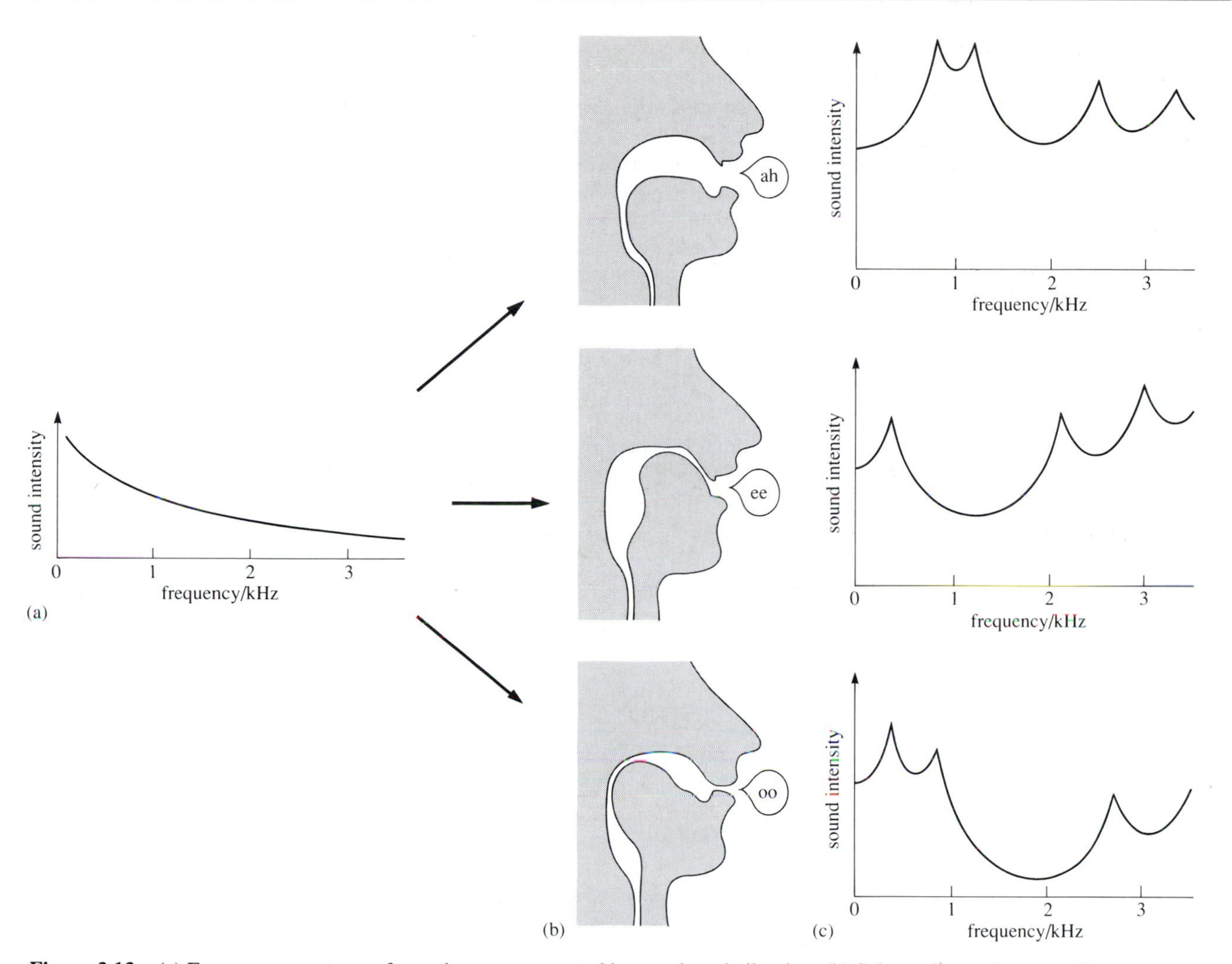

Figure 3.13 (a) Frequency spectrum of sound energy generated by vocal cord vibration. (b) Schematic vocal tract outlines. (c) Frequency spectra of the output speech waves.

The consonant sounds can be sub-divided. The main types we are concerned with here are the **stop consonants** [b, p, d, t, g, k] and the **fricatives** (e.g. [s] and [f]). Stops are produced by closing the vocal tract completely, building up pressure behind the closure and then suddenly opening the tract again. Try saying [p]. Your lips are closed (this is known as a bilabial closure) and as you open them the pressure behind them produces the [p] sound. In fricative sounds the vocal tract is only partly closed and, as the air is forced through, the fricative sound is produced.

The foregoing provides an outline of the general principles governing the relationship between vocal tract articulatory configurations and the kinds of sounds they can produce. The rather large repertoire of vowels that adult humans are capable of producing is apparently a direct consequence of the specific and rather specialized properties of the mature human vocal tract. Recordings of the vocal output of non-human primates and of human neonates suggest that sounds like the vowels [ee] and [oo] are rare in non-humans, reflecting the large anatomical differences that exist between their vocal tracts and those of adult humans. The particular difference that appears most important is the presence in the mature human vocal tract of a well-defined bend in the acoustic tube formed by the tract, plus a mobile tongue body forming part of the wall of the tube, and a vertically mobile larynx. Together these features permit large changes in vocal tract cavity size, leading to the stable and relatively extreme values of formant frequencies found for the vowels [ee], [ah] and [oo].

3.4.2 Descriptions of speech

An important assumption in conventional descriptions of speech is that it can be represented as a sequence of discrete segments, corresponding very roughly to the letters used in writing. These segments are called **phonemes**. Phonemes form the basic units of speech, from which words are constructed. We can define phonemes for a given language in terms of the minimal sound change that will alter the meaning. Thus the English word 'bat' consists of three phonemes, since we can form the words 'pat', 'bit' and 'bag' for example by changing the first, second and third segments, respectively. Each segment thus corresponds to a phoneme.

□ How many phonemes is the word 'shark' composed of?

■ There are three phonemes, not five as implied by the number of letters in the word as written, since we can form e.g. 'bark', 'shock' and 'sharp' by changing the first, second and third segments.

Various phonetic alphabets exist which provide symbols to represent the range of distinctive speech sounds in the world's languages. Languages differ in the number and range of phonemes that are distinctive. English uses a relatively large number—about 40, depending on the dialect—whereas some languages use many fewer. New Zealand Maori, for example, is said to use only 15 distinctive phonemes.

As well as differing in the total number of distinctive phonemes, languages also differ in the way in which their phonemes are used. For example, although in English the sounds at the ends of the words 'ice' and 'eyes' are distinctive, those two sounds in similar positions in Spanish words are not distinctive. Similarly, in Arabic a distinction is made between the sounds at the beginning of the English words 'keel' and 'call', which for English would be represented by the same symbol in the phonetic alphabet.

3.4.3 Vocal tract action

A number of general principles of vocal tract action can be described, based on the position of the articulations (Section 3.4.1). Such classifications of speech elements are economical, and have historical respectability—they were employed by Sanskrit grammarians roughly 2 600 years ago. A relatively small number of articulatory dimensions is sufficient to describe linguistically significant contrasts. These are the degree of constriction of the vocal tract and the position in the tract where maximum constriction occurs (often as a result of the position of the tongue and the presence or absence of vocal cord vibration).

Vowels, semi-vowels ([w] as in 'way'), fricative consonants ([s] as in 'say'), and interrupted or 'stop' consonants ([d] as in 'day', [b] as in 'beef') form a natural ranking of articulations with increasingly narrow constriction of the vocal tract. The initial consonants in 'gay', 'day' and 'bay' involve maximum constriction of the vocal tract at increasingly more forward locations, towards the front of the mouth. The vowels can be divided similarly in terms of the characteristic position of the tongue in the mouth, so the vowels in 'heed' and 'hard' are described as high-front and low-back, respectively.

The two dimensions, position of maximum constriction and position of tongue, correspond roughly to those known to phoneticians as manner and place of articulation. The voicing contrast, referring to the presence or absence of vocal cord vibration at some point in the phoneme, provides a further subdivision of phonemes with given manner and place.

□ Say the words 'fast' and 'vast'. What is the difference between the way that you say them?

■ The [f] in 'fast' does not require vocal cord action, but the [v] in 'vast' does.

So the fricatives [f] (in 'fast') and [v] (in 'vast') differ only in the presence of voicing in the latter. The stop consonants [b] (in 'bay') and [p] (in 'pay') differ in the synchronization of vocal cord vibration with the moment of release of the constriction at the lips: they are roughly simultaneous in [b], but voicing is delayed by a few hundredths of a second in [p].

The classification outlined above allows the phonemes of a language to be represented as an intersecting set of features, and hence allows utterances to be represented as articulatorily-defined segments arrayed serially in time.

□ Say the word 'beef'. How many phonemes is it composed of?

■ Three phonemes:

1 [b] uses the vocal cords, both lips and it is stopped. It is a voiced, bilabial stop consonant.

2 [ee], a high-front vowel.

3 [f].

□ The third phoneme [f] is an unvoiced fricative. As you spoke the word, what position were your teeth and lips in at the end of the third phoneme?

■ As you finished saying the word you would have been aware that your lip and teeth were in contact. This phoneme is a labio- dental (lips-teeth) fricative.

It is important to note, however, that although phonetic segments and features are convenient descriptors of speech, this does *not* imply that speech is necessarily planned or recognized by processes in human brains in which phonemes or phonetic features are the actual functional elements, i.e. phonemes do not necessarily have *psychological reality*. The psychological reality of such elements can only be established by suitable experimentation with human talkers and listeners. The results of such experiments provide some support for a phonemic representation during the planning of speech production, but only equivocal support for the psychological reality of phonemes during speech perception.

It is clear that the various means for describing phonetic information considered so far do not represent the full range of information in speech. There is a wealth of other information which is carried in the way a particular string of words is uttered. Variations in emphasis and emotional content, which can have profound effects on the meaning of an utterance, are achieved by variation in pitch, loudness and rhythm. Similar variables are employed to carry other cues such as those which facilitate turn-taking in fluent conversation. The context in which the words are spoken is also important in determining meaning, as you will read in Section 3.7.1.

3.4.4 Speech as a dynamic activity

In the previous sections it has been convenient to discuss speech as though it consisted of a sequence of relatively steady-state configurations of the vocal tract. However, a moment's thought about the antics of your vocal tract articulators as you read this sentence aloud is sufficient to show that steady-state vocal tract configurations are not characteristic of normal, fluent speech. Instead you move smoothly from one vocal tract configuration to the next, with the consequence that the distinctions between individual configurations become blurred.

In this section focus is on the fact that speech is a sequence of actions that unfolds in time. Note that some speech movements can also have visual consequences which may be important for the listener when the talker's face can be seen, particularly when there is ambiguity in interpretation of the acoustic signal.

Inspection of the sonagram shown in Figure 3.14 reveals, acoustically, several of the articulatory processes discussed above. The formants resulting from vocal tract cavity resonances are seen clearly as horizontal dark bands, and the presence of vocal cord vibration (for example during the [ee] vowel in 'The') as vertical striations on the pattern. Fricative consonants ([th] in 'The' and [s] in 'University') show up as sudden switches to high-frequency energy as shown by the dark bands at higher frequencies. Most dramatically, however, the display confirms our intuition that fluent speech involves highly skilled, carefully synchronized and continuous movements of the major articulators. So if you track the movements of the second formant, for example (F2 in Figure 3.14), in the early part of the utterance you see that it rises smoothly in frequency as the vocal tract configuration shifts from that appropriate for the [th] fricative, passes through a peak associated with the [ee] vowel, and then descends in frequency during the [o] of 'open' towards the configuration appropriate for the stop consonant [p] that is produced by lip closure. Notice that even when the vocal tract is closed for [p] it is not stationary, as the second formant frequency is higher after the closure for [p] than it was before closure as the lips came together.

Box 3.3 Observing the vocal tract and recording the sounds

Obtaining data on the details of vocal tract movements is difficult without disrupting the movements under scrutiny, so some fairly elaborate experimental procedures have been used to avoid such problems. The activity of the larynx can be observed directly using narrow fibre-optic bundles inserted through the nose and positioned in the pharynx. The movements of the major articulators can be measured by fixing small pellets to the relevant structures. The pellets reflect X-rays and allow the experimenter to track their movements with computer-controlled X-ray microbeams. The activity of the vocal tract musculature can be measured by recording and averaging electro-myographic (EMG) signals from electrodes inserted into appropriate muscles. All these techniques have provided valuable information about the way in which vocal tract movements are controlled by the brain, although scientists are far from having a complete understanding of how skilled motor behaviour like speech is planned and executed.

Recording and representing the acoustic characteristics of running speech is relatively easier. The first instrument to be able to do this conveniently was the sound spectrograph, a spectrum analyser that displays a plot of the frequencies of speech arrayed over time, with the amplitude at a given time and frequency shown by the darkness of the plot at that point. This display is called a sonagram or spectrogram (or voiceprint in North America), and an example for an utterance of the phrase 'The Open University' is shown in Figure 3.14.

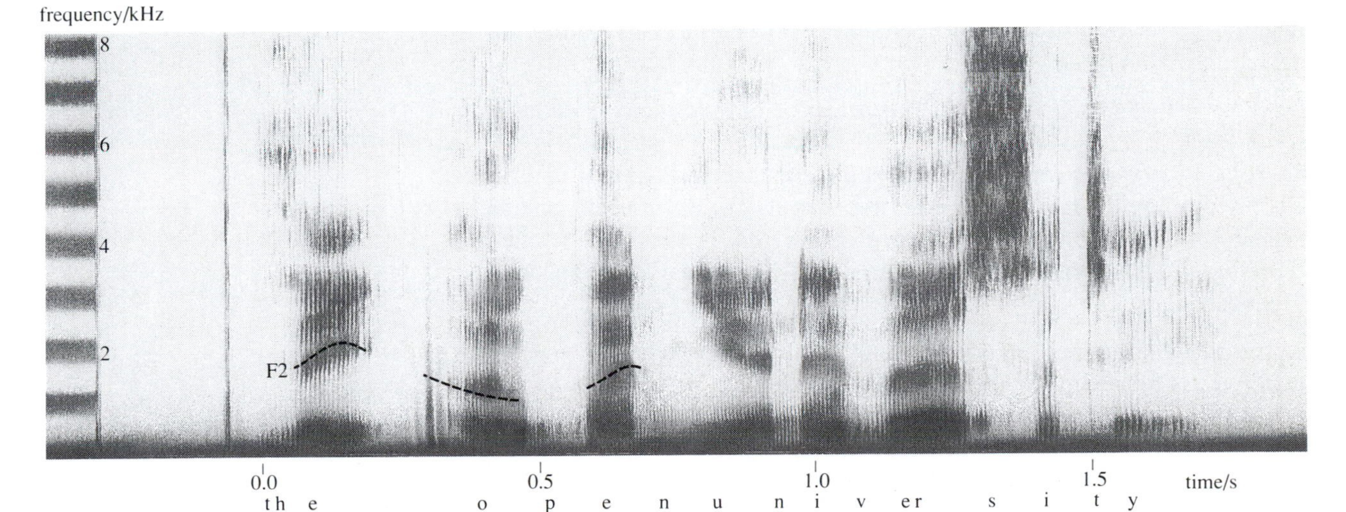

Figure 3.14 Sonagram of the phrase 'The Open University' spoken by an adult male. The axes are frequency and time, with amplitude represented by darkness on the grey scale. The letters of the phrase as written have been aligned roughly with the signal, although in fact information about the individual speech sounds is 'smeared out' in time. Note that the acoustic signal is not clearly segmented in linguistically relevant ways.

A further observation that can be made from Figure 3.14 is that segments corresponding to phonemes in the message are not clearly delimited in the sonagram. There are significant acoustic discontinuities visible, for example the relatively long period of low energy associated with the [p] closure, but they are not well correlated with the boundaries between any linguistic unit, a phoneme, syllable, or word. It is not easy to say exactly where the [p] begins or ends, since it is part of a smooth, extended pattern of movement. As a result one cannot say that a phoneme has a single physical moment of occurrence.

In the context of speech, the smooth transition between successive phonemic segments is called **co-articulation**. It implies a high degree of context dependence in the articulatory, and hence acoustic form of a given phoneme. In other words, the same articulatory manoeuvres will not always be executed to produce a specific phoneme, rather the form of the manoeuvres will depend on those that precede and follow them. Co-articulatory effects that operate forward in time—that is when the movements for a particular phoneme are influenced by the articulatory requirements for a coming phoneme—are called anticipatory. Backward co-articulation—in which a movement is influenced by the preceding articulatory activity—is usually the result of lag in articulatory movement due to inertia, and as such is of less theoretical interest than anticipatory co-articulation, which has been closely studied because it provides a window on the style and scope of articulatory pre-planning.

Anticipatory co-articulation can span several phonemes, but the actual extent of co-articulation that is observed depends on several factors. For example, watch the lips of a talker uttering the two words 'consume' and 'conceit' or watch yourself in a mirror—lip rounding to anticipate that required for the final vowel in 'consume' is usually seen early in the utterance. Note that the lip-rounding movement is not only visible, but will endow the acoustic signal with information that potentially signals to a listener the imminent arrival of a rounded vowel. In the two words 'compute' and 'compete', by contrast, the scope of anticipatory lip rounding is restricted because lip rounding conflicts with the requirements of bilabial closure for the [p].

This does not mean that a talker will always indulge in anticipatory co-articulation over as great a range as possible. It is important to remember that the talker's objective is successful communication, and the fundamental task of the speech planning mechanism must be to make the articulatory movements maximally efficient, consistent with maximizing the distinctiveness of the speech sounds that are produced. So the talker must sometimes trade smoothness for communicative effectiveness.

Of possible relevance here are the results of experiments which found a tendency for anticipatory lip rounding prior to a rounded vowel to begin earlier for talkers of Swedish—a language with a relatively large vowel inventory—than for Americans speaking English—a language with relatively fewer vowels. A plausible although unconfirmed rationalization of this result is that, when the vowel inventory is large, there is a greater need for articulatory precision to avoid auditory confusions among the vowels, and greater articulatory precision requires, where possible, earlier anticipatory co-articulation to get the right articulators into the right place at the right time.

3.4.5 Modelling the speech planning process

Despite a wealth of experimentation on speech production, there is still no fully successful model of the planning process for speech. The simplest model one might contemplate is one in which the talker, having determined the words to be uttered and their sequence, looks up the entry for each in an 'internal dictionary' which gives, among other things, a specification of the phonemes required for each word. The fact that many speech errors are of the 'Spoonerism' variety—that is, involve phoneme transpositions—lends some support to this view. (The term 'Spoonerism' is derived from the Reverend Spooner, who became famous for transposing parts of words—'town drain' instead of 'down train'. Most of the other examples associated with Spooner are later fabrications.) However, it is not sufficient to string a number of phoneme specifications together to produce a string of phonemes that make up a complete sentence. There are production rules that talkers use when joining certain phonemes together which change the speech sounds that are actually produced. Familiar instances of this are when the [v] of 'have' becomes [f] in the sequence 'have to' and when 'facts' is produced without the [t] or with only a barely audible one.

After applying the rules the talker has a specification of the string of phonemes that must be uttered. It is temptingly simple to suppose that each phoneme in a language is associated with a particular set of muscle commands. Producing an utterance would then consist of running off in sequence the muscle commands that would produce the movements specified by the sequence of phonemes in the utterance, rather in the way that sea-anemone behaviour is the result of a particular sequence of identifiable muscle contractions (Book 2, Section 7.2).

☐ What drawback does this model have when applied to fluent speech?

■ It does not take full account of co-articulation. The context in which a phoneme is uttered will vary and, as a result, so will the muscular movements before and after the phoneme.

Co-articulatory effects could arise in such a model as a result of the temporal overlapping of commands to the articulatory musculature. However, there is fairly clear evidence against this simple view of speech motor control. Most directly, when electrical signals recorded from the articulatory musculature were scrutinized for common properties associated with a particular phoneme in all phonemic contexts, no systematic common properties were found.

Summary of Section 3.4

Humans produce a wide range of sounds but generally only a small portion of the sounds that can be produced by the vocal cords are used to convey linguistic information. The main parts of the vocal apparatus are the vocal tract and the major articulators and the speaker can control both the overall shape of the vocal tract and the sound energy fed into it. The sound produced by changes in the air flow through the larynx is modified by the filtering action of the vocal tract. Certain frequencies are accentuated, others attenuated. The accentuated frequencies produce peaks in the sound spectrum called formants.

Words are formed from basic units of speech called phonemes. Different languages may have different numbers of phonemes. However, although phonemes and phonetic features are useful ways to describe speech, it is not necessarily planned and produced by the brain from equivalent elements. Speech is a dynamic activity and phonemes follow each other smoothly. Thus it is not always possible to define the boundaries of each phoneme. The smooth transition between phonemes means that a particular phoneme will not always be produced by precisely the same sequence of movements of the vocal tract articulators. A phoneme will be influenced by both the one that follows it and the one that preceded it. This makes it more difficult to develop models of the speech planning process.

3.5 Perception of speech

Humans detect, and can discriminate, patterns of acoustic energy over a wide range of frequencies and intensities, as you read in Section 3.1. Large numbers of these patterns can also be *identified*—that is, perceived, recognized and labelled as similar to some previously encountered pattern. These abilities are central to the process of speech recognition, which will now be considered.

3.5.1 Auditory processing of speech

You have seen a little of how the talker's articulatory movements structure the acoustic signal in ways that are specific to those movements, and how some of the acoustic structure of speech can be revealed by representations such as the sonagram. It is important to realize, however, that the representation of speech upon which the brain goes to work is not the same as that produced by a sonagram. Auditory and spectrographic representations have certain broad similarities, but also enough differences to make some consideration of the nature of human auditory processing necessary.

Auditory sensitivity is not constant as a function of frequency, as you read in Section 3.1. This is seen most clearly in a plot of the absolute threshold for simple tones against tone frequency (Figure 3.3). The threshold is roughly U-shaped, with a broad maximum in sensitivity in the frequency region between 800 Hz and 3.5 kHz, deriving substantially from the resonance of the external auditory canal.

□ What is the importance of this region of relatively greater sensitivity?

■ This frequency response coincides with the frequency region in which the bulk of the information for phoneme identity lies (Section 3.1.2).

Thus information-rich parts of the speech signal are given relatively greater prominence in the auditory representation.

The frequency-analysing capabilities of the auditory system vary in resolving power with frequency. At low frequencies, auditory frequency resolution (the ability to discriminate between different frequencies) is relatively good. At higher frequencies resolution is not as good, but the resolving power is still sufficient to preserve the general shape of the frequency spectrum.

Associated with the variation in frequency resolving power is an inverse variation in temporal resolving power (the ability to discriminate between auditory events that are very close together). Temporal resolution increases with increasing frequency up to about 2 kHz. By this arrangement the auditory system can simultaneously achieve relatively good frequency and temporal resolution, albeit in different frequency regions. In this respect the auditory system is particularly well suited to the processing of speech, in which phonetic distinctions depend on differences both in frequency and temporal detail.

3.5.2 Resolution and the hearing-impaired

It should be noted in passing that the importance of resolution in the discrimination and identification of speech sounds gives an insight into the difficulties experienced by hearing-impaired people when listening to speech. A common observation by people with hearing impairment of cochlear origin (probably the most common type) is that, even when using their hearing aid, speech tends to sound indistinct and words are difficult to identify. Studies using hearing-impaired listeners have shown not only that their thresholds are raised, but also that their auditory filters tend to be broader than normal. Broad auditory filters are less able to resolve the details of frequency and timing in speech, and thus lead to poorer speech identification performance. Wearing a conventional hearing aid may go some way to correcting the *threshold* elevation, but can do little to ameliorate impaired *resolution*.

3.5.3 Adaptation, inhibition and suppression

There are two other properties of auditory processing that probably have a significant influence on the clarity of phonetic information detectable in speech. First, and apparently common to all sensory systems, is a phenomenon known as lateral inhibition (for visual processing) or **lateral suppression** (for auditory processing), in which strongly-stimulated sensory receptors inhibit the responses of their less-strongly-stimulated neighbours (Book 2, Section 9.5.3). Although the mechanism underlying this effect appears to be different for different senses, the general advantage for all perceptual processes is similar: it will tend to enhance *discontinuities* in the stimulus and so sharpen the edges. In visual processing these may be contours created by relatively abrupt changes in lightness or colour; in auditory processing they may be formant peaks in the frequency spectrum, and as you have seen, the characteristics of formant peaks are perceptually important because they correspond directly to the articulatory behaviour of the talker.

A second property of auditory processing that may enhance perception is frequency-specific **short-term adaptation**. Whereas lateral suppression may make details of the frequency spectrum clearer, short-term adaptation will enhance responses to changes in frequency over time. When stimulated, the responses of auditory sensory cells tuned to specific frequencies tend to increase sharply in magnitude and then rapidly decrease. If the energy in a stimulus moves from one frequency region to an adjacent region, a very common property of speech, short-term adaptation will have reduced the response of receptors to the original frequency while the response to the new frequency is at a maximum. This will accentuate the differences between the responses of receptors tuned to the original frequency and those tuned to the new frequency.

Summary of Section 3.5

Perception of speech depends upon both the frequency sensitivity and the amplitude sensitivity of the ear. The maximum sensitivity of the ear to incoming sound is over the frequency range that contains the most important information in speech. The ability to distinguish between frequencies (frequency resolution) is relatively good at low frequencies but less good at high frequencies. Auditory processing at the receptor level in the nervous system enhances discontinuities in the stimulating sound, for example the formant peaks in the frequency spectrum. Short-term adaptation in the receptors accentuates differences between frequencies when there is a transition from one to another.

3.6 Phonetic processing of speech

The previous section identified some of the processes that make the auditory representation of phonetic information clearer than it would otherwise be. This section turns to the processes by which the listener extracts and integrates this phonetic information. There is still some debate about whether recognizing words in the incoming stream of speech must necessarily involve perception of phonemes, or a different intermediate representation, perhaps involving the syllable as a processing unit, or even whether word recognition might proceed directly from the auditory representation itself. Of course, none of these possibilities excludes the others and, in view of the remarkable ease with which listeners understand speech even in adverse conditions, it is perhaps appropriate to suppose that the speech recognition system has the flexibility to use whatever types of information and whichever processing strategies yield the most accurate and plausible interpretation of the talker's message. Most models of speech understanding incorporate an implicit, if not an explicit, level of phonemic representation, particularly for the acquisition of new words.

Although for the listener the perception of speech seems an effortless process, a more detailed analysis of what must be involved makes the task appear formidable. The fact that it has not yet proved possible to design a machine to recognize speech with capabilities even approximating to human performance lends support to this view. You saw in Section 3.4 something of the complexities of the acoustic consequences of vocal tract configurations and of the movements of the major articulators during speech production. The more important implications of those complexities for the listener as potential perceiver may be summarized as follows:

1. Because of co-articulatory influences, the acoustic form of a given phoneme will vary with the context in which it is uttered.
2. Because of the articulatory restructuring required for speech at different rates, the acoustic form of a given phoneme will vary with changes in speech rate.
3. Because fluent speech also carries other information, the acoustic form of a given phoneme will vary with its assigned stress and its position in an utterance.

4 Because talkers' vocal tracts differ in size, and vocal tract cavity size is a primary determinant of formant frequencies, the acoustic form of a given phoneme will vary between different talkers. Related to this is the fact that talkers tend to have dialectal and idiosyncratic differences in speech style which will generally have acoustical consequences.

Notice that, although the points made above refer to variability in the acoustic form of phonemes, similar comments apply to the acoustic form of syllables and words.

These various effects mean that any simple perceptual strategy involving matching unknown incoming auditory speech patterns against stored representations of patterns for all phonemes in the language is clearly unworkable. However, it is neither necessary nor appropriate to view variability in the context in which it is uttered as a perceptual problem. On the contrary, it contributes significantly to the redundancy of information which allows speech to be recognized relatively easily even when spoken rapidly or in adverse listening conditions. It is the multiplicity of information in speech that is one of its great strengths as a medium of communication.

It has not proved easy for perceptual theories to describe how the speech perception system registers acoustic cues. One approach, stimulated initially by theories of visual perception (which will be discussed in Chapter 4), has been to argue for the existence of discrete phonetic feature detectors, which respond when the specific type of acoustic pattern corresponding to their phonetic feature is present in the input. Evidence for this view has come from adaptation experiments in which subjects are presented repeatedly with a clear example of, say, a voiced consonant (e.g. [ba]), and then required to identify a consonant that has been synthesized to be ambiguous between voiced and voiceless (e.g. intermediate between [ba] and [pa]). It is usually found that listeners experience adaptation, so after hearing [ba] repeatedly, an ambiguous syllable is more likely to be heard as [pa].

This type of result has been taken to demonstrate the existence of 'detectors' for the phonetic feature voicing which are fatigued by adaptation, but care is needed to rule out the possibility that at least some of the effect should be attributed to adaptation at the receptor level (auditory adaptation), as distinct from phonetic adaptation. When the appropriate experimental controls are included it can be shown that both auditory and phonetic adaptation occur. In any case, doubts have been expressed concerning the explanatory power of the feature detector concept—some have argued that it is not an explanation of phonetic perception but rather a redescription of it.

A striking feature of the results of these identification experiments is that they demonstrate sensitivity to several cues for each phonetic contrast. For example, it has been claimed that for the voicing contrast in English there are at least 17 distinct, identifiable cues which can carry information about voicing in initial stop consonants. Similarly, numerous cues have been identified for place and manner of articulation. Indeed, it is plausible to claim that adult listeners come to be sensitive to every auditory difference that exists between two utterances that contrast minimally on some phonetic dimension.

3.6.1 Heterogeneity of acoustic cues

One of the notable properties of the acoustic cues is their heterogeneity and a correspondingly remarkable property of phonetic perception is its capacity to integrate this acoustically diverse information. To take one example that has been a popular research tool, the presence or absence of a stop consonant after [s]—the distinction between, say, 'slit' and 'split'—is dependent primarily on two acoustic properties. One is the presence and duration of a period of silence or low acoustical energy (a 'gap') after the [s] and before the [lit] portion of the word, and the other is the presence and extent of a transition in the first formant frequency at the onset of the [lit] portion of the word. Natural productions of 'split' as distinct from 'slit' tend to have a relatively long gap (approximately 100–180 ms) and a relatively large first formant frequency transition (approximately 200 Hz over 30 ms). Perceptual experiments using synthetic versions of the words confirm that presence of a long gap and a large formant frequency transition will predispose a perception of 'split'. So, it appears that these two very different acoustic properties—a temporal interval and a frequency change—have the same phonetic 'meaning'. Furthermore, it appears that these cues can enter into a phonetic trading relationship, by which is meant that a 'weak' setting in one cue can be offset by a 'strong' setting of the other. In other words, if the gap duration is decreased the extent of the formant transition must be increased to maintain a certain probability that a [p] will be heard.

It is worth dwelling briefly on the integrative abilities of the speech perception system. The ability to integrate spectrally and temporally distributed segmental information from the acoustic signal has been discussed above. It was observed in Section 3.3.3 that visual information from the talker's face also carries segmental information. It has been shown fairly recently that when the talker's face can be seen this visual information is integrated with acoustic information in the phonetic perception process. A rather compelling audiovisual illusion illustrates this kind of integrative perceptual processing. A video recording of a talker uttering the word 'goes' has the audio signal of the word 'bows' (as in 'bows and arrows') dubbed on in synchrony with the talker's facial movements. As expected, when listening to the soundtrack alone the word 'bows' is recognized. When listening to the soundtrack while watching the screen, the most commonly recognized words are 'doze' or 'those', corresponding neither to the acoustic *nor* the visual signal but to a rational combination of the information from both. The visual specification rules out bilabial place of articulation since the lips are clearly not involved in articulation of 'goes', so the word is recognized as being most consistent with the combined visual and acoustic information.

As a final observation, it must be acknowledged that there is still only an incomplete understanding of the way phonetic information is represented in the acoustic signal, and how that information is used in perception. Furthermore, a complete account of speech perception will need to include stages that tend to be overlooked until one attempts to build a working perceptual model for speech. For example, speech is usually heard against a background of other sounds, often other speech sounds, which unless de-emphasized in the perceptual process, will introduce spurious information into the listener's auditory representation. An early step in the perceptual process must therefore be to group together all the acoustic information derived from the talker whose speech is to be perceived. The mechanism by which this is achieved is currently the focus of considerable research effort.

Summary of Section 3.6

Most models of the speech understanding process assume some form of phoneme representation, though this does not exclude other possibilities such as recognition of syllables or whole words. Any model of the recognition process for words has to account for the variability that results from variation in the size of vocal tracts between individuals, the consequences of co-articulation, variation in speech rate and the importance of non-auditory information. Of course, the context in which words are produced affects their interpretation by the listener, as the next section describes.

3.7 Language understanding

The objective of speech is successful communication, that is for the listener to understand the talker's message. This final brief section considers the processes required for understanding. You have an intuitive grasp of what understanding involves, and you know when you, or those with whom you attempt to communicate, have achieved it, but it is not easy to define, or always to establish that it has occurred.

In general it is accepted that someone has understood something when his or her behaviour (linguistic or other) is consistent with the message. What intervenes between word recognition and this appropriate behaviour?

3.7.1 Word meanings and sentence meanings

If you hear a simple utterance like 'The girl threw the ball', you are likely to be able to identify the individual words without difficulty. You can probably say something about the form-class of each of the words, so you may say of 'girl' and 'ball' that they are nouns, and of 'threw' that it is a verb. You can usually say something about the meaning of each of the words, so 'girl' denotes [young female human], 'ball' denotes [small spherical physical object] and 'threw' denotes [action launching projectile using arm]. For a specific utterance the words will have a context: 'girl' would be the particular girl who was the subject of the utterance. In addition, words often have connotations for individual listeners, so the meaning of 'ball' for a particular listener might be [nasty dangerous projectile to be avoided while walking in the park]. The nature of word meaning is complicated further by the issue of ambiguity, as many words have more than one meaning. So 'ball', for example, could denote [large assembly for the purpose of social dancing], in which case 'threw' would denote [made arrangements to hold].

The form of the representation of these various types of word meanings in the brain has proved elusive, and is not known in detail. This is as much a question about the structure of memory as about the representation of word meaning. It has been proposed that storage of information about concepts is hierarchical in organization, with properties associated with concepts at different levels in the hierarchy, after the fashion of the network in Figure 3.15 (*overleaf*).

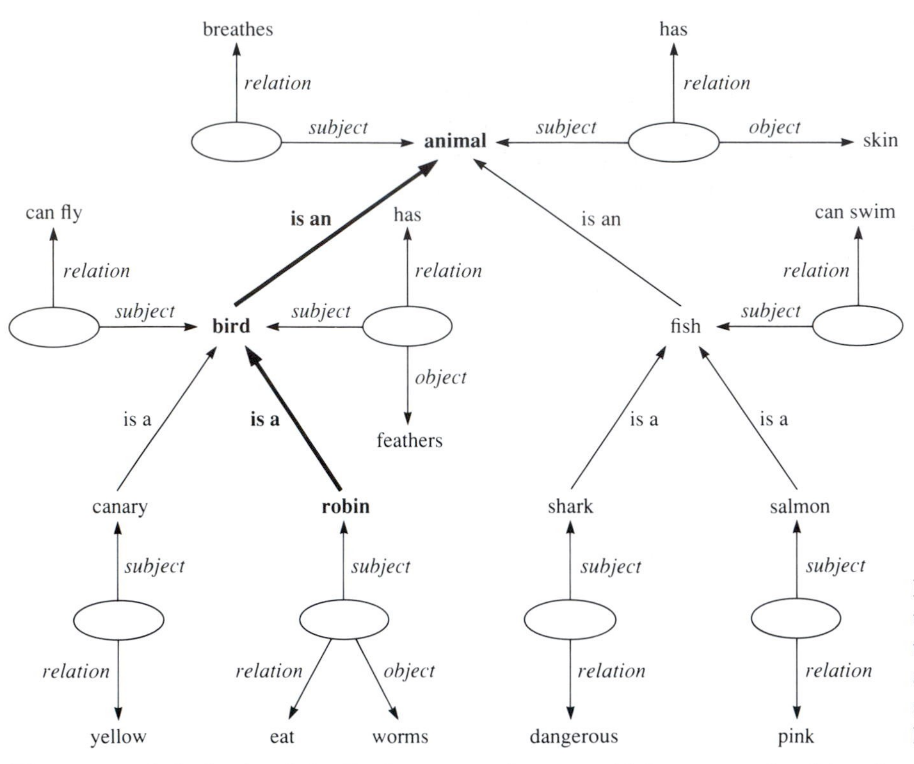

Figure 3.15 The network of concepts used in experiments to compare reaction times in making true/false judgements about statements. The pathway for the statement 'A robin is a bird' is highlighted.

The proposal derived originally from experiments which measured subjects' latencies (reaction times) in a task which required them to identify a statement as true or false. A relationship between latency and the number of links in the network separating the particular concept and property was found. Thus, it took subjects longer to agree that 'A robin is an animal' than 'A robin is a bird' because of the greater number of links between 'robin' and 'animal' (Figure 3.15). However, more recent experiments have shown that latency is affected by other factors, such as the frequency with which the concept is encountered, so it may be that 'A robin is an animal' takes longer because it is a proposition that is encountered relatively infrequently.

The processes by which sentence meanings are derived and represented are similarly obscure and inaccessible. There have been several attempts, primarily by linguists, to devise theoretical frameworks within which sentence understanding can be characterized, but they have tended to be based on linguistic theories of limited durability. The relationship between formal linguistic theories and psychological theories has tended to be an uneasy one. Although plausible linguistic models may seem like good candidates for psychological models, attempts to establish the psychological reality of particular linguistic theories have tended to founder. One reason for this is the difference in approach by those who have developed linguistic and psychological models.

One of the important stimulants for thinking about language understanding has been the concern with artificial intelligence, that is the design and creation of machines that exhibit 'intelligent' behaviour. The creation of a working system of language understanding requires that all of one's assumptions about the representation and processing of linguistic information be made explicit, and thus forces a thoroughness that is often lacking in theories expressed in less rigorous natural-language terms.

Although some impressive language understanding systems have been demonstrated, their capabilities do not match those of human listeners. One of the many reasons for this is the difficulty of representing in formal and tractable ways the several kinds of linguistically-relevant knowledge essential for language understanding. Some insight into the diversity and wealth of this knowledge can be gained by reflecting on the examples of ambiguity in the sentences listed in Table 3.1. Resolution of each type of ambiguity involves linguistic knowledge of a specific kind, and although some of the examples may appear contrived, they serve to focus attention on processes that most normal language use must require.

Table 3.1 Examples of ambiguous sentences.

They took the plane to London.
They watched her duck.
The fish were too old to eat.
She took the sweet from the wrapper and ate it.

Summary of Chapter 3

This chapter has given a very brief introduction to a fascinating area of brain and behaviour research—human hearing and human speech. This is an area of collaborative research since it spans psychology, physiology, neuroscience and anatomy. The processes of production and reception of sound are complex and there is a match between the acoustic properties of the ear and the sound content of speech. However, for speech to be recognized as such and for information transfer between individuals to take place there has to be processing of the incoming sound. There is processing at the level of the receptors in the ear that sharpens distinctions between linguistically important parts of the the incoming sound. The processing that must occur within the brain is difficult to study at the neural level but much information can be obtained from psychological studies. It is possible to construct models for speech perception and understanding based on the phoneme. The phoneme is a unit that can readily be defined and used in experiments but it is not possible to be certain that all the analysis of incoming sound that is carried out by the nervous system is actually based on the phoneme. Recognition of words is a stage in successful communication, but the words need to be assigned a meaning. Meaning is not an absolute term since the meaning of a word can be different in different contexts. This course is about the links between neurons, brains and behaviour. In the area of understanding language there is much information available from the psychologists but the neuroscientists are not yet in a position to provide much experimental data since speech is a process in humans that is not explicable in terms of animal models.

Objectives for Chapter 3

When you have completed this chapter, you should be able to:

3.1 Define and use, or recognize definitions and applications of each of the terms printed in **bold** in the text. (*Question 3.6*)

3.2 Give a description of the main structures in the human ear and explain how the external sound source is coupled to the fluid in the cochlea. (*Question 3.1*)

3.3 Describe the role of the hair cells in hearing. (*Question 3.2*)

3.4 Outline the major acoustic pathways leading to the auditory cortex in the brain. (*Question 3.3*)

3.5 Give a description of the main structures in the human vocal tract and explain how speech sounds are produced. (*Questions 3.4 and 3.5*)

3.6 Explain how speech is classified in terms of phonemes. (*Question 3.6*)

3.7 Give a brief description of speech perception, and the subsequent processing of speech within the ear and brain.

Questions for Chapter 3

Question 3.1 (*Objective 3.2*)
Does the human ear hear all frequencies equally well?

Question 3.2 (*Objective 3.3*)
The auditory hair cells are the receptors in the ear. How do they 'detect' acoustic signals?

Question 3.3 (*Objective 3.4*)
The auditory cortex has a tonotopic organization. What does this mean?

Question 3.4 (*Objective 3.5*)
What are formant frequencies and where are they produced?

Question 3.5 (*Objective 3.5*)
Classify each of the following sounds as voiced or voiceless:

(a) [sh]

(b) [z]

(c) [s]

Question 3.6 (*Objectives 3.1 and 3.6*)
Define *phoneme* and give an example of how words are described in terms of phonemes.

Further reading

Carpenter, R. H. S. (1996) *Neurophysiology*, 3rd edn, Physiological Principles of Medicine Series, Oxford University Press Inc. (Edward Arnold UK). Chapter 6 gives a good account of the neurophysiology of hearing.

Denes, P. B. and Pinson, E. N. (1993) *The Speech Chain: The Physics and Biology of Spoken Language*, 2nd edn, W. H. Freeman and Co. Ltd.

Griffin, D. R. (1958) *Listening in the Dark*, Cornell University Press. (Reprinted 1986.)

Liebermann, P. and Blumstein, S. E. (1988) *Speech Physiology, Speech Perception and Acoustic Phonetics*, Cambridge University Press.

Moore, B. C. J. (1997) *An Introduction to the Psychology of Hearing*, 4th edn, Academic Press.

CHAPTER 4
VISION

4.1 Introduction

Chapter 1 used the analogy of stones falling into a pond to make the point that perception is difficult because contact with the external world relies upon proximal stimuli which are quite unlike the distal objects and events which produce them. This may seem reasonable for hearing, where the proximal stimulus is a complex pattern of pressure variation, but you may feel that the situation is rather simpler for vision, where the proximal stimulus is just a sequence of images. After all, images seem easy to deal with and, through pictures and television, are used as powerful and effective symbols in society. This line of thinking is understandable but mistaken: images are just as indirect and difficult to deal with as patterns of pressure variation. They just *seem* easy, indeed they have become so important to society, because the human visual system is very, very good at making sense of them.

To convince yourself of this, try the following exercise. Write down a simple description of a cube in terms of edges, corners and surfaces. Now draw a simple sketch of a cube. Do both of these things before reading the next paragraph.

If perceiving the distal object (a cube) were just a matter of describing the proximal stimulus (an image of a cube) then your written description and your sketch should match each other pretty closely. But how do you compare these two very different things? Clearly there are equivalences between the object and its image: for example, edges of the object correspond to lines in the image, corners to junctions between lines, and surfaces to closed regions of the image. So try writing down a description of your sketch in terms of lines, junctions and regions.

Does your description of the cube in terms of edges, corners and surfaces match your description of the sketch in terms of lines, junctions and regions? A cube has six surfaces. How many regions does your sketch have? A cube has eight corners. How many junctions does your sketch have? A cube has 12 edges. How many lines does your sketch have? More important, the *defining* characteristics of a cube are that its surfaces should be square and that they should all be joined at right angles. The sketch shown in Figure 4.1 does not contain *any* of these defining characteristics—there are no squares and no right angles—and yet you have no difficulty in recognizing it as a cube. The message should be clear: you *can't* perceive visual objects (distal objects) simply by *describing* their images (proximal stimuli)—because the descriptions just don't match. (It is conventional in the study of vision to refer to an object as a *distal stimulus*, in contrast to the *proximal stimulus*, which is the image formed by that object on the retina. Because this terminology is potentially confusing, the terms *distal object* and *proximal stimulus* are used in this chapter.)

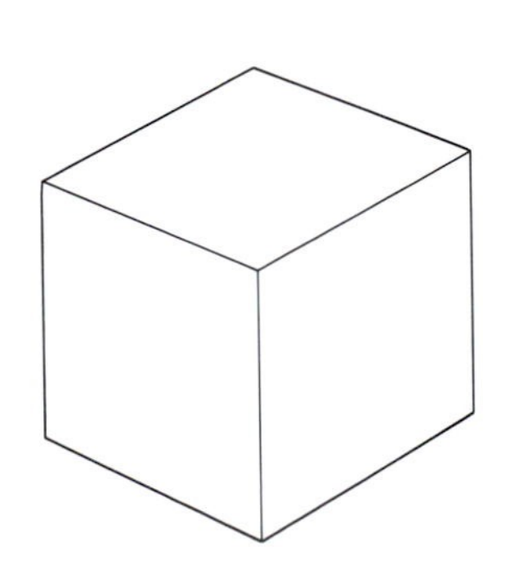

Figure 4.1 A sketch of a cube. Compare the properties of the image with the properties of the object it depicts.

The exercise above aims to convince you that, for vision as well as hearing, there is an important distinction between distal objects and proximal stimuli and that images are not direct, complete or even very good representations of the external world. Because of this, throughout this chapter vision will be regarded as an indirect and active process which begins with a description of the image but which must also solve the formidable problem of working out what object produced the image. The chapter consists of three main sections. Section 4.2 briefly describes how the proximal stimulus—an image—is produced from the complex pattern of light which falls upon the eyes. Sections 4.3 to 4.8 describe in some detail how the informative features of the proximal stimulus are encoded in the first stages of vision. These first stages of visual processing are essentially *descriptive*. Finally, Section 4.9 briefly outlines the subsequent *interpretive* stages which, given a description of the proximal stimulus, must work out what things in the world could have produced it.

Vision is such a large topic that it is impossible to provide more than an overview of a few themes in just one chapter—the most recently commissioned review of current vision research runs to 16 substantial volumes. This chapter concentrates on the descriptive stages of vision because these are the most relevant to a broadly-based biological approach. Even given this restriction, it is not possible to provide a comprehensive picture and what follows is intended to be a coherent but very selective review.

A good strategy for studying this chapter is first to skim through it quickly to get an overall picture of the topics and issues that it covers. You should then start studying it more carefully, paying particular attention to Section 4.3, which examines how the visual system codes for brightness. Sections 4.4 to 4.7 examine other, equally important, aspects of vision, including colour vision, movement detection and stereoscopic vision.

One of the first things to do is to make sure that you are familiar with the gross anatomy of the visual system, as illustrated in Figure 4.2. You are not expected to learn this; it is intended simply as a map to allow you to place the detailed discussions into a proper anatomical context.

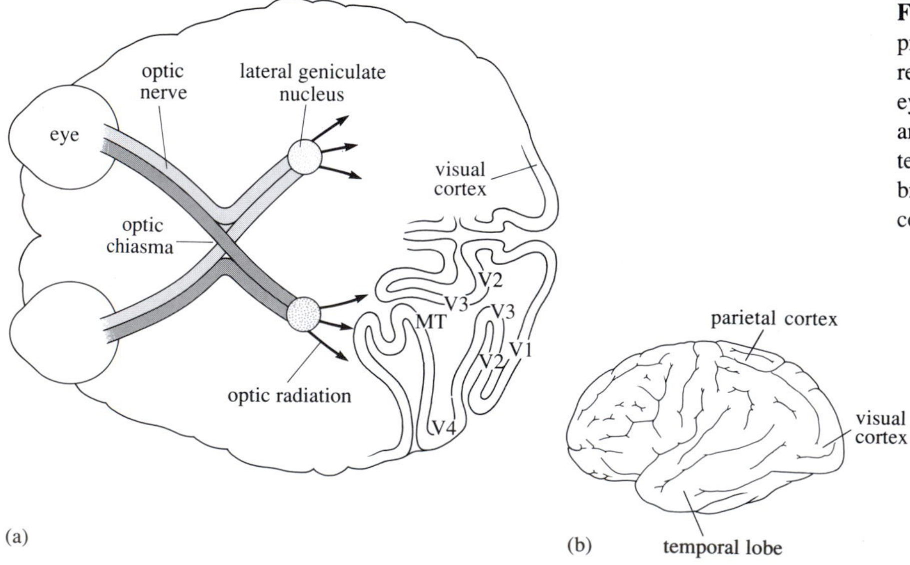

Figure 4.2 Gross anatomy of the primate visual system. (a) A schematic representation of the pathways from the eyes to the visual cortex. V1–V4 are areas of the visual cortex; MT: middle temporal cortex. (b) Lateral view of the brain, showing position of the visual cortex.

4.2 Image formation

Light travels extremely fast, for the most part in straight lines, and so it is an excellent medium for carrying information about the layout of the world. The easiest way to represent this information is to form an image. An image is simply a two-dimensional map of the three-dimensional world and it has just two essential properties. First, each point in the world is represented by one, and only one, point in the image. Second, points in the image preserve the spatial relationships between points in the world. The simplest way to form an image is to use a box with a hole in it—a pinhole camera.

Figure 4.3a shows a flat, reflecting surface. Light hits each point on this surface from all possible directions and is reflected in all possible directions. The purpose of the hole in the camera is simply to stop most of the light and, ideally, to ensure that only a single ray from each point gets into the box. Thus, light from point A on the surface only reaches the image at point A′—a single point in the world projects to one, and only one, point in the image. Making the hole bigger would cause problems, as shown in Figure 4.3b, because a cone of light rays from each point could get into the box and each point in the world would project to a considerable area in the image—the image would thus be blurred.

Figure 4.3c shows the same idea extended to several points on the surface. Notice that the image is upside-down—but this is unimportant. The important thing is that point B in the world is halfway between points A and C, and that the same is true of the corresponding points, A′, B′ and C′, in the image—spatial relationships are preserved.

□ Take two pieces of paper and make a very small hole in one of them. Hold them vertical and parallel about 10 cm apart, like the front and back surfaces of a box. According to the geometry explained above, there should be an image of the world on the sheet of paper without the hole. Why isn't there?

■ In fact there is—but you can't see it because there is too much 'noise': the image is swamped by all the other rays of light which can reach the surface without going through the hole. You have to enclose the hole and the screen in a lightproof box to eliminate this noise.

The pinhole camera is actually a very good device for forming images because it has one great advantage: if the hole is small enough, the image will always be sharp for objects at any distance and so the camera will never need to be focused.

□ Pinhole cameras have one serious disadvantage. Can you identify what it is?

■ Pinhole cameras work by excluding most of the light from each point and so the image is inevitably dim. Old cameras worked on the pinhole principle and had to make up for the dimness of the image by collecting light over a considerable period of time. The subject of the photograph had to keep perfectly still during this time or it would appear in the photograph, if it appeared at all, as a streaky blur.

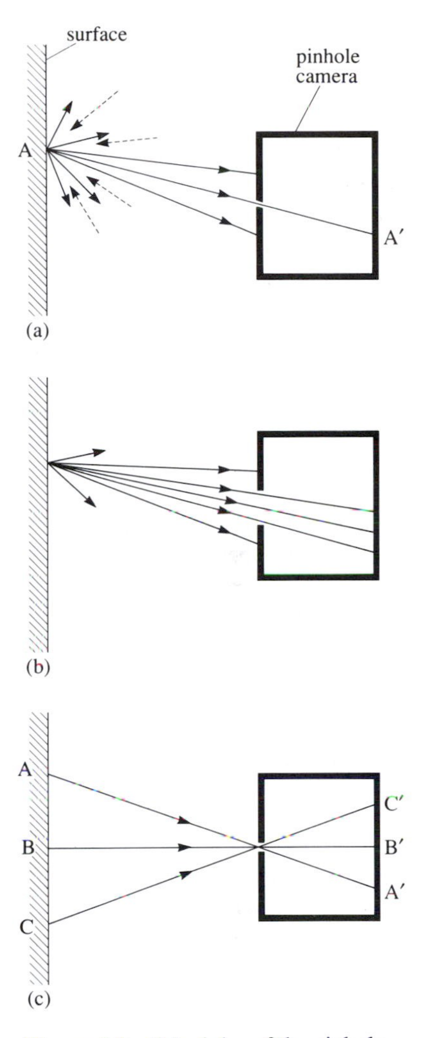

Figure 4.3 Principles of the pinhole camera. The small size of the pinhole ensures that light from each point on the surface projects to only one point in the image. (a) Pinhole camera with small pinhole. (b) Pinhole camera with large hole. (c) Formation of an image in a pinhole camera. See text for further details.

Unlike cameras, animals cannot solve the problem of image dimness by collecting light over considerable periods of time, because one of the main functions of vision is to detect movements in the world. Moving things are much more potentially dangerous or appetizing than stationary things. A better solution would be to make use of more of the rays of light coming from each point in the world. One way to do this is to make use of *refraction*—the fact that light bends as it travels from a medium with one optical density (e.g. air) to a medium with a different optical density (e.g. water). To make use of refraction, you would have to fill the box with water, make the front surface transparent, and then—and this is the difficult part—make sure that this surface is precisely the right shape to bend more of the rays of light from each point in the world so that they project to one, and only one, point in the image. This principle is illustrated in Figure 4.4a, which shows a simple refracting eye. Unlike the pinhole camera which admits only one ray of light from each point, a whole cone of rays from each point is focused as the light enters the eye so that they project to a single point in the image.

The image is now brighter, but solving one problem has caused another. Unfortunately, the simple system depicted in Figure 4.4a only works for objects at one particular distance. If the object is closer or further than this, as shown in Figure 4.4b, the cone of light from each point no longer projects to a single point in the image, and the image is consequently blurred. This presents a dilemma: a smaller hole ensures that the image remains in focus over a greater range of distances but produces a dimmer image; a larger hole produces a brighter image but it is only in focus for objects at a limited range of distances.

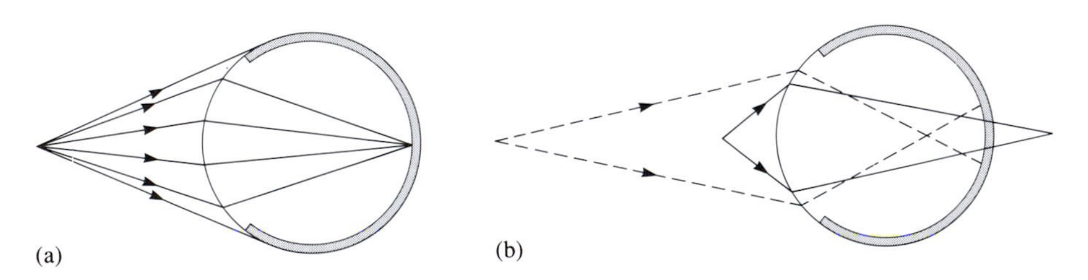

Figure 4.4 A simple refracting eye. (a) A cone of light rays from each point on the surface is bent so that they project to a single point in the image. (b) A cone of light rays from a point nearer or farther away does not project to a single point in the image, which as a result is blurred.

The vertebrate eye depicted in Figure 4.5 goes some way to resolving this dilemma by introducing two refinements. First, it has a variable sized hole, the pupil, through which light is admitted. In dim conditions, the pupil opens up to admit as much light as possible, while in bright conditions, when light is more plentiful, it closes down to a pinhole so that objects remain in focus over a greater range of distances. You can demonstrate this to yourself, with a friend or a mirror, by using your hand to shade one eye and watching its pupil change size.

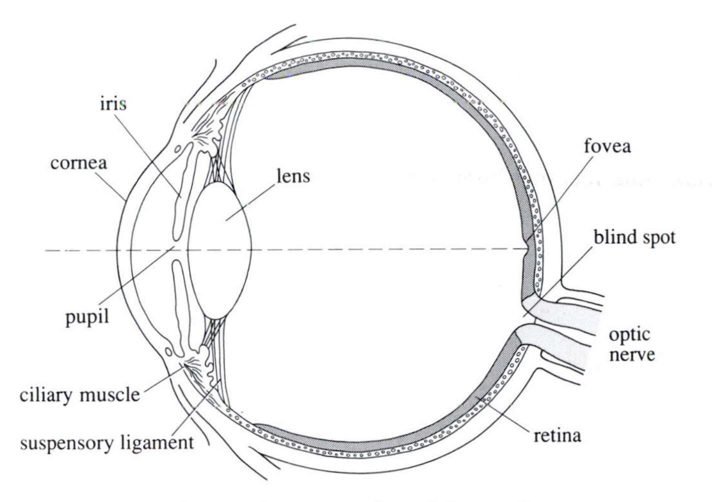

Figure 4.5 Schematic cross-section of the vertebrate eye.

The second refinement is the introduction of the lens. This is a refinement, rather than a strict necessity, because most of the refraction takes place at the cornea, the membrane covering the eye, as light moves from one medium (air) to another (in effect water). The function of the lens is to allow some flexibility in focusing and it does this in many mammals, including humans, in a very clever way. The amount that light rays are bent by a lens depends upon the curvature of the lens—the more curved it is, the greater the bending. The lens in the human eye is an elastic structure which can vary its curvature. In the resting state, when the ciliary muscle is relaxed, residual tension in the suspensory ligament pulls the lens flat and so focuses light from distant objects (Figure 4.6a). To bring closer objects into focus, the lens needs to be more curved (Figure 4.6b), so the ciliary muscle contracts, releasing tension in the suspensory ligament and allowing the lens to bulge into its natural shape. This process of changing the lens shape to focus on objects at different distances is called **accommodation**.

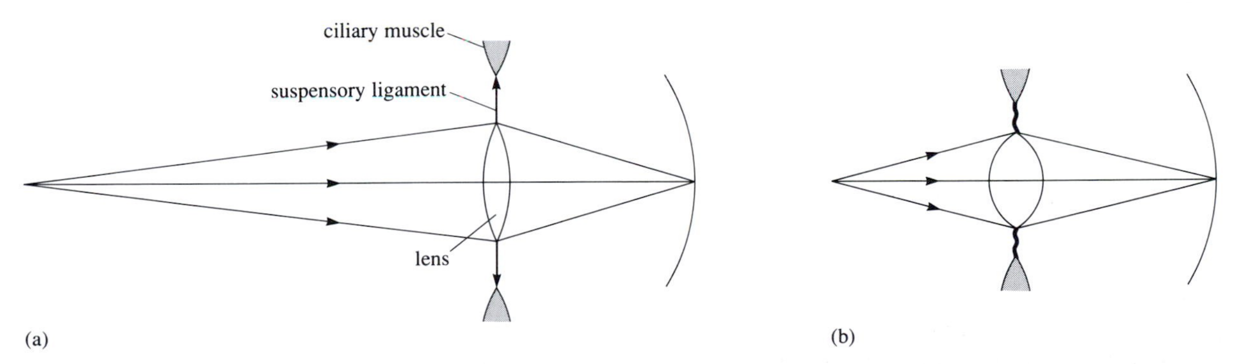

Figure 4.6 Accommodation in the vertebrate eye. (a) Tension in the suspensory ligament pulls the lens flat so that its curvature decreases, light is bent less and distant objects are in focus. (b) Tension is released in the suspensory ligament so that the lens bulges to its natural shape, curvature increases, light is bent more and near objects are in focus.

□ Many people have optical defects and cannot accommodate over the normal range of distances. Why should they find their problems most noticeable under dim lighting conditions?

■ Under bright conditions the pupil closes down, so that the eye functions more like a pinhole camera and does not need to accommodate. In dim conditions, the pupil opens up to allow in as much light as possible and this is where accommodation is most needed.

□ The image formed in a human eye is upside-down, as demonstrated in Figure 4.4. Why is this, as claimed above, unimportant?

■ The fact that the image is upside-down only presents a problem if you believe that visual perception is just a matter of describing the image. It was argued earlier (Sections 1.3 and 4.1) that description is not enough and that visual perception must also involve some *interpretation* of the image. All that is needed for such an interpretation is a *consistent relationship* between the image and the world. It does not matter what the relationship is, and an upside-down relationship is, in principle, no more difficult to interpret than a right-way-up relationship.

4.3 Image description: luminance

The first neural stage of visual processing is to transduce the pattern of light which constitutes the image into a pattern of neural activity. Once this has been accomplished, further neural processing can extract and process the information provided by the proximal stimulus. Before reading further, you should pause for a moment to consider what kind of description of the image would be most useful. This question is best addressed by considering the functions that vision serves because, if you know what vision is for, it is easier to identify the information needed to make it possible.

□ Write down a list of the main functions which vision serves in everyday life.

■ A complete list will be very long indeed but it should be possible to recognize several distinct classes of function including, for example, the recognition of distant objects, the direction of skilled hand movements, and the direction of movements of the whole body about the world.

The point of this exercise is to emphasize that vision serves several rather different functions, that each of these functions may require rather different types of information, and that, consequently, it may not be possible to derive a single description of the image which is suited to all the different tasks. In fact, the image provides a number of different types of information: the overall pattern of light and dark, the distribution of colours, the speed and direction of motion, and so forth. The early stages of visual processing appear to deal with these different types of information separately, and it is useful to think of several descriptions of the image being developed in parallel, one dealing with light and dark, one with colour, one with motion, and so forth. This section takes the same approach, dealing at first only with the overall pattern of light and dark and treating the proximal stimulus as though it were a single, stationary, black-and-white image, rather than two streams of coloured images in continuous motion. The processing of light and dark is treated in some detail because it illustrates many of the main principles of visual coding and introduces the relevant parts of the visual system. Colour, movement and binocular vision are then introduced much more briefly and the section ends by considering the anatomical and neurophysiological evidence that each of these different types of information is analysed separately in the early stages of the process.

4.3.1 Black-and-white vision: encoding information about image luminance

Figure 4.7 is a photograph of a cross-section of human retina, and Figure 4.8 is a simplified schematic diagram which makes the structure clearer. Familiarize yourself with the overall structure and locate the receptors, which are the first neural stage of proper processing. There are two types of receptor—rods and cones—which have rather different roles, but this distinction need not concern you yet and you can assume, for the moment, that all receptors are essentially the same.

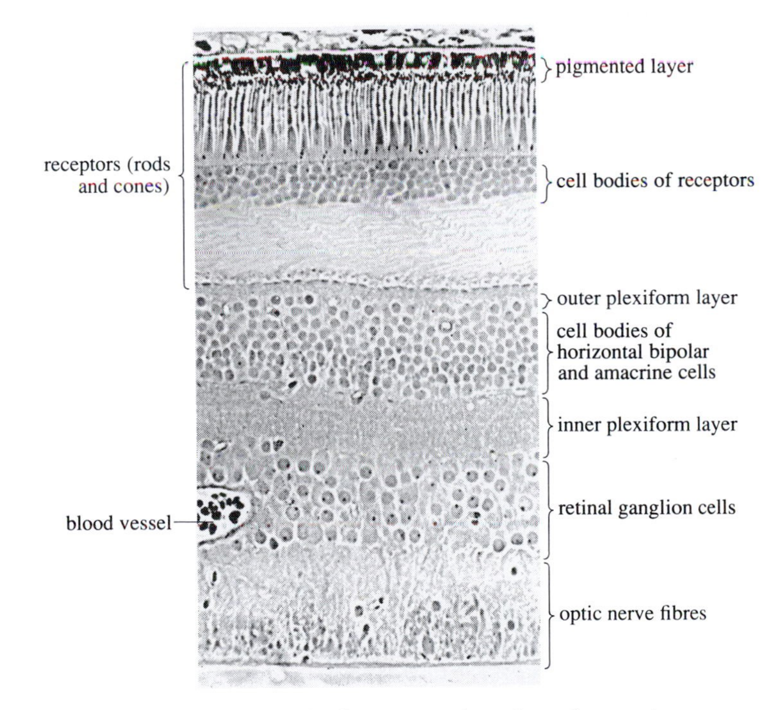

Figure 4.7 Photograph of a cross-section of vertebrate retina.

Retinal receptors are specialized mechanisms for transducing light energy into the electrochemical energy that is the first stage of the neuronal response. Even though, as Figure 4.8 (*overleaf*) shows, light passes through several interneurons before reaching the receptors, these interneurons are not specialized for transduction and so do not respond to light. They only respond indirectly as a result of the neural activity which the light causes in the receptors.

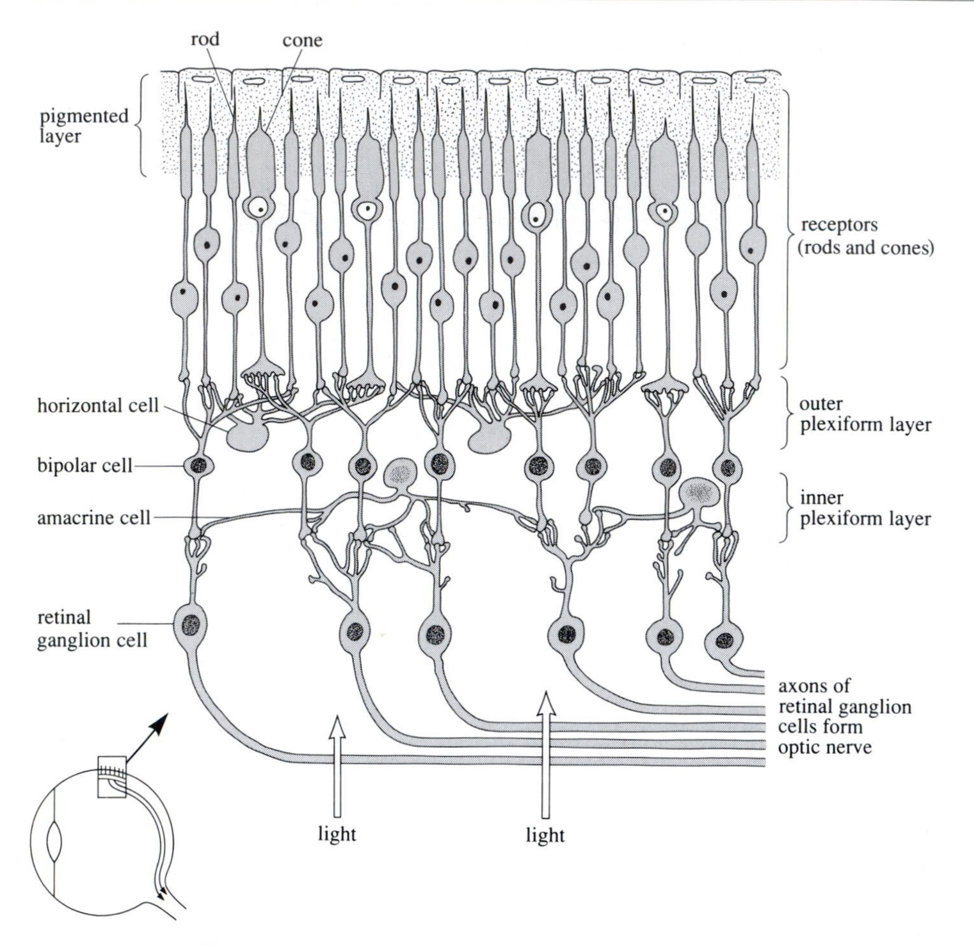

Figure 4.8 Schematic cross-section of vertebrate retina.

Each receptor measures the amount of light at one point in the image, so the pattern of neural activity over many receptors forms a kind of **neural image**, which directly mirrors the pattern of light in the image. All that has changed is the units in which the pattern is measured—from units of light to units of neural response.

Retinal receptors, as illustrated in Figure 4.9, have an outer segment made up of a pile of stacked membranes, and an inner segment containing the nucleus and synaptic contacts. The membrane stacks contain the pigment *rhodopsin*, which is broken down, or bleached, by light. This breakdown changes the cell's permeability to ions, thus producing a receptor potential (Section 1.2.1). Once bleached, rhodopsin is automatically regenerated so that the process can be repeated.

The molecular events involved in transduction are similar across a wide variety of species, and appear to have been conserved during evolution from single-celled organisms sensitive to light. Recordings from receptors in the dark reveal that there is a **dark current** (potential present in the absence of light), produced by sodium ions, Na^+, entering the cell to maintain its depolarization. The bleaching of rhodopsin causes the Na^+ channels to close, so that Na^+ can no longer enter and the membrane hyperpolarizes. This process is exquisitely sensitive—a very small amount of light produces a measurable receptor potential.

You may recall from Book 2, that depolarization generally causes an increase in transmitter release, while hyperpolarization causes a decrease. Thus, retinal receptors actually release more transmitter in the dark than in the light. This may seem strange—but remember that, to a simple animal, a darkening (e.g. a shadow) is often a more relevant stimulus than a lightening.

So far, visual processes are quite straightforward. Transduction just converts a pattern of light into a corresponding pattern of neural response and all the spatial relationships in the original image are preserved in the resulting neural image. Rather like a photograph in a newspaper, the neural image is made up of small points, each point being the response of a single receptor. Since the receptors are very small and closely packed, there are many small points and the neural image provides a faithful representation of the original pattern of light.

□ The point-like nature of the image as represented by the responses of retinal receptors was likened to a photograph in a newspaper. In what important respect does the neural response differ from the photograph?

■ In terms of the amount of transmitter released, retinal receptors respond more to dark regions of the image and less to light regions, thus their pattern of response is more like a photographic negative than the photograph itself.

Although the notion of a neural image is useful, it can be misleading if taken too far. A faithful neural representation of the image is a good place to *start* visual processing, but the purpose of visual processing is not just to take some sort of neural photograph. Remember that vision is not only about describing the image, but also about working out what might have produced it. The function of the early stages in visual processing is thus to extract information that is particularly relevant to that interpretive task, rather than simply to preserve the original image. This point is brought home by the very next steps in the visual process, which also take place in the retina.

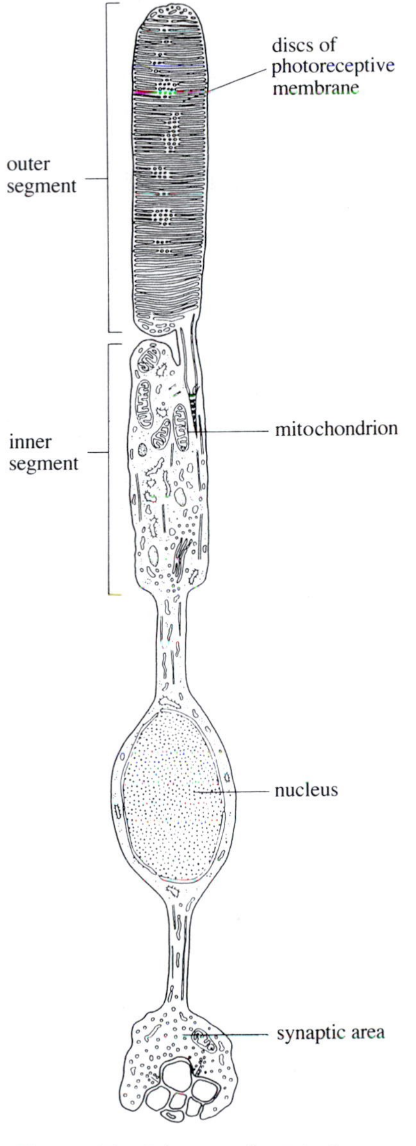

Figure 4.9 Diagram of a retinal receptor.

4.3.2 Retinal processes for encoding spatial structure

Refer back to Figure 4.8. Clearly the retina is a complicated structure and is equipped to do some fairly complex visual processing. Although the precise details vary from species to species, a number of common features are worth noting. First, the 'output' of the retina is via retinal ganglion cells. These are the first stage in the visual pathway where action potentials, as opposed to graded potentials, appear. The axons of the retinal ganglion cells form the optic nerve, which leaves the eyeball at the blind spot (Figure 4.5). Between the receptors and the retinal ganglion cells there are three layers of interneurons. The responses in these interneurons are graded potentials; there are no action potentials. Two layers of interneurons, the horizontal and the amacrine cells, link horizontally along the retina, allowing cells in different positions in the retina to communicate with each other. The third layer, the bipolar cells, links vertically through the retina allowing information to flow from the receptors towards the retinal ganglion cells.

Note particularly that there are many more receptors than there are retinal ganglion cells. In primates the ratio is about 100 to 1. This means that there must be considerable convergence (Book 2, Section 2.3.3) of information in the retina, with many receptors contributing to the response of each retinal ganglion cell.

Basic retinal neurophysiology

The effects of these convergent inputs are revealed when the responses of retinal ganglion cells in cats or monkeys are examined using the set-up shown in Figure 4.10. A microelectrode is placed either directly into a retinal ganglion cell or, equivalently, into a single fibre (the axon of a retinal ganglion cell) of the optic nerve. The animal's eyes are then immobilized with suitable drugs so that each point on the screen in front of the animal always projects to exactly the same point on the retina. The screen is moderately illuminated so that it is possible to make small regions lighter or darker than their immediate surroundings. Then, by moving light or dark spots about the screen, it is possible to study what kind of stimuli will cause the selected retinal ganglion cell to respond.

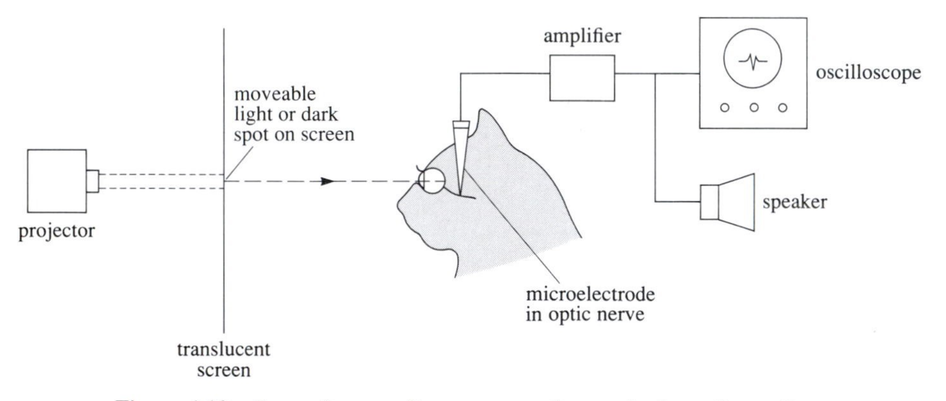

Figure 4.10 Set-up for recording responses from retinal ganglion cells.

When this is done, the first point to emerge is that retinal ganglion cells respond even without direct stimulation by light: they generate action potentials spontaneously and are said to have a **background firing rate**. Thus they have the potential either to increase or to decrease their firing rate in response to light.

The next point to emerge is that each retinal ganglion cell responds to light not just at one point on the retina but over a circular area. This circular area is termed the cell's receptive field (Book 2, Section 9.5.2) and is simply the region of the retina which, when stimulated by light or dark, causes a change in the response of the cell. Receptive fields are just what you would expect from retinal anatomy: many receptors are connected to each ganglion cell, so illumination of any of these receptors will affect the ganglion cell's response.

The final point to emerge is a little more unexpected. Depending upon its position in the receptive field, a spot of light may cause either an increase or a decrease in the firing rate of the retinal ganglion cell (this is why the receptive field is defined in terms of a *change* in response). Moreover, as the spot of light is moved around within the cell's receptive field, some regions cause an increase in firing rate while others cause a decrease. For convenience, the two types of region are called **on-**

regions (+) and **off-regions** (–). An increase in the light falling upon the on-region of a receptive field causes an increase in the retinal ganglion cell's response, while a decrease in light causes a decrease in response. For off-regions, the exact reverse applies. An increase in the light falling upon the off-region of a receptive field causes a decrease in the retinal ganglion cell's response, whereas a decrease in light causes an increase in response.

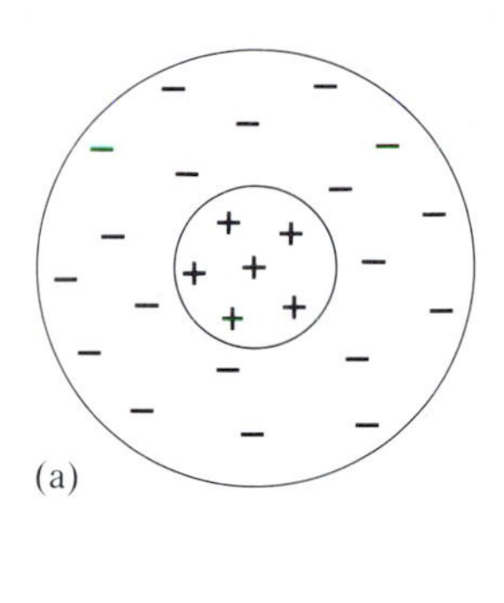

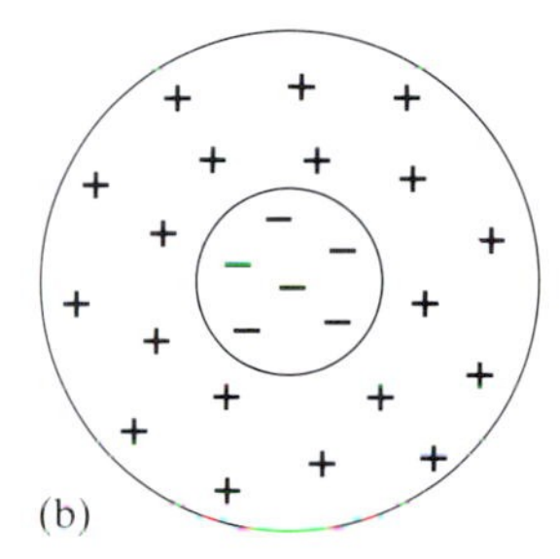

Figure 4.11 Receptive fields of retinal ganglion cells. (a) For an on-centre cell. (b) For an off-centre cell.

The receptive fields of most retinal ganglion cells contain both on- and off-regions, and they are usually arranged, as shown in Figure 4.11, into concentric, antagonistic regions. Some cells have on-centre/off-surround receptive fields, and these are termed **on-centre cells** for short. An equal number of cells have off-centre/on-surround receptive fields and these are termed **off-centre cells**. The arrangement of receptive fields into neighbouring antagonistic regions is an example of lateral inhibition (Section 3.5.3)—so-called because the antagonistic regions are the result of inhibitory influences between cells in neighbouring (lateral) regions of the retina.

Having mapped out the receptive fields of retinal ganglion cells using small spots of light and dark, their responses to larger stimuli can be predicted. According to Figure 4.11, a good stimulus for an on-centre cell should be a spot of light that exactly covers the whole central sub-region of the receptive field: this would increase the response to well above the background firing rate. Conversely, a good stimulus for an off-centre cell would be a spot of darkness which exactly covers the centre of the receptive field.

□ What would happen if the spot is made any larger than this?

■ A larger spot will begin to invade the antagonistic surround and the response of the ganglion cell will decrease. Indeed, the receptive fields of retinal ganglion cells are arranged so that the two antagonistic regions have roughly equal but opposite effect. If the spot is big enough to cover the *whole* receptive field, then both the centre and the surround will be fully activated and the equal and opposite responses will cancel out: the ganglion cell will not change its response from its normal background level.

□ What is the optimal stimulus for an on-centre cell?

■ A bright circle, which exactly covers the central on-region, with a dark ring around it. The increase in light in the central on-region and the decrease in light in the surrounding off-region will both cause the retinal ganglion cell to increase its response.

Figure 4.12 (*overleaf*) summarizes the responses of retinal ganglion cells to spots of light and dark of various sizes.

Suddenly, visual processing is not quite so straightforward. The receptive fields of retinal ganglion cells have two very strange, and at first sight counterproductive, properties. First, having started from a set of small, closely-packed receptors which provide a fine-grained neural representation of the original image, ganglion cell receptive fields sum the light over a region of the image, thus introducing a kind of neural blur and producing a much coarser-grained neural representation. Second, having started with a sensitive mechanism of transduction which can detect very small amounts of light, if all the individual receptors making up a receptive field respond, then lateral inhibition will ensure that the ganglion cell does not change its response at all.

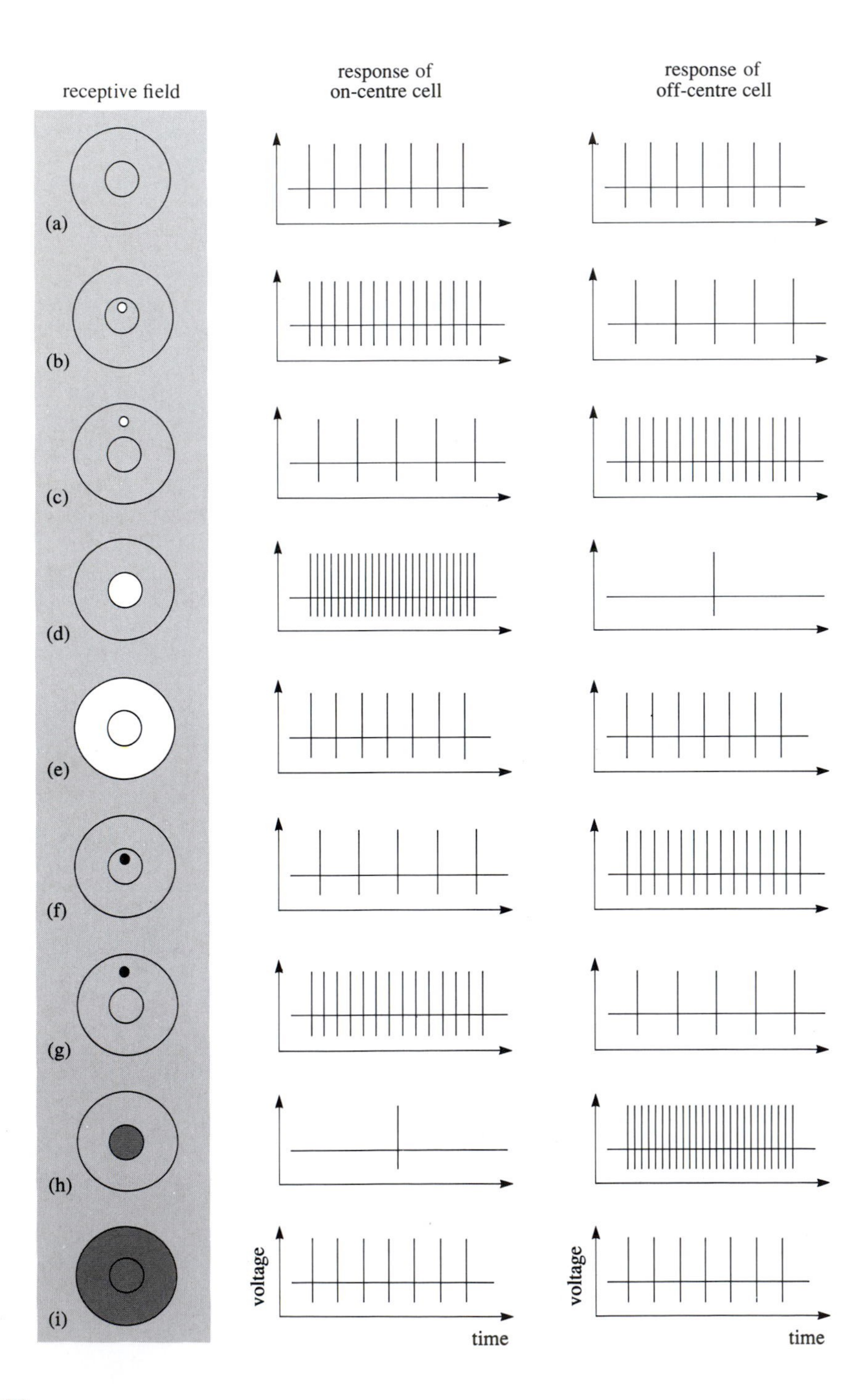

Figure 4.12 Responses of retinal ganglion cells to spots of light or dark of various sizes at different positions in the receptive field. (a) Neutral illumination over whole field: on- and off-centre cells at background firing rate. (b) Small light spot in centre sub-region: on-centre cell—slight increase in firing rate; off-centre cell—slight decrease in firing rate. (c) Small spot of light in surround sub-region: on-centre cell—slight decrease in firing rate; off-centre cell—slight increase in firing rate. (d) Light covers whole centre sub-region: on-centre cell fires at high rate; off-centre cell fires at low rate. (e) Light covers whole receptive field: centre and surround cancel, on- and off-centre cells at background firing rate. (f) Small dark spot in centre sub-region: on-centre cell—slight decrease in firing rate; off-centre cell—slight increase in firing rate. (g) Small dark spot in surround sub-region: on-centre cell—slight increase in firing rate; off-centre cell—slight decrease in firing rate. (h) Dark covers whole centre sub-region: on-centre cell fires at low rate; off-centre cell fires at high rate. (i) Dark covers whole receptive field: centre and surround cancel, on- and off-centre cells at background firing rate.

Yet receptive fields and lateral inhibition are found not only in the visual systems of almost all animals, but also in almost all other sensory modalities (Book 2, Section 9.5.3). Although they seem strange at first, they actually perform an extremely useful function. To understand this function, it is important to understand the problem that the early stages of processing must solve.

The purpose of retinal processing

If you were asked to draw a picture of your immediate surroundings, you would probably begin by sketching down a few lines rather than filling in solid blocks of shading. These lines would represent the boundaries of objects in the visual scene, and you would choose them because object boundaries are the most informative bits of the world (as in Figure 4.1). If you know where the boundaries are, you know where things are and what shape they have. The early stages of visual processing do something very similar—they pick up object boundaries and ignore everything else. Just as you can think of transduction as producing a kind of neural photograph, so you can think of the next visual stage as reducing this to a rather sophisticated *neural line-drawing*.

How do object boundaries show up in the retinal image? Figure 4.13 shows a simple black-and-white image. The line below it shows the amount of light coming from the image, measured at each point along the line AB. This measure of the amount of light projected by a surface is called **luminance** and the graph plotted out in Figure 4.13 is called a **luminance profile** of the image. The boundaries of an object show up in the luminance profile as places where the luminance changes quite abruptly from one position to the next—these are called **luminance discontinuities** (e.g. point d in Figure 4.13). So, if you want to know about object boundaries and you have to deal with images, a good strategy would be to begin by locating luminance discontinuities.

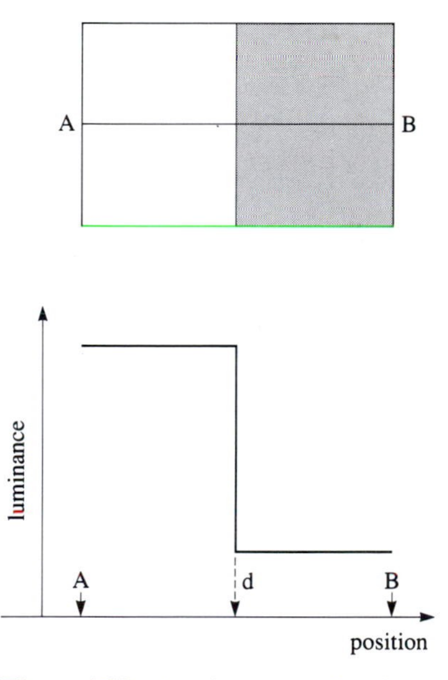

Figure 4.13 Luminance profile of a simple image, measured along the line AB. There is a luminance discontinuity at point d.

However, this mechanism is not foolproof. For instance, if two surfaces of an object are evenly illuminated they will produce more or less the same amount of luminance in the image. In this situation there will be no change in luminance in the image and so not all object boundaries produce luminance discontinuities. The reverse is also true, that is, a luminance discontinuity may be produced even when there is no object boundary.

- □ Can you think of a situation where there will be a luminance discontinuity in the image that does not correspond to an object boundary in the external world?
- ■ A shadow is a good example. Shadows produce luminance discontinuities which do not correspond to object boundaries.

What is the best way to locate luminance discontinuities in images? First, measure the light at each point in the image—just as retinal receptors do. Next, *compare the amount of light at each point with its immediate neighbour*: if the two values are the same then there is no change in luminance but, if they are different, you have potentially found a luminance discontinuity. A simple way to do this kind of comparison would be to connect adjacent pairs of receptors together in an antagonistic way, as shown in Figure 4.14 (*overleaf*). This in effect subtracts the responses of neighbouring receptors. If the light at each point is the same, then both receptors will respond similarly, and the result of subtracting two equal amounts is zero. But if there is a luminance discontinuity in the image, the receptors on each side of it will respond differently, and the result of the subtraction will not be zero. Put another way, this simple process of *spatial comparison* (comparing the responses of adjacent receptors) will produce responses only where there are luminance discontinuities, and thus produce a description of the image containing only things of interest—such as object boundaries.

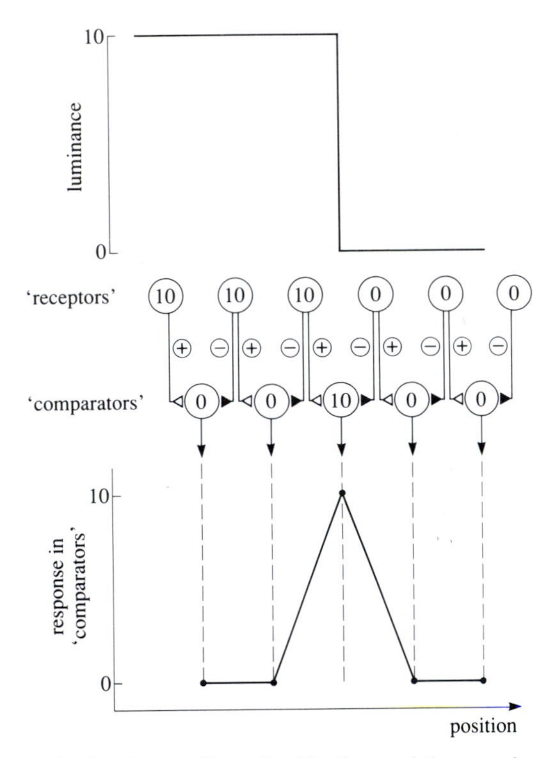

Figure 4.14 Detecting luminance discontinuities by spatial comparisons. The numbers in the 'receptors' and 'comparators' represent the response of each unit. The 'comparators' simply subtract the inputs which each receives from two 'receptors' and respond relative to the result of the subtraction.

Before relating this process back to the receptive fields of retinal ganglion cells, there are three important refinements to add.

First, comparing receptor responses at adjacent positions will detect sharp changes in luminance but will be rather insensitive to the more gradual changes which also occur in natural images. This is because, as shown in Figure 4.15a, there isn't much change in luminance between the two points at which the measurements are made.

- □ How might you solve this problem?
- ■ Doing the comparison between more distant points would solve this problem, as shown in Figure 4.15b.

However, this causes another problem, shown in Figure 4.15c. In complex images, where luminance discontinuities can be closely packed together, comparisons over large distances would be very confusing because they would misrepresent much of the important detail. Unfortunately, there is no perfect solution to this problem: comparisons over small ranges will inevitably be insensitive to gradual luminance changes, while comparisons over large ranges will inevitably misrepresent complex images. The best that can be done is a compromise: work out the comparisons simultaneously at several different scales and then interpret the results of all of the comparisons.

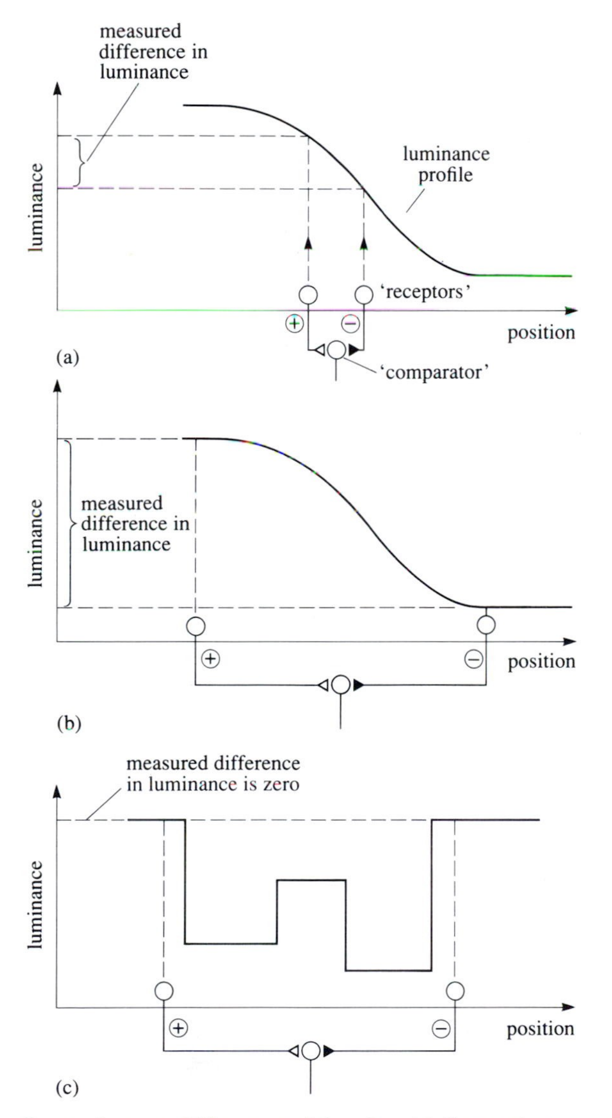

Figure 4.15 Comparisons at different spatial scales. (a) Comparisons at small scales are relatively insensitive to gradual luminance changes. (b) Comparisons at large scales are more sensitive to gradual luminance changes. (c) Comparisons at large scales may give misleading results in complex images.

Second, comparing responses along a horizontal line will detect vertical luminance discontinuities but not horizontal ones.

☐ How might you solve this problem?

■ The comparison must actually be done in all directions so that, no matter what the orientation of a discontinuity, one pair of receptors will always register a difference.

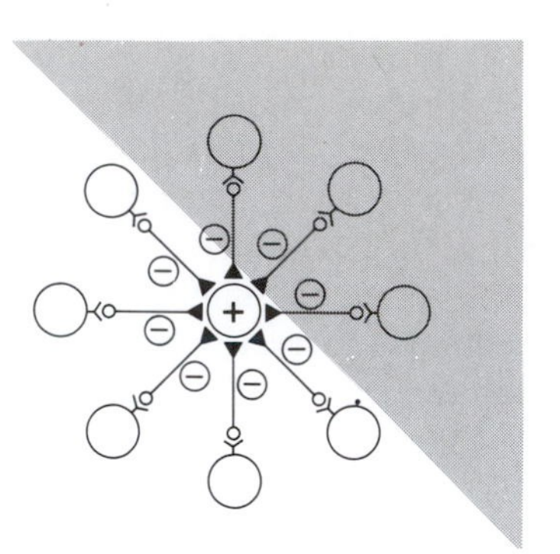

Figure 4.16 Comparisons at all orientations using a ring of receptors.

This refinement is illustrated in Figure 4.16, which uses a ring of receptors to make spatial comparisons at all possible orientations.

Finally, retinal images are far from perfect. The amount of light reaching each point in the image actually fluctuates considerably over time and biological systems, like receptors, will also introduce a certain amount of variation in response. This means that the luminance profile of an object will actually be very variable and might look, for example, like that shown in Figure 4.17a. This is important because the amount of light will vary from point to point in the image, even in the absence of a genuine luminance discontinuity and, since the process of comparison is very sensitive to precisely this type of variation, it is liable to signal many 'spurious' edges.

☐ How might you solve this problem?

■ In order to avoid this, the outputs of several receptors can be combined to produce a more reliable, *average* estimate of response before the comparison is done.

The result of this type of averaging is shown in Figures 4.17b and c, where each point used at the comparison stage was obtained by averaging together the responses of the 3 receptors immediately above it.

You are now in a position to understand just how beautiful retinal processing really is. Retinal ganglion cells function to pick up luminance discontinuities in the image because these usually correspond to object boundaries in the real world. They do this by making spatial comparisons of the amount of light at neighbouring points in the image—thus their receptive fields contain adjacent, antagonistic sub-regions. The comparisons are done simultaneously at all orientations—so receptive fields are circular and consist of a central sub-region surrounded by an annular antagonistic sub-region. In order to pick up luminance discontinuities of different scales, at each point on the retina there are retinal ganglion cells with receptive fields of different sizes: cells with small receptive fields do small-scale comparisons, while cells with large receptive fields do comparisons at a larger scale. Finally, retinal ganglion cell receptive fields do not just make use of a few receptors. In general, many receptors contribute to each sub-region so that their outputs are averaged before the spatial comparison is done.

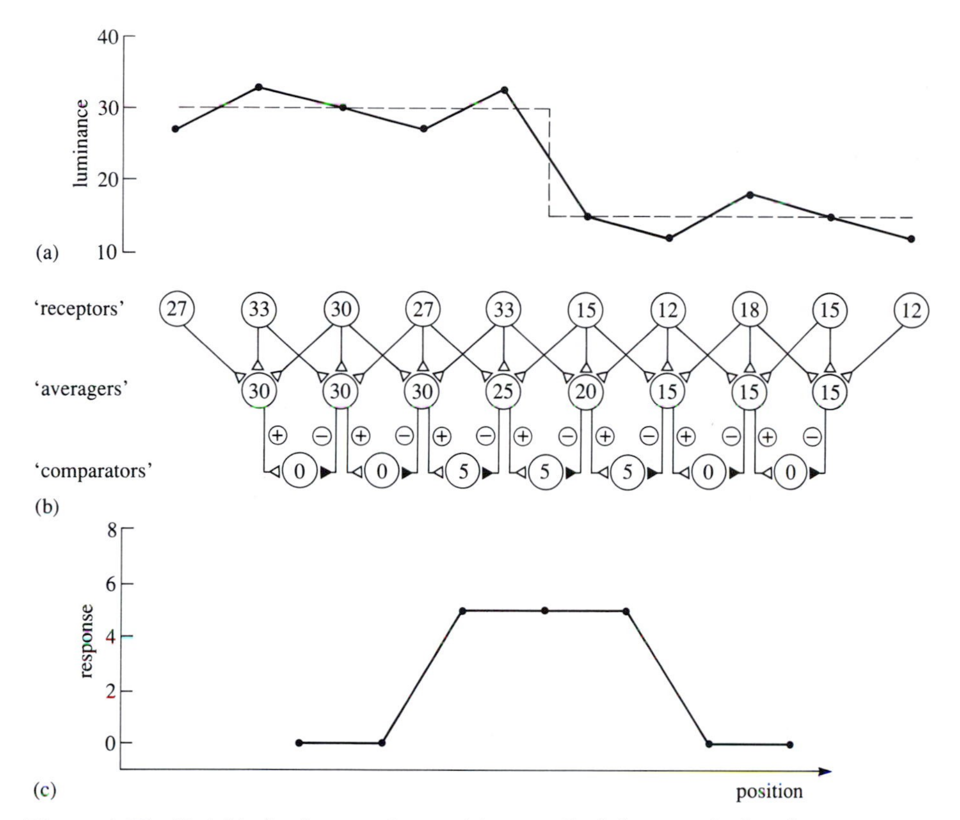

Figure 4.17 Variable luminance of natural images. Each 'averager' takes the average response of the three 'receptors' immediately above it. Each 'comparator' subtracts the inputs from the two 'averagers' immediately above it. The result is a less variable pattern of response in the 'comparators' than would be obtained by comparing the outputs of individual 'receptors'.

The final result of retinal processing is shown in Figure 4.18 (*overleaf*). The top line represents a realistic profile of the kind of luminance discontinuity commonly found in retinal images. The bottom line represents the pattern of response in an array of retinal ganglion cells with receptive fields of the same size. Each point on this line represents the response of a single ganglion cell. The characteristic 'peak–trough' response emphasizes just how efficient retinal processing is. Only a few ganglion cells respond above or below the background firing rate because the receptive fields of most of them are evenly illuminated, yet these few cells convey all the important information about the discontinuity. The *existence* of a discontinuity is signalled by the very fact that some ganglion cells are responding above or below their background rate. Its *position* is given by the point at which the peak turns into a trough, its *amplitude* by the height of the peak or the depth of the trough, its *spatial scale* by the width of the peak or the trough, and its *polarity* (light to dark or dark to light) by the arrangement of the peak and trough. This is very efficient: when you look at a uniform surface, millions of receptors respond and all of them say more or less the same thing. Rather than sending this repetitive hubbub to the brain, just a few responses in a few ganglion cells are all that are needed because, if you know where the changes are and how big they are, you can (if you wish) reconstruct the original image.

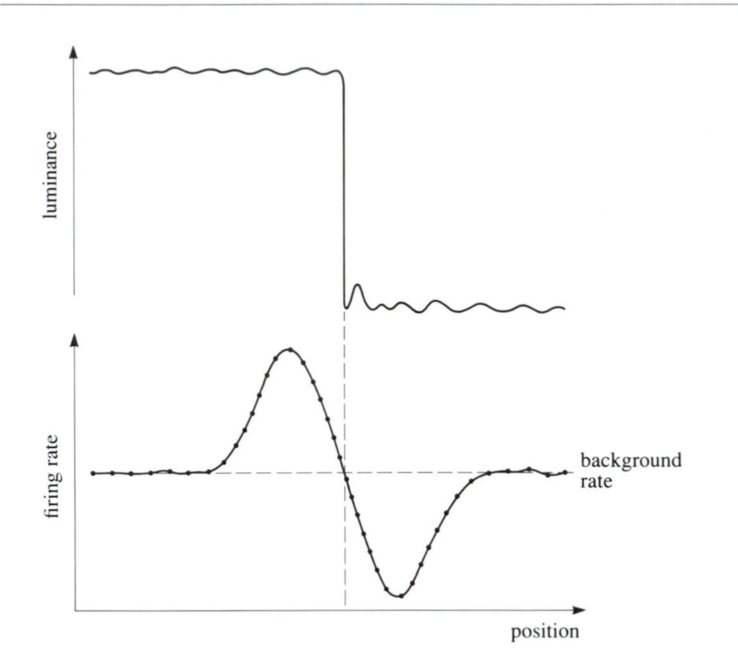

Figure 4.18 The response of an array of on-centre retinal ganglion cells to a luminance discontinuity in a natural image.

Detailed retinal anatomy and physiology

Now that you understand the function of receptive fields, you can consider exactly how retinal cells are 'wired up' to do the job. In fact, the details of retinal anatomy vary widely from species to species but the following principles hold across various animal groups. Most of these details have been worked out on whole retinas which are carefully dissected out and kept alive in a suitable oxygenated nutrient solution. Detailed anatomy and neurophysiology can then be studied using standard dye-labelling and electrical recording techniques.

The basic functional anatomy of the retinal receptive field is illustrated in Figure 4.19. Work your way through this Figure as you read the text that follows.

The central sub-region of each receptive field is produced by the bipolar cells (B), which synapse directly with several neighbouring retinal receptor cells (R). The antagonistic surround region is produced by the horizontal cells (H), which pool the responses from a group of retinal receptor cells and synapse onto the bipolars.

The four important things to remember are:

1 Retinal receptors hyperpolarize in response to light and so release less transmitter when they are illuminated.

2 Horizontal cells are excited (depolarized) by receptor transmitter.

3 Retinal ganglion cells are excited (depolarized) by bipolar transmitter.

4 There are two different types of bipolar cell, those connected to on-centre retinal ganglion cells, which are termed on-centre bipolars, and those connected to off-centre cells, which are termed off-centre bipolars. The differences lie in the detailed organization of the postsynaptic membrane of the bipolar cell, so that the two types of bipolar cell respond differently to the transmitters released by retinal receptors and horizontal cells.

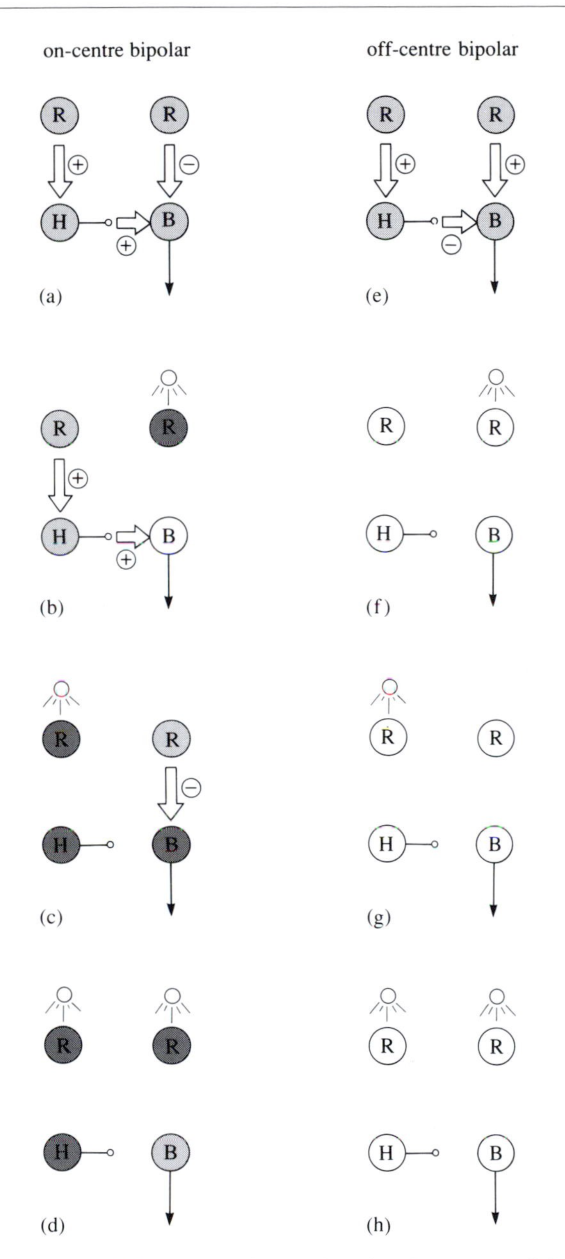

Figure 4.19 Schematic illustration of basic functional anatomy of the retinal receptive field. R: receptor cell; H: horizontal cell; B: bipolar cell. White cells are depolarized (excited), dark cells are hyperpolarized (inhibited) and grey cells are in a neutral state of polarization. See text for further details.

Figures 4.19a–d refer to an on-centre bipolar cell. This type of cell is inhibited (hyperpolarized) by transmitter from receptors and excited (depolarized) by transmitter from horizontal cells.

Look first at Figure 4.19a. This shows two receptors, one connected to a horizontal cell and the other to an on-centre bipolar cell. The horizontal cell is in turn connected to the bipolar cell. The arrows symbolize the release of transmitter and the associated symbols indicate whether the transmitter has an excitatory (+) or inhibitory (–) effect on the postsynaptic cell. The shading of the cells indicates their state of polarization: white indicates depolarization (excitation) and dark grey indicates hyperpolarization (inhibition). When neither receptor is illuminated, there is a background release of transmitter from the receptors which results in both direct inhibition of the bipolar cell and indirect excitation via the horizontal cell. The bipolar cell thus receives a balanced input and is consequently in a neutral state of polarization (shown by light grey shading).

What happens when a light is turned on only in the on-centre of the receptive field of the retinal ganglion cell to which the on-centre bipolar cell is connected? This situation is illustrated in Figure 4.19b by illuminating only the receptor connected directly to the bipolar cell. The illuminated receptor hyperpolarizes and its release of transmitter declines. The bipolar cell thus receives less inhibition (hyperpolarization) from the receptor but the same excitation (depolarization) from the horizontal as it did before. It therefore has a net excitatory input and consequently depolarizes. The retinal ganglion cell to which it is connected now receives a greater excitatory (depolarizing) input and so its firing rate increases.

What happens when a light is turned on only in the off-surround of the receptive field? This situation is illustrated in Figure 4.19c by illuminating only the receptor connected to the horizontal cell. This receptor hyperpolarizes and so releases less transmitter. The horizontal cell receives less excitatory input, hyperpolarizes, and consequently releases less of *its* transmitter. The bipolar cell now receives the same amount of inhibition (hyperpolarization) directly from the receptor but less excitation (depolarization) from the horizontal cell. It has a net inhibitory input and so hyperpolarizes. The retinal ganglion cell to which it is connected is hyperpolarized and its firing rate decreases.

Finally what happens when the whole of the receptive field is illuminated? This situation is illustrated in Figure 4.19d by illuminating both receptors. Both receptors are hyperpolarized and release less transmitter. The bipolar cell thus receives less inhibition (hyperpolarization) directly from the receptor. But it also receives less excitation (depolarization) indirectly from the horizontal cell. Although both excitatory (depolarizing) and inhibitory (hyperpolarizing) inputs decline, they remain in balance and so the bipolar cell remains in a neutral state of polarization. The retinal ganglion cell to which it is connected does not change its response from the background firing rate.

While this is complicated, it is also beautiful. Check your understanding of it by following through the same logic for an off-centre bipolar cell, using Figures 4.19e–h, but this time filling in the shading, arrows and symbols for yourself in Figures 4.19f–h. Use the same convention as before (white for depolarization, dark grey for hyperpolarization, light grey for neutral). The only differences between the on-centre and off-centre cells are that the bipolar is excited (depolarized) by receptor transmitter and inhibited (hyperpolarized) by horizontal cell transmitter. The answer is shown in Figure 4.20 (p. 98).

☐ Do the receptor cells in the on- and off-regions of a retinal ganglion cell's receptive field respond in different ways?

- ■ No, all retinal receptors in vertebrates respond to light with a graded hyperpolarization. Different receptors can exert different effects on a given retinal ganglion cell because they are connected to it by different interneurons. Some interneurons exert an excitatory effect upon the retinal ganglion cell, so that an increase in their response produces an increased firing rate in the retinal ganglion cell. Others exert an inhibitory effect, so that an increase in their response produces a decrease in the retinal ganglion cell's firing rate.

- □ Two adjacent retinal ganglion cells will have receptive fields at more or less the same position on the retina, In fact their receptive fields will almost certainly overlap. How might this arise?

- ■ Each receptor can contribute to many different receptive fields—being connected to many different ganglion cells by many different interneurons. Indeed, because interneurons may exert either an excitatory or an inhibitory effect, the same receptor may form part of the on-region of the receptive field of one retinal ganglion cell, and the off-region of the receptive field of an adjacent retinal ganglion cell.

Psychological demonstrations of retinal spatial coding

The study of retinal processes provides a good example of an interdisciplinary approach: a theoretical consideration of the 'task' of vision suggests a method of extracting an efficient description of the image, while anatomists and neurophysiologists reveal that this method is actually implemented in the retina. Psychologists too have contributed to this area, strengthening understanding even further by devising simple displays which appear to reveal retinal processes actually functioning in humans.

Figure 4.21 (*overleaf*) shows a blurred luminance discontinuity, shading gradually down from a uniformly light region to a uniformly dark region. The luminance profile is shown below the stimulus. Your perception does not correspond with this physical profile—you should see a fuzzy dark stripe on the dark side of the edge and a fuzzy bright stripe on the light side of the edge. These illusory stripes are called Mach bands. The bottom line in Figure 4.21 shows the pattern of response which the stimulus would produce in an array of retinal ganglion cells. Notice that this response profile is distorted in just the same way as your perception—there is a peak on the light side of the edge and a trough on the dark side. Here, then, it appears that what you see is directly related to the output of retinal ganglion cells.

Figure 4.22 (p. 99) shows a slightly more subtle stimulus called the Craik–Cornsweet–O'Brien illusion. The luminance profile, shown below the stimulus, consists of two smooth ramps surrounding a sharp luminance discontinuity, the whole flanked by two regions of the *same* uniform lightness. The illusory perception is of a sharp edge between two roughly uniform fields of *different* lightness. You can confirm that this perception is illusory by using a thin strip of paper to cover the central region of the stimulus—the two regions should now look the same.

☐ Why should this rather strange stimulus look like a sharp light–dark edge?

■ Remember that retinal ganglion cells respond only to luminance changes. They will thus respond to the sharp luminance discontinuity in the centre of the stimulus but not to the smooth ramps on either side, which have been carefully designed so that the change in luminance is too gradual to cause any response. Thus the output of the retina says that there is a sharp change from light to dark in the centre of the display and no evidence of further change in the flanking regions, so that is what you see. Put another way, the pattern of response which the stimulus evokes in retinal ganglion cells, as shown in the bottom line of Figure 4.22, has the same characteristic peak–trough form as would be evoked by a normal light–dark edge (Figure 4.18). This pattern of response is therefore *interpreted* by higher visual processes as a normal light–dark edge.

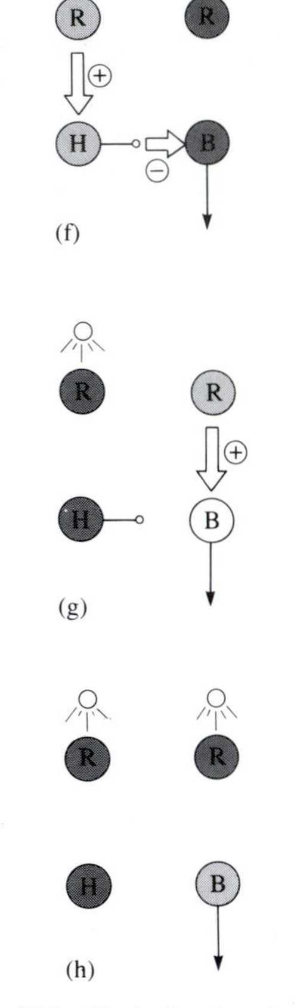

Figure 4.20 Basic functional anatomy of the retinal receptive field of an off-centre bipolar cell. This is what (f) to (h) of Figure 4.19 should look like when you have filled them in.

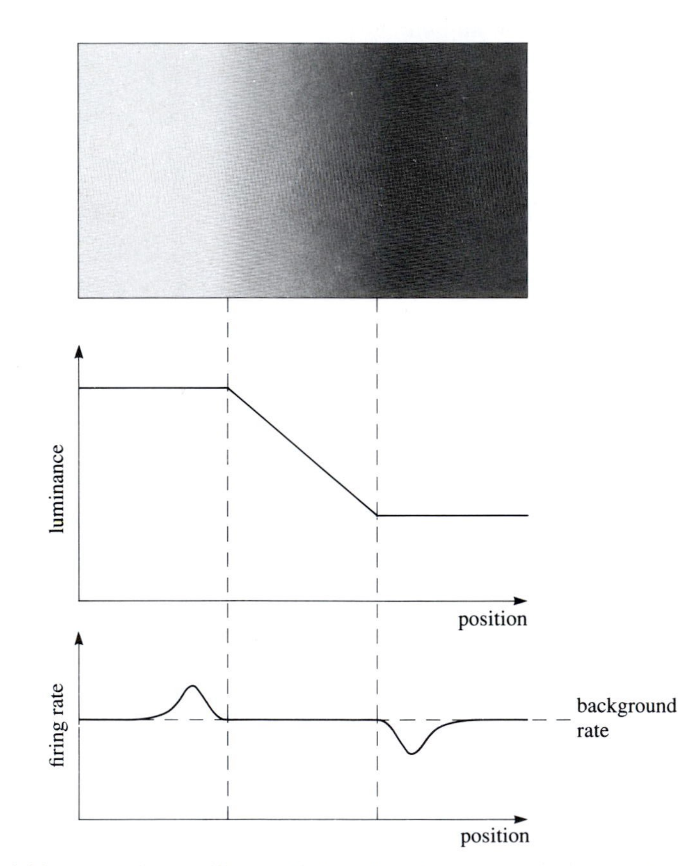

Figure 4.21 Mach bands: illusory light and dark stripes appear on either side of the boundary.

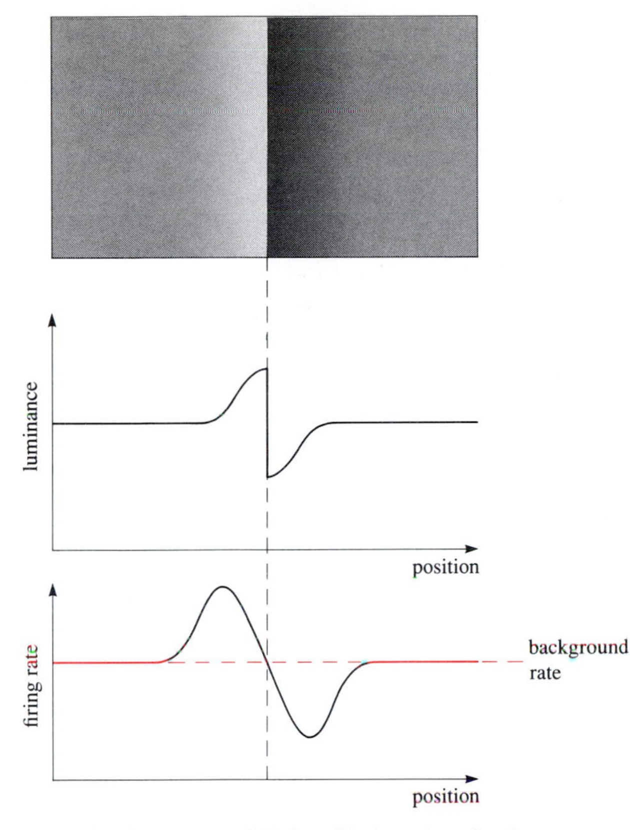

Figure 4.22 The Craik–Cornsweet–O'Brien illusion: the stimulus appears as an edge between two regions of different lightness.

- □ How do you need to refine the explanation of Mach bands to fit in with the explanation of the Craik–Cornsweet–O'Brien illusion?

- ■ The discussion of Mach bands suggested that the output from the retina directly determined our perception. The discussion of the Craik–Cornsweet–O'Brien illusion was a little more sophisticated because it recognized that the output of the retina has to be interpreted by subsequent processes. So the explanation of Mach bands should really have said that the pattern of output in the ganglion cells, shown in the bottom line in Figure 4.17, is *interpreted* as a pair of light and dark fuzzy bars flanking a blurred edge.

4.3.3 Retinal coding of brightness and lightness

The discussion so far has only been concerned with retinal processing of the spatial structure of the image—detecting and locating luminance discontinuities. What about the *amount* of light in the image? This may seem relatively straightforward but, in fact, it presents the visual system with a terrible problem. The brightest light encountered in normal life is physically about 100 million million (100 000 000 000 000) times brighter than the dimmest. How can the visual system possibly function efficiently over such an unimaginable range?

Fortunately, the full range of brightnesses never occurs simultaneously: the brightest object in a given scene is unlikely to be more than a few thousand times brighter than the dimmest. The visual system is thus able to function in a rather ingenious way. The processes which encode brightness are only sensitive to a relatively limited range of variation, but the visual system automatically adjusts its sensitivity to bring the prevailing lighting conditions into this range. This process of adjusting to the prevailing lighting conditions, called **dark adaptation**, is rather like sliding a small magnifying glass up and down a very long ruler to make accurate measurements at whatever point they are needed.

The processes of dark adaptation are complex but two aspects are worthy of brief mention.

There are actually two types of retinal receptor: rods, which are very sensitive and function best under dim conditions; and cones, which are less sensitive and function best under bright conditions. In effect, there are two separate retinal systems—one based on rods which is well adapted to night (scotopic) conditions, and the other based on cones which is well adapted to daylight (photopic) conditions. The two systems have complementary properties. Cones predominate in and around the fovea where retinal receptive fields are rather small, so the photopic system can resolve the fine detail of objects looked at directly. Rods predominate in the periphery of the retina where receptive fields are much larger, so the scotopic system maximizes sensitivity by pooling the responses of many receptors but is consequently less able to resolve spatial detail because it produces a much coarser-grained representation of the image.

□ You have already encountered a second mechanism of dark adaptation. What is it?

■ The pupil can change its size to control the amount of light entering the eye.

This ability actually allows only relatively fine adjustments because the range of possible pupil sizes is limited.

The coding of luminance intensity: the retinal basis of lightness constancy

Dark adaptation functions to adjust the sensitivity of the retina to the prevailing lighting conditions. Once properly adapted, how does the retina encode the amount of light within a given scene? You already know that retinal receptive fields are insensitive to uniform regions of the image and that they respond only at places where the luminance changes. This means that receptive fields signal only *change* in luminance, rather than the absolute amount of light. Although this may seem strange, it is actually a very good solution to the problem that the visual system has to solve.

Most objects reflect, rather than emit light and the amount of light that a surface reflects depends upon the amount of light striking the surface (the **illumination**) and upon the proportion of this light that the surface reflects (the **reflectance** of the surface). If the visual system simply measured the amount of light at each point in the image, it would inevitably confound these two factors and you would not be able to distinguish between a well-lit unreflective surface and a badly-lit reflective one. This problem is illustrated in Figure 4.23.

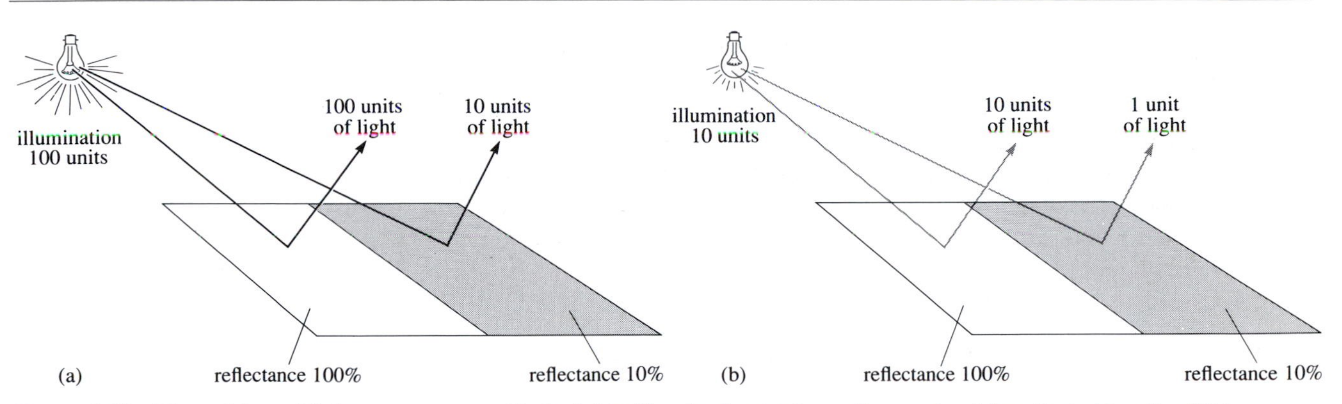

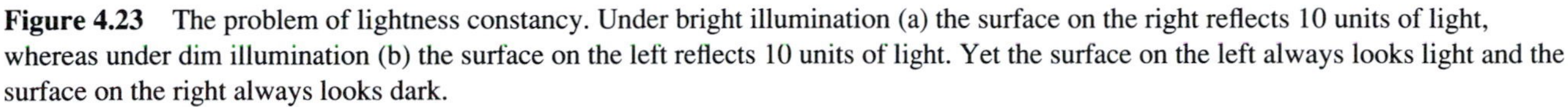

Figure 4.23 The problem of lightness constancy. Under bright illumination (a) the surface on the right reflects 10 units of light, whereas under dim illumination (b) the surface on the left reflects 10 units of light. Yet the surface on the left always looks light and the surface on the right always looks dark.

In fact, of course, the visual system is not fooled in this way. **Lightness constancy** refers to the fact that a grey (i.e. neutrally reflective) surface continues to look grey over a huge range of lighting conditions: it does not look white (strongly reflective) when well-lit, or black (weakly reflective) when badly-lit. This is a remarkable and extremely useful ability. It means that, even though the visual system has access to only a single value (image luminance), it can somehow disentangle the two underlying factors to extract an unvarying and informative property of objects (surface reflectance) despite potentially confusing variations in illumination. It is this remarkable ability which allows you to make quite separate comments about the reflectance of a surface and its illumination. The perceptual correlate of surface reflectance is called **lightness**, while the perceptual correlate of illumination is called **brightness**. Thus it makes sense to talk of a light surface which is brightly-lit.

How is lightness constancy achieved? Look again at Figure 4.23. Although the amount of light coming from each surface varies along with the illumination, there is something about the amount of light coming from them that always stays the same.

□ Can you see what it is?

■ The *ratio* of the light from the two surfaces always remains the same: the surface on the right always reflects 10 times as much light as the surface on the left, because it is 10 times as reflective.

You already know that retinal receptive fields compare the amount of light in adjacent parts of the image. So far, the comparisons have been described as *subtracting* the response in one region from the response in the next and so working out the *difference* in response.

□ How would you modify the process to calculate the *ratio* of responses rather than the difference?

■ You would calculate the ratio by *dividing* the response in one region by the response in the next. In fact, the effect of lateral inhibition is more like a division than a subtraction, so retinal receptive fields do perform the appropriate calculation.

By calculating the *ratio* of the amount of light in neighbouring regions of the image, the output of a retinal ganglion cell signals the *relative* reflectance of adjacent surfaces in the world and this will be completely unaffected by variations in the illumination of the scene.

The psychology of lightness perception

Television is actually based on a lightness illusion. Television images often seem to contain black regions which are much darker than the screen appears when the set is turned off. But a television screen cannot actually be physically darker than when it is turned off, because cathode rays can only brighten the screen: dark regions are just brightened less than light regions. You see apparent blacks because the visual system is relatively insensitive to *absolute* luminance, and more concerned with *relative* luminance. Relatively dark regions appear black in the television image because they are much darker than the lighter regions of the image and because, within a given context, the visual system interprets 'relatively dark' as 'black', and 'relatively light' as 'white'.

This phenomenon is called **lightness contrast** and is illustrated in a simple form by Figure 4.24. Here all of the central squares have the same physical luminance, but those with a dark surround appear lighter than those with a light surround. Since retinal ganglion cells calculate the ratio of adjacent luminances, perceived lightness will depend not on the *absolute* luminance of the central square, but on its *relationship* to the surrounding luminance: the bigger the difference between these two luminances, the lighter (or darker) the square appears in relation to its surround.

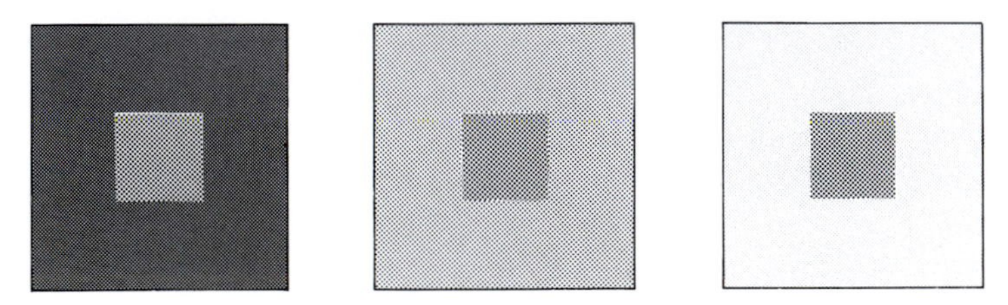

Figure 4.24 Lightness contrast: the central squares appear to have different lightness even though they are physically identical.

These examples support the idea that the responses of retinal ganglion cells are important in signalling relative lightness. But you should not assume that the response of a retinal ganglion cell is the *only* thing that determines how you see lightness.

☐ You have already encountered one example where the response in retinal ganglion cells is *not* interpreted as a difference in lightness. Can you work out what it is?

■ A shadow will generally produce a luminance discontinuity in the image and, consequently, a response in retinal ganglion cells. But the visual system correctly interprets a shadow as a change in illumination, rather than a change in surface reflectance.

An even more compelling demonstration of the role of more sophisticated processes in lightness perception is provided by the Mach card. Take a blank sheet of white paper or card, fold it in half, and stand it on a table like a tent. Arrange the lighting so that one side of the tent is better illuminated than the other. Your retinal image now contains a luminance discontinuity corresponding to the ridge of the tent. The very fact that you continue to see both sides of the tent as white is good evidence that you do not always interpret the responses of retinal ganglion cells as signalling changes in surface reflectance. But there is more.

Close one eye, keep your head still and gaze at the ridge of the tent. With a little perseverance you can reverse the apparent depth of your tent so that it looks like an open book with the pages sticking out towards you. When this happens, your perception of surface lightness will change dramatically and the surface of the card will appear shimmery and unreal. Nothing whatsoever has changed in the image—the pattern of luminance is just the same—yet your interpretation of that pattern is very different.

This is compelling evidence that the perception of surface lightness depends not only upon retinal responses, but also upon the perceived spatial arrangement of surfaces in the world. The interaction between perceived depth and perceived lightness is not really so surprising: if you have ever tried your hand at drawing you will know that, while line drawings are quite effective, shading (i.e. the pattern of image luminance) is very important in conveying an impression of depth.

Summary of the retinal processing of luminance

The vertebrate retina is specialized for the detection of object boundaries. Retinal ganglion cells have circular receptive fields composed of concentric antagonistic sub-regions which compare the amount of light in adjacent parts of the image and so detect luminance discontinuities. The comparison is done in the form of a ratio so that the output of the cells is unaffected by changes in overall illumination. Retinal output can thus be thought of as a sophisticated neural line-drawing, in which the boldness of the lines is closely related to the relative reflectance of neighbouring surfaces. You must remember, however, that retinal output is not the same as your perception of an image—it is just a very efficient description of that image and provides the basis for subsequent interpretive processes.

4.3.4 Central processing of luminance

From Figure 4.2 (p. 78), and from Book 2, Section 8.8.3, you will know that, in mammals, the optic nerves from each eye meet at the optic chiasma and then project via the lateral geniculate nucleus (LGN) to the visual cortex. This pathway links the next stages in the process of visual recognition. The output from the retina is a rich but general description of the image, which emphasizes the spatial changes in image luminance. What happens next?

The lateral geniculate nucleus

The axons of retinal ganglion cells synapse directly onto cells in the lateral geniculate nucleus (LGN), which organizes and then relays the information to the cortex. Cells in the LGN have very similar properties to those in the retina: they have circular receptive fields organized into concentric antagonistic sub-regions. Rather than performing any new types of processing, the LGN seems to be more concerned with controlling which signals reach the cortex. The responses of LGN cells are very much affected by wakefulness, through interactions with the reticular system, and there is an important input from the cortex itself, suggesting a powerful feedback system.

The primary visual cortex

Axons from the LGN reach the cortex via the optic radiation and terminate in Area 17. Rather confusingly, this region is also called the striate cortex, the primary visual cortex or, in primates only, V1 (Figure 4.2). The projection is **retinotopically mapped,** which means that the spatial arrangement of the image is preserved: retinal ganglion cells with receptive fields which are close together in the image project to cells which are close together in the cortex. However, receptive fields at and near the fovea are small, so many more retinal ganglion cells are found here than in the retinal periphery, where receptive fields are much larger. This unequal representation is preserved in the cortex, where up to 30% of the neural tissue is devoted to processing just the input from the fovea. Consequently, the cortical map of the image is distorted as though viewed through a wide-angle ('fish-eye') camera lens: the small central region of the fovea projects to a large area of the cortex, while the large area of the retinal periphery projects to a relatively small cortical area.

Basic cortical anatomy and neurophysiology

Some cells in V1 have the familiar circular receptive field organization found in the retina and LGN, but most cells have receptive fields with rather different properties. The visual properties of cortical cells were first investigated by the Nobel laureates David Hubel and Torsten Wiesel. Hubel and Wiesel used a set-up very like that shown in Figure 4.10 (p. 86), the only difference being that the recording electrode was placed in the cortex rather than the retina. Remember that, even though the recorded cells are located in the cortex, they are connected, via other cells, to retinal cells and their receptive fields are on the retina.

The most striking property of most cortical cells is that their receptive fields are elongated, rather than circular, as shown in Figure 4.25. Hubel and Wiesel further distinguished several different types of cortical cell on the basis of their receptive field properties:

Simple cells have elongated receptive fields divided into distinct antagonistic sub-regions.

- □ What would be the most effective stimuli for simple cells with receptive fields like those shown in Figures 4.25a?
- ■ A cell with a receptive field like that shown Figure 4.25a would respond best to a vertically-oriented bright bar of light which covers the entire on-region without invading the off-region. In Figure 4.25b, the best stimulus would be a horizontally-oriented luminance discontinuity which exactly coincides with the boundary between the on- and off-regions of the receptive field.

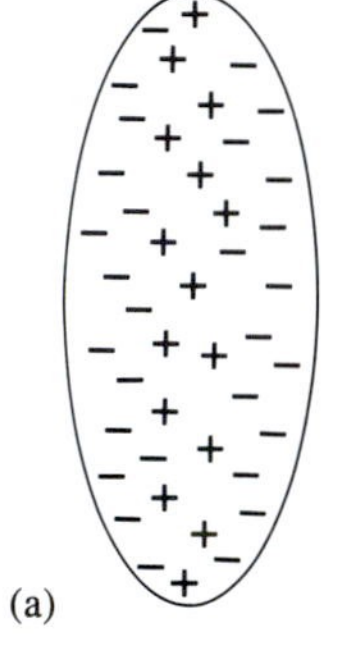

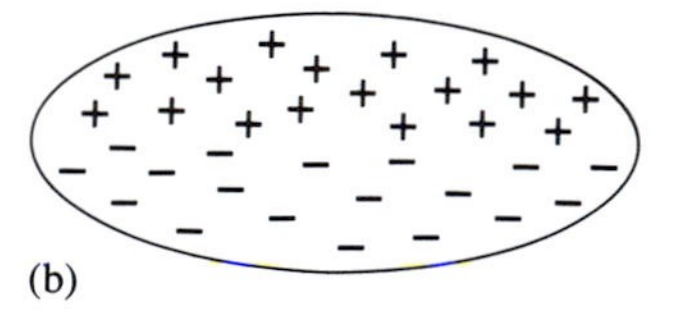

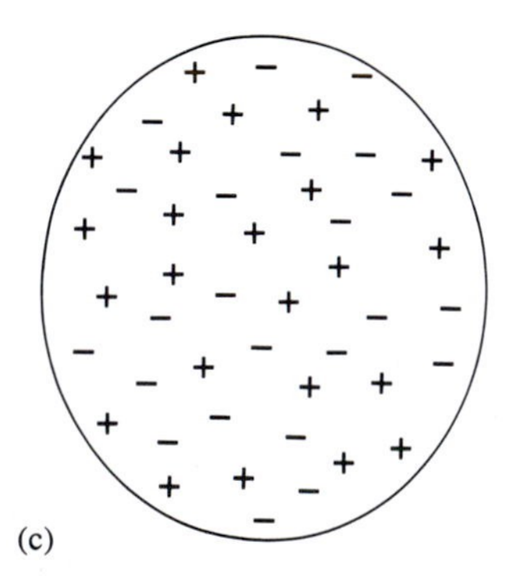

Figure 4.25 Typical receptive fields of cortical cells. (a) and (b) Simple cells. (c) A complex cell.

Complex cells have elongated receptive fields but the fields are larger than those of simple cells and are not divided into distinct sub-regions. These cells also respond well to a bar or luminance discontinuity at a specific orientation, but the stimulus can be positioned anywhere in the receptive field.

Hypercomplex cells are similar to complex cells, but respond best to oriented bars or luminance discontinuities which end within the receptive field. These cells are nowadays often called **end-stopped cells**. Hubel and Wiesel originally distinguished between lower-order hypercomplex cells, which have one preferred stimulus orientation, and higher-order hypercomplex cells, which have two preferred orientations at right angles to each other.

These neurophysiological findings almost inevitably lead to a theory of visual perception based on the related concepts of **feature detection** and **hierarchical wiring**. According to this theory, simple cells would function as feature detectors, signalling the presence of object boundaries with particular orientations. These cells would then feed into complex cells which, because of their larger receptive fields, would be able to detect these important features irrespective of their retinal position. Complex cells would in turn feed into hypercomplex cells, which would combine object boundaries and so act as corner detectors.

It is easy to imagine the next stages in this hierarchical process, with each stage becoming progressively more selective for stimulus properties and less selective for stimulus position, combining corner detectors to make shape detectors, shape detectors into object detectors, and so forth.

This theory, although simple and attractive, remains controversial. The next sections deal with the notions of feature detection and hierarchical processing in turn.

Feature detectors: The coding of stimulus orientation

The original idea that an individual simple cell can act as a feature detector is wrong, at least in detail, because simple cells are just not selective enough to fill this role. The problem is illustrated in Figure 4.26: an individual simple cell will respond to a considerable range of stimulus orientations and so its output cannot be taken to signify the presence of a stimulus with one particular orientation. Moreover, an optimally-oriented but fairly dim stimulus produces exactly the same response in the cell as a non-optimally oriented but bright stimulus: an individual cell cannot distinguish between the orientation and the brightness of the stimulus and so cannot unambiguously signal the orientation of a feature.

Modern theories of visual coding have shifted the emphasis away from the outputs of individual cells towards the *pattern* of output across a population of cells. This principle is illustrated in Figure 4.27 (*overleaf*). Imagine a population of cells with receptive fields at the same retinal position but with slightly different orientations. A stimulus with a particular orientation evokes a characteristic pattern of activity in these cells: the cell whose receptive field lines up exactly with the stimulus responds most, those with slightly misaligned receptive fields rather less, and so forth. Changing the brightness of the stimulus changes the amount of response in each cell, but the pattern of response will be undisturbed: the same cell will still be responding most.

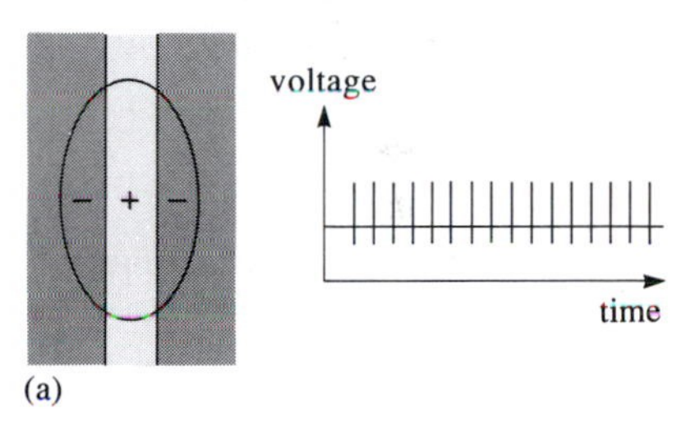

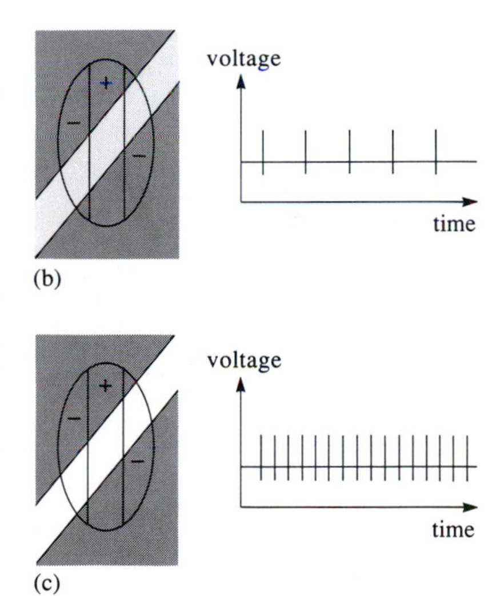

Figure 4.26 Ambiguity of responses of individual cells. (a) A light vertical bar covers the excitatory sub-region of the receptive field and produces a strong response from the cell. (b) When the bar is rotated so that it no longer exactly covers the central sub-region, the response declines. (c) The response can be returned to its previous high level by increasing the brightness of the bar.

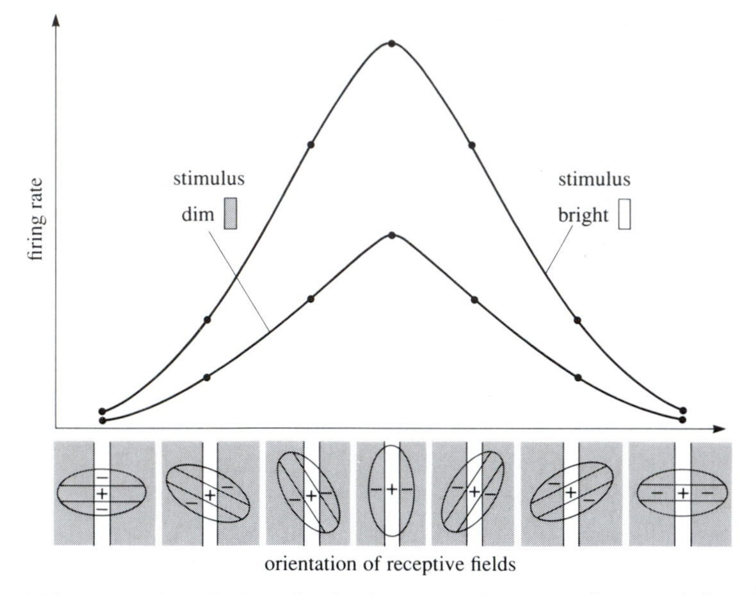

Figure 4.27 The coding of orientation by the pattern of responses in a population of cells, each selective for a slightly different stimulus orientation. Irrespective of the brightness of the stimulus, the same cell always responds most.

There is good anatomical evidence that the visual cortex is equipped to perform this more sophisticated type of coding, and good psychological evidence that it actually does so. Hubel and Wiesel found that, for each retinal location, there is a set of cortical cells with receptive fields at all orientations, and that these cells are beautifully arranged anatomically, as shown in Figure 4.28. Cells in the visual cortex, like those in the rest of the cortex, are arranged into columns at right angles to the surface of the cortex. A microelectrode passing down one of these columns will encounter many cells of different types as it passes from layer to layer. Hubel and Wiesel found that, irrespective of type, all the cells in a given column had receptive fields with the same orientation. They termed this arrangement an **orientation column**. Moreover, as the electrode shifted from

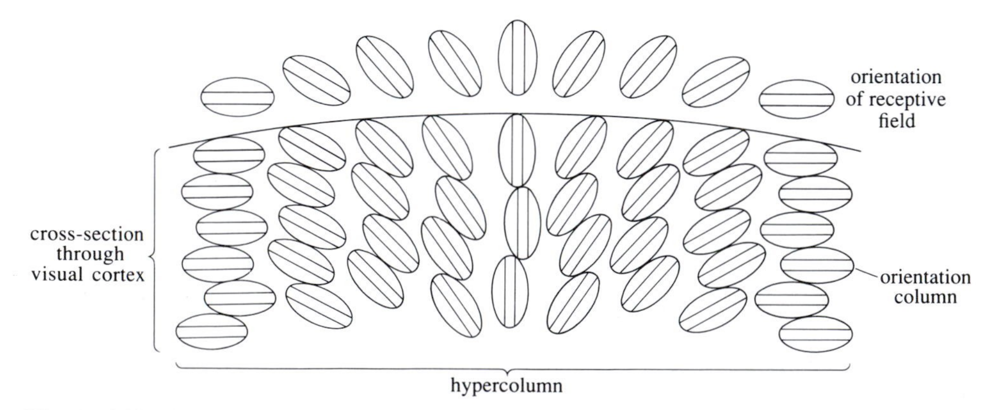

Figure 4.28 Schematic drawing showing the arrangement of orientation columns in a hypercolumn. Cells with the same preferred stimulus orientation are arranged into columns. A complete set of columns, representing all the possible stimulus orientations, is called a hypercolumn.

column to column, the preferred orientation of the cells gradually shifted around the clock. Within each small region of the cortex, consisting of a fairly small set of orientation columns, all the cells have receptive fields in roughly the same retinal position, each column of cells is selective for one orientation, and all orientations are represented by the different columns. The complete set of cells 'looking at' each region of the retinal image is called a **hypercolumn**.

Box 4.1 Autoradiography

Once a radioactively-labelled substance has been incorporated into living cells, they can be recognized by a technique called autoradiography. The basis of autoradiography is explained in Figure 4.29. The tissue of interest is removed from an animal previously injected with some kind of radiolabelled substance (one that will be incorporated into the particular cells that are of interest, e.g. radiolabelled 2-deoxyglucose (2-DG) which will accumulate in active cortical cells) and cut into thin slices. Each slice is mounted on a glass slide and coated with a photographic emulsion. After a period of time in the dark, the emulsion is processed using normal photographic techniques. The emulsion contains silver salts which react to radioactivity to become silver ions. Only those silver salts immediately next to radiolabel in the tissue become silver ions. The processing turns the silver ions into grains of silver metal which can be seen under the microscope. Silver grains will be seen in the emulsion in a pattern which exactly corresponds to the pattern of radiolabel in the tissue slice.

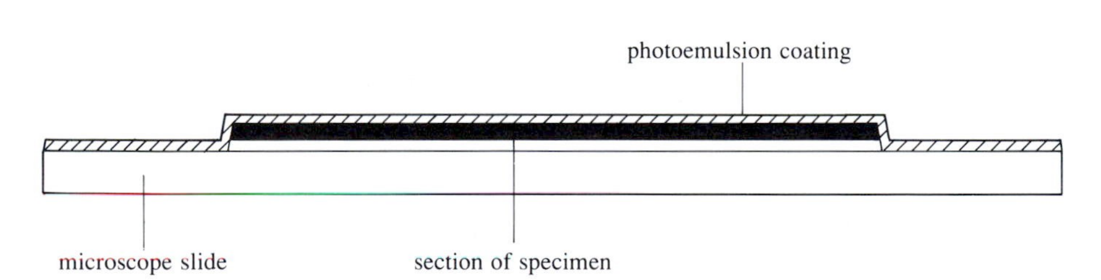

Figure 4.29 Diagram of a glass slide on which is a tissue slice coated with photographic emulsion.

The beautiful anatomical arrangement revealed by single-cell recording has since been confirmed using radioactively labelled 2-deoxyglucose (2-DG), which accumulates in actively responding cells. An animal is exposed to a striped pattern at a specific orientation while labelled 2-DG is present in the bloodstream supplying the brain. Subsequent *autoradiograms* (see Box 4.1) of sections down through the cortex (parallel to the columns) show a characteristic pattern of stripes—each stripe corresponding to the small set of columns which responded to the stimulus and which thus took up the radioactive 2-DG (Figure 4.30a, *overleaf*). Tangential sections (at right angles to the columns) show a complex pattern of 'whorls'—tracing out how a single orientation is represented across the surface of the cortex so that each hypercolumn receives its full complement of orientations (Figure 4.30b, *overleaf*).

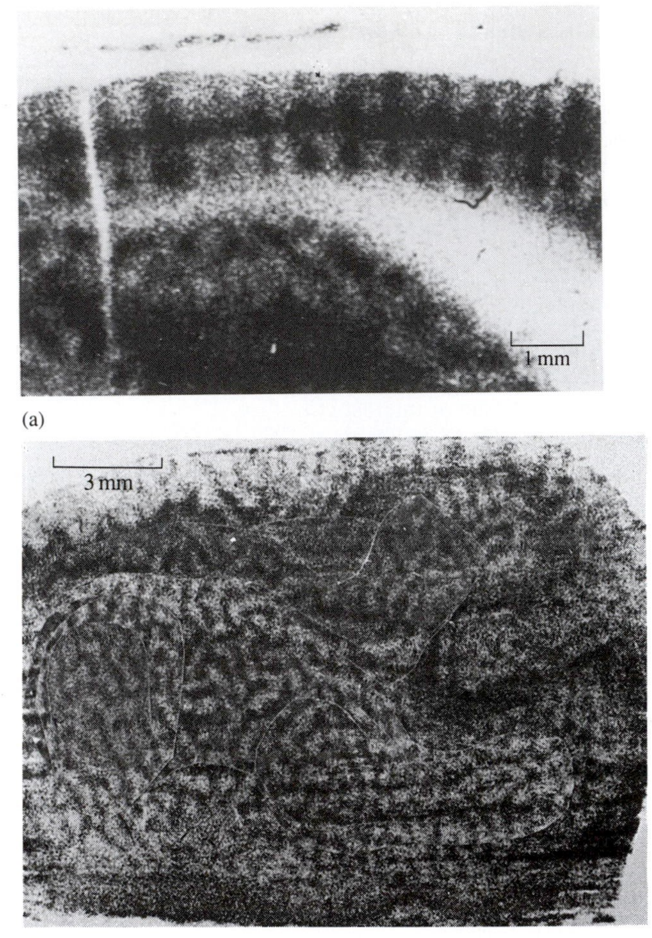

Figure 4.30 2-DG autoradiograms showing cortical arrangement of cells responding to a vertical stimulus, shown (a) in cross-section, and (b) in tangential section.

Psychological evidence for orientation coding by cell populations is provided by the *tilt after-effect*, illustrated in Figure 4.31. Briefly fixate (look directly at) point T to confirm that the two gratings on either side of it appear vertical. Then allow your eyes to move slowly along the bar labelled A for about a minute. Quickly move your eyes back to point T.

The grids on either side should for a short time appear tilted away from the vertical: the one on the left should appear to be tilted anticlockwise, and the one on the right clockwise.

In general, adaptation to a stimulus of one orientation shifts the perceived orientation of a subsequently-viewed test stimulus *away* from the adapting orientation. The population theory of orientation coding can explain this quite simply, because the adapting stimulus will distort the pattern of response evoked by the test, as shown in Figure 4.32 (p. 110). During adaptation, cells responding to the adapting stimulus become fatigued so that the pattern of response subsequently produced by the test stimulus is distorted, with the peak shifted away from the adapting orientation.

Although these arguments throw considerable doubt on the role of individual cells, you do not need completely to abandon the notion of feature detection. You just need to think in terms of populations of cells, rather than individual cells. It would be more appropriate, for example, to think of the hypercolumn, rather than a simple cell, as a feature detector.

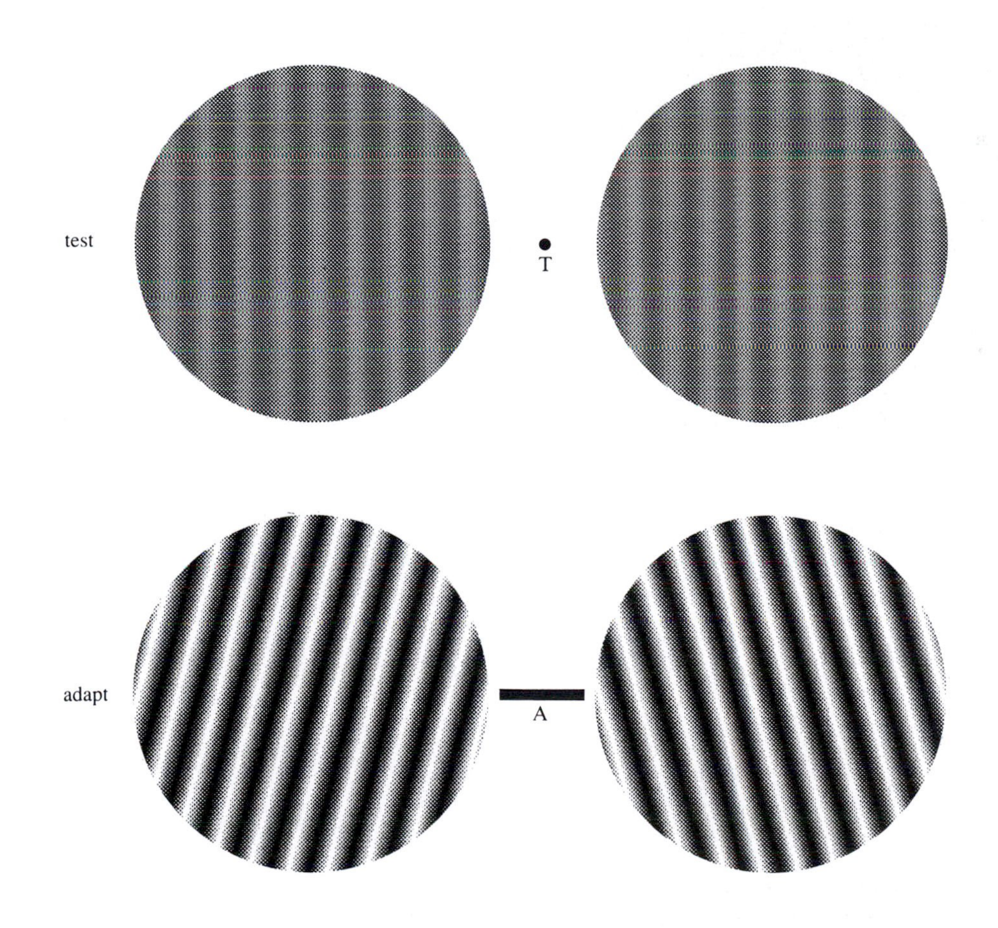

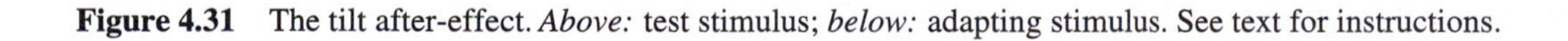

Figure 4.31 The tilt after-effect. *Above:* test stimulus; *below:* adapting stimulus. See text for instructions.

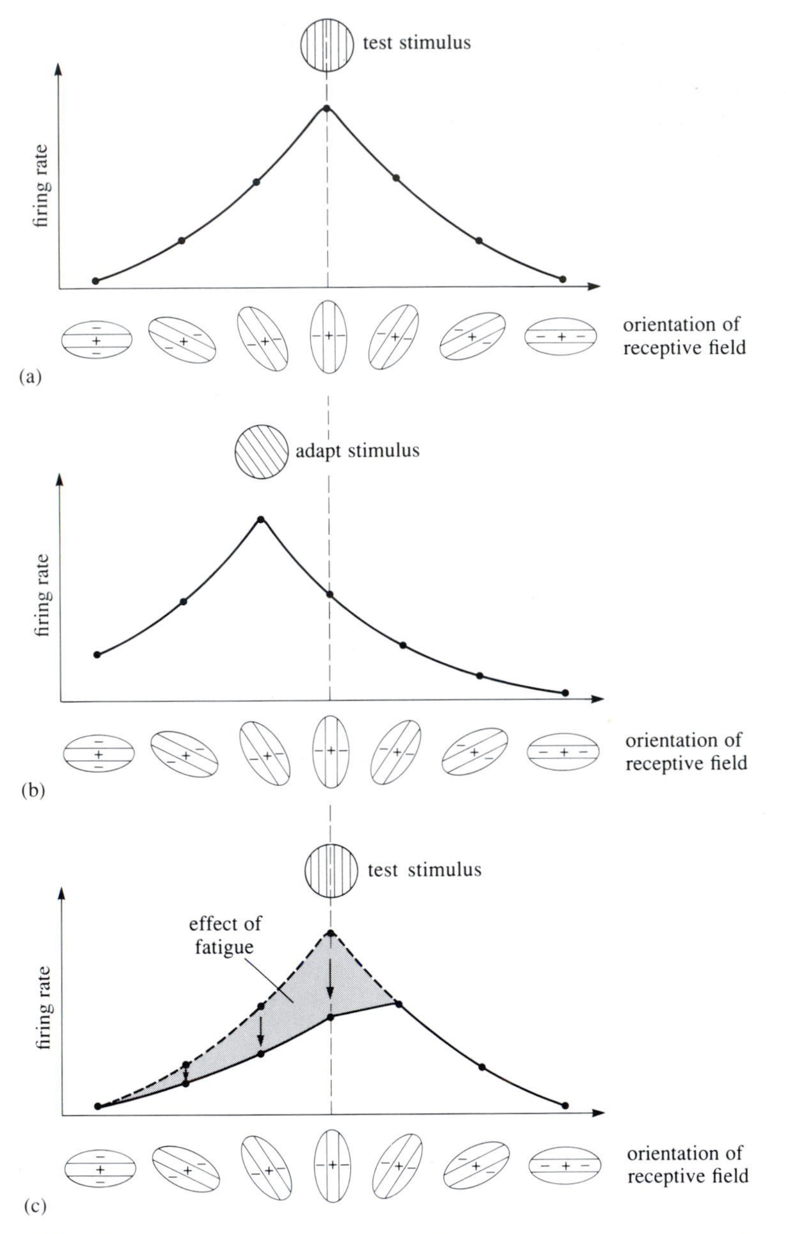

Figure 4.32 Explanation of the tilt after-effect. The adapting stimulus fatigues the cells which respond to it so that they respond less to the test stimulus than they would if not fatigued, thus distorting the pattern of response produced by the test stimulus so that the peak response shifts away from the adapting orientation.

Hierarchical processing

The first stages of cortical processing are almost certainly hierarchical: simple cells do combine the output from a group of retinal ganglion cells to produce oriented receptive fields, as shown in Figure 4.33. However, the notion of a similar hierarchy from simple to complex cells, and from complex cells to hypercomplex cells, is unfortunately wrong. When cells in the LGN are directly stimulated electrically, for example, complex cells tend to respond *before* simple cells—though the hierarchical model would require precisely the opposite.

Subsequent work has given much more emphasis to the *parallel* organization of functionally distinct cortical systems, rather than to *sequential* hierarchies. Parallel organization is considered in more detail in Section 4.7, after the description of the other visual sub-modalities. The remainder of this section considers hierarchical visual theories from a more general perspective.

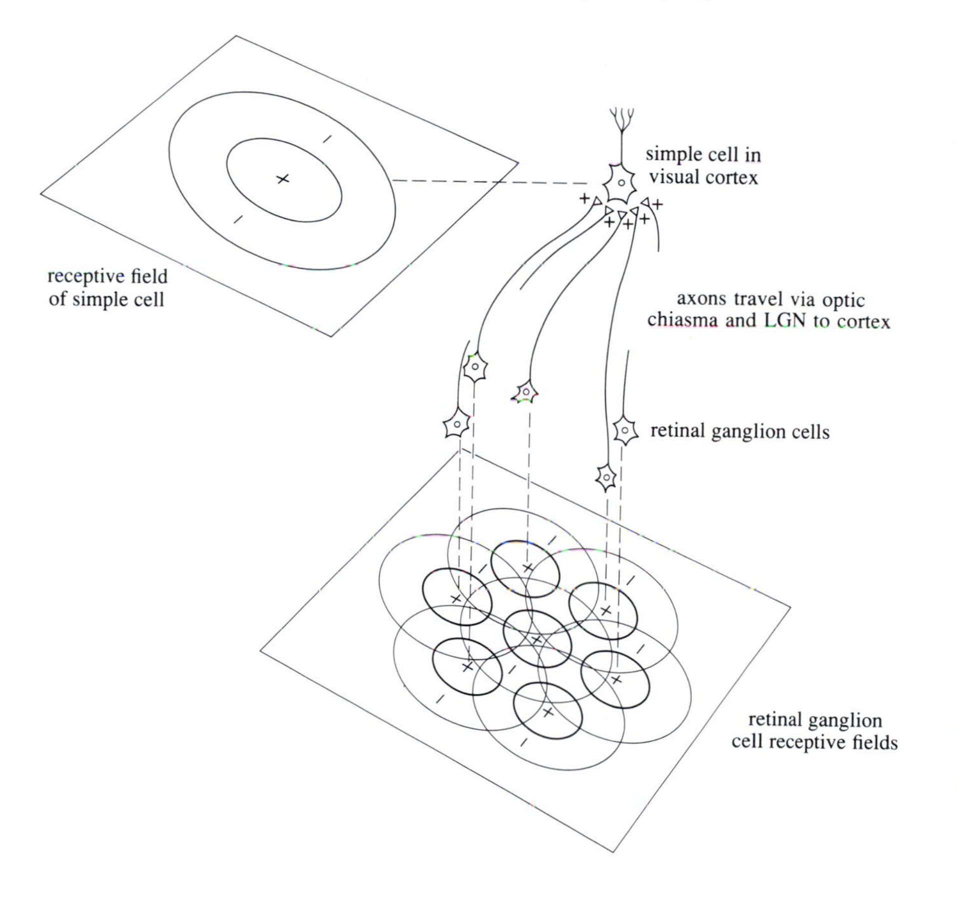

Figure 4.33 The receptive field of a simple cell produced by connecting together retinal ganglion cells with appropriately positioned receptive fields.

The notion of a strict hierarchy of visual processing was originally criticized on a number of grounds. It was often argued throughout the 1970s and early 1980s, for example, that a hierarchy was impractical because, at the top of the hierarchy, there would have to be a cell for every recognizable object—a grandmother detector, a yellow Volkswagen detector, and so forth. Taken in this extreme form, there surely would not be enough cells in the brain, let alone the visual system, to deal with all recognizable things in all their possible variations.

In the late 1980s, however, theories of hierarchical processing began to enjoy a resurgence. It was pointed out, for example, that hierarchies need not be as inefficient as the opponents of 'grandmother detectors' had argued. The basic idea of a hierarchy is that all images can be broken down into a common set of simple descriptive 'primitives'—such as lines and, perhaps, even a limited set of characteristic shapes. The lower levels of the hierarchy would be designed to detect these primitives and so these cells would respond to all stimuli. Only at the upper levels in the hierarchy would there need to be specialized cells which respond to some stimuli but not others—and even here you would not need one cell for each recognizable object. All faces, for example, consist of eyes, nose and mouth in more or less the same spatial arrangement. Suppose you could isolate just ten important facial features: length of nose, separation of eyes, and so forth. Suppose further that you could wire up one cell to measure each of these characteristics and that each cell was capable of distinguishing just two levels of what it was measuring (short or long noses, for example). It could signal its measurement by changing its firing rate (slow rate for short noses, fast rate for long noses, and so forth). These ten cells would, in fact, be able to signal over 1 000 (2^{10}) different facial patterns and thus, potentially, recognize over 1 000 different faces. If the cells could each distinguish ten levels of what they were measuring, then just ten cells could potentially recognize 10 billion (10^{10} or 10 000 000 000) different faces.

Face recognition and the middle temporal cortex

Throughout the 1970s and early 1980s, there emerged very little evidence of the higher order 'object recognizer' cells required by early hierarchical models. This is perhaps not surprising because, in looking for an object detector, it is very difficult to know where to begin—the potential range of stimuli is, by definition, almost infinite.

In the middle and late 1980s, however, an area of the middle temporal cortex (Figure 4.2, p. 78) was found in monkeys which did seem to be concerned with the detection of complex visual stimuli and, in particular, with the recognition of faces. Many cells were found that were selective for eyes, noses, full face views, profiles, and so forth, and for specific constellations of these features. There were even reports of individual cells which would respond only to specific faces—plausible candidates for the often-ridiculed 'grandmother detectors'.

At the beginning of the 1990s, these findings offer an exciting prospect for future research and promise the answers to some quite fundamental questions. How is the system wired up to produce cells selective for specific facial features from the more general feature encoders found in V1? Are there really individual face recognizer cells (grandmother detectors) or is each face represented by a pattern of response distributed across several sub-populations of cells? How does the system learn to recognize new faces?

The answers to these questions will provide a giant step towards the understanding of visual object recognition. The only note of caution is that the processes described above are concerned entirely with deriving a *structural description* of the image. Faces may be well-suited to recognition on these grounds but it is much more difficult to derive a structural description of, for example, all recognizable chairs. A few minutes in any modern furniture store should convince you of this. Chairs seem better defined in terms of *functional potential*: they are designed to be sat upon, and there are many different structures which meet this requirement and which are, in the appropriate context, acceptable as chairs.

☐ How does this discussion of face recognition relate to the argument about proximal stimuli and distal objects at the beginning of this chapter and in Chapter 1?

- In general, distal objects cannot be recognized simply by describing proximal stimuli (images). Faces may be an exception, because they are distinguishable *only* in terms of spatial structure and because they are also particularly important. But many objects require much more flexible interpretation of an initial structural description. This simply supports the idea, also discussed in Chapter 1, that the balance between bottom-up (descriptive) and top-down (interpretive) processes will vary with different types of stimulus.

Summary of Section 4.3

Retinal processes are specialized to respond only at places where luminance changes in the image, and so to signal the presence of object boundaries in the world. These processes make use of lateral inhibition to compare the luminance in adjacent regions of the image. The comparison is in the form of a ratio, so that the response of the cells is unaffected by changes in overall illumination. Information about luminance changes is further processed within hypercolumns in the primary visual cortex where, for example, orientation is encoded by the pattern of response across many cells organized into orientation columns. Subsequent processes in the middle temporal cortex refine the structural description of the image even further and may be specialized for the recognition of specific objects such as faces.

4.4 Image description: spectral composition

Since it is possible to make sense of black-and-white photographs, the processes described in the previous section must be adequate for visual object recognition. However, in normal life, knowing about the colour of objects can help in at least three different ways.

First, colour is useful in *image segmentation*—the process of deciding which regions of the image belong together. A blue car standing behind a tree, for example, produces two quite distinct regions in the image. The fact that both regions have the same colour is useful in deciding that these two, spatially separate, regions in fact relate to the same object. The natural importance of colour in this role is highlighted by camouflage: animals which need to remain hidden often have many irregular patches of different colours and so they do not stand out as a coherent object against the background.

Second, many natural objects have a characteristic colour which can provide an important cue for recognition and for distinguishing between, say, a nutritious and a poisonous berry, or an apple that is ready to eat and one that should be left to ripen.

Finally, colour has an important signalling role in nature. In many species, it is essential in distinguishing gender or to indicate sexual maturity as in sticklebacks (Book 1, Chapter 2). More subtly, in people for example, it can provide a useful clue to mood or even intention.

4.4.1 The physical basis of perceived colour

Vision provides a small window onto a much broader spectrum of electromagnetic radiation which ranges from long wavelength (low frequency) radio and TV waves, up through infrared (heat) to visible light and on through ultraviolet to shorter wavelength (higher frequency) X-rays and gamma rays. The sun emits all of these wavelengths but only the relatively narrow band between about 400 and 700 nm is available to humans as visible light (Plate 1).

Sunlight contains all visible wavelengths in roughly equal proportions and is perceived by humans as white, or uncoloured. However, **monochromatic light**, consisting of a single wavelength, is seen as coloured: long wavelengths are perceived as red, medium as green, and short wavelengths as blue. Monochromatic light is very useful in probing the mechanisms of colour vision and in demonstrating that the physical correlate of perceived colour is wavelength. But in nature you almost never encounter monochromatic light. Instead, as shown in Plate 2, light consisting of all wavelengths is typically reflected by a surface, and surfaces tend to reflect some wavelengths and absorb others. The light reaching your eyes from the surface thus consists of a complex *spectrum*, in which some wavelengths are more predominant than others. You see this complex spectrum as the colour of the surface. Some surfaces, for example, reflect more long wavelengths than medium, and more medium wavelengths than short—and are perceived as red or orange, others absorb both long and short wavelengths and reflect only medium wavelengths—and are perceived as green or yellow.

The task of colour vision is, therefore, to describe the *spectral reflectance* of surfaces (the relative amount of each wavelength that each surface reflects), and is really just an extension of luminance coding. Whereas black-and-white vision is only concerned with the overall reflectance of a surface—a single number representing the total amount of light reflected, irrespective of wavelength—colour vision has to break this down into a more detailed description, giving the relative amounts of each wavelength reflected. To do this, colour vision needs to be equipped with at least two things: a way of breaking light down into separate bands of wavelengths, and a way of making comparisons between these bands. The retina of many animals, including primates, is equipped to do both these things.

4.4.2 Retinal coding of colour

Cone mechanisms and trichromaticity

Fish such as goldfish have comparatively large cone receptors and it is relatively easy to measure their sensitivity to different wavelengths by recording the responses from a single cone while stimulating it with monochromatic light. When this is done, three distinct types of cone receptor are found, each with a different spectral sensitivity (Figure 4.34). One type (L cones) responds best to long wavelengths, a second (M cones) to medium wavelengths, and the third (S cones) to short wavelengths. The different spectral sensitivities are due to slight variations in the visual pigment contained within the cones.

☐ Does each type of cone respond to just one wavelength of light?

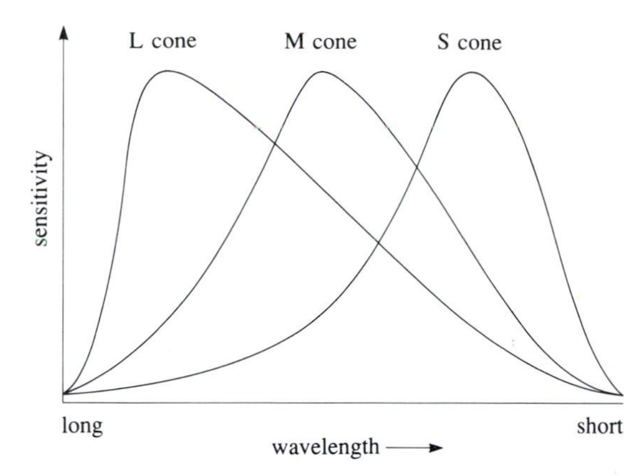

Figure 4.34 Spectral sensitivities of different cone types. For clarity, the sensitivities of the three cone types have been adjusted to be roughly equal. In reality, the S cones are very much less sensitive than the other two types.

■ No, each cone responds to a broad band of wavelengths, but different types of cone are most sensitive at different wavelengths.

Thus, the three different types of cones provide the visual system with a way to break down light into three separate bands of wavelengths, and form the basis of 'three-colour' or **trichromatic vision**. Light with a particular spectral composition evokes a particular pattern of response across the three different types of cone, and each distinguishable colour evokes a different pattern of response.

Trichromatic vision is not universal throughout the animal kingdom—some animals are monochromats (effectively colour blind), others dichromats (two-colour vision) and some birds are even tetrachromats (four-colour vision), using different coloured oil droplets as filters in their cones rather than different types of pigment. However, humans and most other primates are trichromats and have three cone types similar to those of the goldfish.

The four different types of receptor (rods and three kinds of cone) are not uniformly distributed across the retina. Rods predominate in the retinal periphery, while cones predominate in the fovea. Human colour vision is thus most effective in the fovea and, because there are relatively few cones, almost non-existent in the far retinal periphery.

□ How might you demonstrate the absence of colour vision in the retinal periphery?

■ Keep your eyes fixated on a point straight in front of you and ask a friend to bring an unknown object slowly into your field of view from behind your head. You should be able to detect the object before you can specify its colour.

□ Why do all cats look grey at night?

■ Cones do not function in dim light, so night vision is mediated entirely by rods. All rods have the same spectral sensitivity and so the visual system is actually monochromatic at night. There is no way to distinguish between different wavelengths and the system can therefore distinguish between surfaces that reflect different *total* amounts of light (perceived as different shades of grey) but not between surfaces that reflect the same amount of light but with *different* spectral compositions.

Bipolar cells and opponency

Different cone systems provide a way to split light into separate bands of wavelengths. The other requirement of colour vision is a mechanism for making comparisons between these bands. This is provided at the very next stage in the visual pathway by the bipolar cells.

Recordings from bipolar cells often show their response to vary with wavelength in the interesting way shown in Figure 4.35. One band of wavelengths causes an increase in response, while another causes a decrease in response. This property of **colour opponency** arises because one type of cone (in this case the L cone) excites the bipolar, whereas another (in this case the M cone) inhibits it. In fact, bipolars with most of the possible combinations of excitatory and inhibitory input from the three different cone types have been found.

In practice, these colour opponent bipolar cells compare the amount of light in different bands of wavelengths.

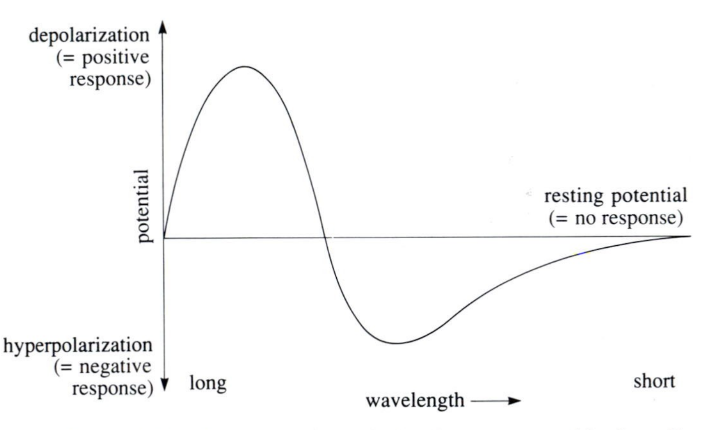

Figure 4.35 Response of a typical colour opponent bipolar cell.

□ How might the output of the cell illustrated in Figure 4.35 be interpreted?

■ A positive response means that there is more long wavelength light than medium wavelength, a negative response signifies that there is more medium wavelength than long, and a zero response signifies that there is an equal amount of long and medium wavelengths.

This type of comparative statement is precisely what is needed to describe the spectral reflectance of a surface.

Colour opponency and complementary after-images

Gaze for half a minute or so at the fixation cross in Plate 3, and remember the arrangement of colours. Then shift your gaze to the small circle. You should see a coloured after-image superimposed on the white surface. Note that where you were looking at red you now see green, and where you were looking at green you now see red.

□ How might colour opponency explain these coloured after-images?

■ Refer back to Figure 4.35. Looking at a red patch will produce a positive response in this bipolar cell. When you shift your gaze to a white field you would normally expect a zero response because the inputs from the L and M cones should balance. But the L cones have been responding and the M cones have not, and so the L cones will be more fatigued. Instead of the normal balance when you look at a white surface, the M cones now predominate over the fatigued L cones, producing a negative response in the bipolar which is interpreted as green.

The colours which result from this kind of adaptation experiment are called **complementary colours**. Every colour has a complementary colour (for example red–green or blue–yellow) which results from the opponent wiring of bipolar cells.

4.4.3 Retinal ganglion cells and early cortical processing

Retinal ganglion cells receive their inputs from bipolar cells and so it is not surprising that research on monkeys has shown that many retinal ganglion cells respond differently to different wavelengths. Some retinal ganglion cells show the centre-surround arrangement described in Section 4.3.2, irrespective of the wavelength of the stimulating light. These cells presumably receive balanced inputs from all three types of cone. A second type of retinal ganglion cell has a normal centre-surround arrangement when stimulated by white light, but also increases its response when one wavelength (for example, red) is presented anywhere in its receptive field, and decreases its response when the complementary wavelength (for example, green) is presented anywhere in its receptive field. These cells have similar properties to colour opponent bipolar cells and seem well equipped to compare the amount of light at different wavelengths. A third and final type of retinal ganglion cell has a central sub-region which increases the cell's response when one wavelength (for example, red) is present, and a surround sub-region which decreases the response when the complementary wavelength (for example, green) is present.

This third type of receptive field arrangement is rather puzzling because it seems to do comparisons both between different wavelengths and between adjacent retinal positions. It is best understood as providing the input to the next stage of colour processing, which takes place in the LGN and visual cortex.

Some cells in the LGN and in area V1 of the visual cortex have circular receptive fields and are called **double opponent cells**. Stimulating the receptive field centre sub-region of a typical double opponent cell with one wavelength (for example, red) increases the cell's response, whereas stimulation with the complementary wavelength (for example, green) decreases the cell's response. The surround sub-region would show precisely the opposite arrangement, causing an increase in the cell's response when stimulated by green wavelengths, and a decrease in the cell's response when stimulated by red wavelengths. These complicated cells may function to detect places in the image where the colour changes.

□ Under what circumstances might colour opponent cells provide an additional clue to object boundaries?

■ Sometimes an object has a different colour from its background but reflects the same total amount of light. Such boundaries would not be detected by the cells described in Section 4.3.2 because they only respond to a change in luminance (the total amount of light).

Plate 4 compares the function of luminance-sensitive retinal ganglion cells with double opponent cortical cells. The regions of the receptive field which would produce an increase in the response of the cell are cross-hatched. For the on-centre retinal ganglion cell illustrated, dark in the off-surround and light in the on-centre would both cause an increase. For the colour opponent cell illustrated, green in the surround and red in the centre would cause an increase. Both cells receive more excitation than inhibition and so respond above the background firing rate.

One possible perceptual consequence of colour opponent cells is called *simultaneous colour contrast* and is shown in Plate 5. The central region of each square is the same neutral grey but the region surrounded by the green patch appears reddish, while that surrounded by the red patch appears greenish.

□ How might double opponent cells account for this?

■ Consider the colour opponent cell shown in Plate 4, positioned near the boundary in Plate 5 so that the centre of its receptive field falls in the grey region and the surround is partly covered by the greenish patch. The green light in the surround region will cause a slight increase in response and there is no colour anywhere else in the receptive field to cancel this out. This cell will therefore respond above the background rate. Such a response can normally be interpreted as signifying a red–green boundary in the image—and that is exactly what you see.

4.4.4 Later cortical processes and colour constancy

When you move from sunlight into artificial light, or through a leafy glade where the light has a distinctly greenish tinge, the spectral composition of the light reflected into your eyes will change because the spectral composition of the illumination is different. Yet the perceived colour of surfaces does not generally change even under very different lighting conditions, a phenomenon known as **colour constancy**. Just as luminance processing faces the problem of changes in the overall amount of light, colour processing faces the problem of changes in the spectral composition of the light.

The solutions to these problems may be very similar. Comparing the overall amount of light at neighbouring positions solves the problem of lightness constancy because changes in the overall amount of light are cancelled out in the comparison. In principle, the same solution should work for colour constancy if the same method of spatial comparison is used *within each cone type*. Thus, for example, a comparison of the response in the L cones at neighbouring positions will be unaffected by an overall change in the amount of red light. The same is true for the M and S cone systems, so the *pattern of response* across the three types of cone, which is the main code for surface colour, will also remain unchanged.

□ Which of the types of cell so far described is most likely to be involved in colour constancy?

■ Double opponent cells seem to have the required properties of comparing the light at neighbouring positions and between different cone systems.

Although the precise details of the solution are not yet known, Semir Zeki, working in London in the early 1980s, reported that area V4 of the monkey cortex, one of the many regions to which V1 projects, contains a very high proportion of cells which are selective for wavelength. When stimulated with monochromatic light, many of these cells respond only to a very narrow band of wavelengths. Most enthralling, however, is that these cells seem to exhibit colour constancy.

Zeki used a complex stimulus made up from many randomly arranged patches of colour, and positioned it so that one uniform patch covered the cell's receptive field. If the cell was selective for long wavelengths, for example, and the stimulating patch looked red when viewed by a human, then the cell would respond. Zeki then drastically varied the spectral composition of the light illuminating the stimulus. Under these conditions, for a human observer, colour constancy holds over quite a range of variation and the patch continues to look red. But if the composition of the illuminating light is very unbalanced, colour constancy does break down and the patch changes in apparent colour. The monkey's cell showed very similar properties—under illuminating conditions where the patch looked red to a human observer, the cell responded, but when the patch did not look red to the observer the cell did not respond. This, for Zeki, must have been an extremely eery experience and it is, perhaps, the only case where the response of a single cell in the brain of an animal has been shown to correlate very strongly with what is perceived by a human.

Summary of Section 4.4

The two main requirements of colour processing appear to be accomplished in the retina by cone receptors, which analyse the light into three different bands of wavelengths, and by bipolar cells, which make comparisons between these different bands. Subsequent processes, in retinal ganglion cells and in double opponent cells in the LGN and visual cortex, build on this to produce cells which are capable of detecting colour boundaries in the image. Later cortical stages reveal cells which show colour constancy: their response is unaffected by changes in the spectral composition of the illumination.

4.5 Image description: motion

Describing static images is difficult enough and allowing them to move seems to make the task even more complicated. For many years, the prevailing view was that human vision worked as though the stimulus were a series of static 'frames'. Each frame would be processed and recognized, and then movement could be *inferred* by comparing successive frames. There is no doubt that vision *can* work like this—a series of static frames rapidly presented in the cinema or on television does produce a compelling impression of movement. But there is much more to movement perception than this. Rather than just being something that is inferred from a static analysis, motion is also one of the primary sub-modalities of vision. Motion in the image can actually *help* perception in a number of ways and most animals possess specialized mechanisms which function to extract and describe it directly.

A description of image motion can help perception in at least three different ways. First, like colour, it can help in segmenting the image and in deciding which regions belong with which. If parts of the image move together in the same direction, this provides a very important clue that they belong to the same object. The importance of movement in nature is, again like colour, demonstrated by animals that need to remain hidden. Such animals typically remain perfectly still, and so do not stand out as a coherent moving object against the background.

Second, though it is less obvious than for colour, many objects have characteristic movements that can aid in their recognition. Indeed human ability to recognize complex scenes only on the basis of movement is quite outstanding, and is demonstrated by a simple technique pioneered in the 1970s by Gunnar Johannson in Sweden. Johannson attached lights to the arms and legs of humans and filmed them in the dark performing complex acts like dancing. When shown a single frame from the film, human observers can make no sense of the stimulus, which just looks like a random collection of dots. But when shown a sequence of a few frames in which the lights trace out complicated trajectories, the observers immediately recognize what is happening. It is even possible, given just a few moving lights, reliably to distinguish the genders of the dancers.

Finally, image motion is extremely useful in aiding balance, in guiding locomotion, and in specifying the three-dimensional lay-out of surfaces in the world. All these properties follow from the fact that, as you move about the world, the image changes smoothly in a characteristic way. For example, as you approach something, its image expands; if you look out to the side while moving forwards, the images of stationary objects move backwards, with more distant objects moving more slowly than near ones. In general, movements by the observer cause smooth **flow patterns** in the retinal image, and these flow patterns contain a wealth of information about the nature of the movement and about the structure of the external world.

Balance is often said to be maintained by organs in the inner ear that detect movements of the head. However, visual flow patterns are also important and this can be demonstrated by an ingenious technique in which an observer is placed in a moveable trolley in a simple 'room' which can also be moved backwards and forwards. When the trolley is moved and the room kept still, balance is maintained. But when the room is moved and the trolley kept still, the observer is thrown off balance and may even fall over.

□ Why does this happen?

■ When the trolley is moved and the room kept still, the movement of the body and head stimulates the organs of the inner ear and also produces a visual flow pattern. The two things signal the same thing so balance can be maintained. However, movement of the room but not of the trolley produces a flow pattern suggesting movement without a matching signal from the inner ear and, since this produces conflicting information, balance is lost. (Recall the discussion of the reafference principle in Book 1, Section 2.3.)

Incidentally, there are more natural situations where information from the visual system and the inner ear are placed into conflict. When travelling by sea, for example, the inner ear detects the movements of the boat but, if you are travelling inside the boat, there is no accompanying motion of the visual field. The unpleasant result of this conflict is called sea sickness. This may be why many people prone to sea sickness prefer to be on deck rather than down below.

Given the brief overview of its possible uses, it may not surprise you that the description of image motion involves at least two stages: first, the motion of individual points in the image is described by simple mechanisms, and then the motion of groups of points is combined by more complex mechanisms sensitive to the types of large-scale motion typical of retinal flow patterns.

4.5.1 Describing the motion of individual points

Retinal processes

The retinas of some mammals, like rabbits, contain specialized mechanisms which appear to function as motion detectors in that they respond best to a stimulus moving in one particular direction. The way that they do this is illustrated schematically in Figure 4.36. The basic motion detector receives input from two receptors in neighbouring retinal positions. One of these receptors exerts an inhibitory influence and its response arrives at the motion detector after a short delay. When a stimulus moves in the 'preferred' direction, in this case right to left (Figure 4.36a), the response from the excitatory receptor arrives at the motion detector and causes a response before the inhibitory response from the other receptor arrives to turn it off. However, when the stimulus moves in the 'null' direction, in this case left to right (Figure 4.36b), the delayed inhibitory response arrives simultaneously with the excitatory response and the two cancel to give no response.

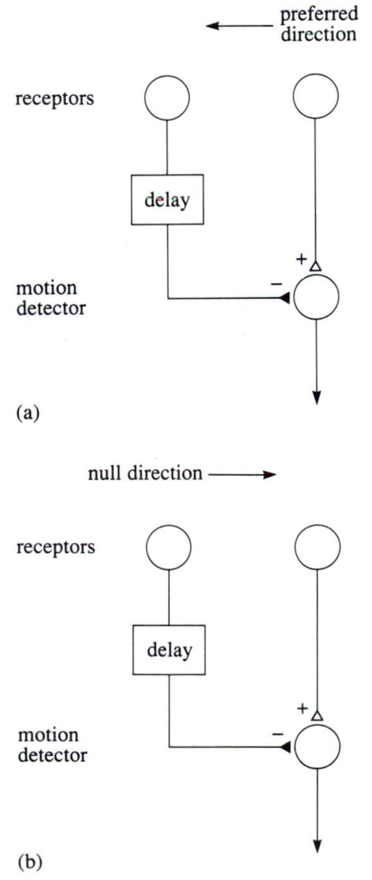

Figure 4.36 Principles of a simple retinal motion detector. See text for explanation.

Although the retinas of other mammals, including primates, do not have specialized motion detectors, they do make use of the above mechanism of **delayed inhibition**. The essential difference is that, whereas in the motion detector shown in Figure 4.36 the excitation and delayed inhibition come from neighbouring regions of the retina, in the primate retina the excitation and delayed inhibition come from the same place. This rather strange arrangement provides a general way to detect changes in the image over time.

Section 4.3.1 described how lateral inhibition provides a way to detect *spatial changes* in the image: if initial responses are the same at adjacent positions in the image then they simply cancel out; only if the responses are different do cells at the next stage of processing respond. Delayed inhibition provides an equivalent way to detect *temporal changes* in the image: if responses at consecutive times are the same then they simply cancel out; only if the responses change over time do the cells at the next stage respond.

□ Figure 4.37a shows the response of a conventional on-centre retinal ganglion cell to a spot of light turned on and off in the central sub-region of its receptive field. This cell does not exhibit delayed inhibition. What would be the pattern of response to the same stimulus if the cell *did* exhibit delayed inhibition? Sketch your answer in the space provided in Figure 4.37b.

■ The cell would respond only at the onset and the offset of the light because these are the only occasions on which the stimulus changes over time. You can compare your drawing with that shown in Figure 4.38 (p. 124).

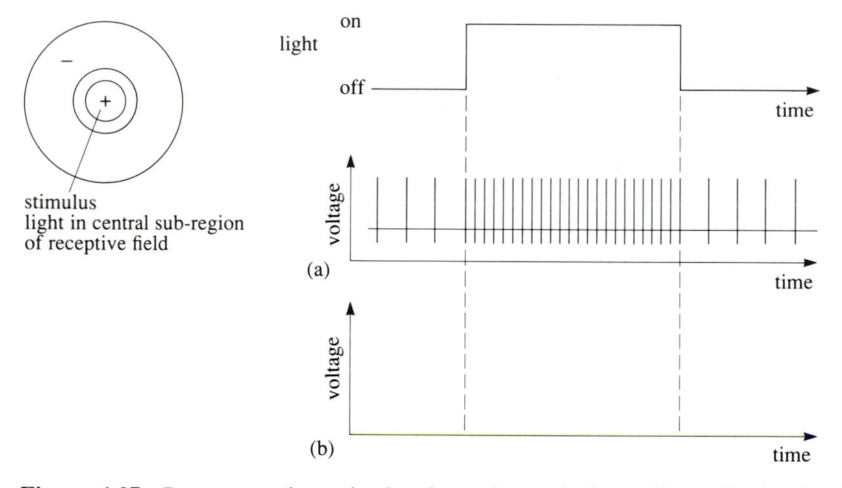

Figure 4.37 Responses of sustained and transient retinal ganglion cells. (a) A cell not showing delayed inhibition (sustained cell). (b) Blank figure for you to sketch the response of a cell that does show delayed inhibition (transient cell).

A retinal ganglion cell of the type shown in Figure 4.37a is sometimes called a **sustained cell** because its response persists throughout the stimulus presentation. A retinal ganglion cell that exhibits delayed inhibition is called a **transient cell** because it responds only briefly at the onset and offset of the stimulus. In fact, just as it is possible to classify retinal ganglion cells into 'on-centre' and 'off-centre' types, or into different groups on the basis of their wavelength sensitivities, it is also possible to classify them into sustained and transient types on the basis of their temporal properties.

Delayed inhibition appears to depend upon the amacrine cells which make up the inner plexiform layer of the retina (see Figure 4.8, p. 84). For example, chemicals that block amacrine cell transmitters abolish the directional selectivity of motion detectors in the rabbit retina. Moreover, in the primate retina, receptors, horizontal cells and bipolar cells all give sustained responses, whereas amacrine cell

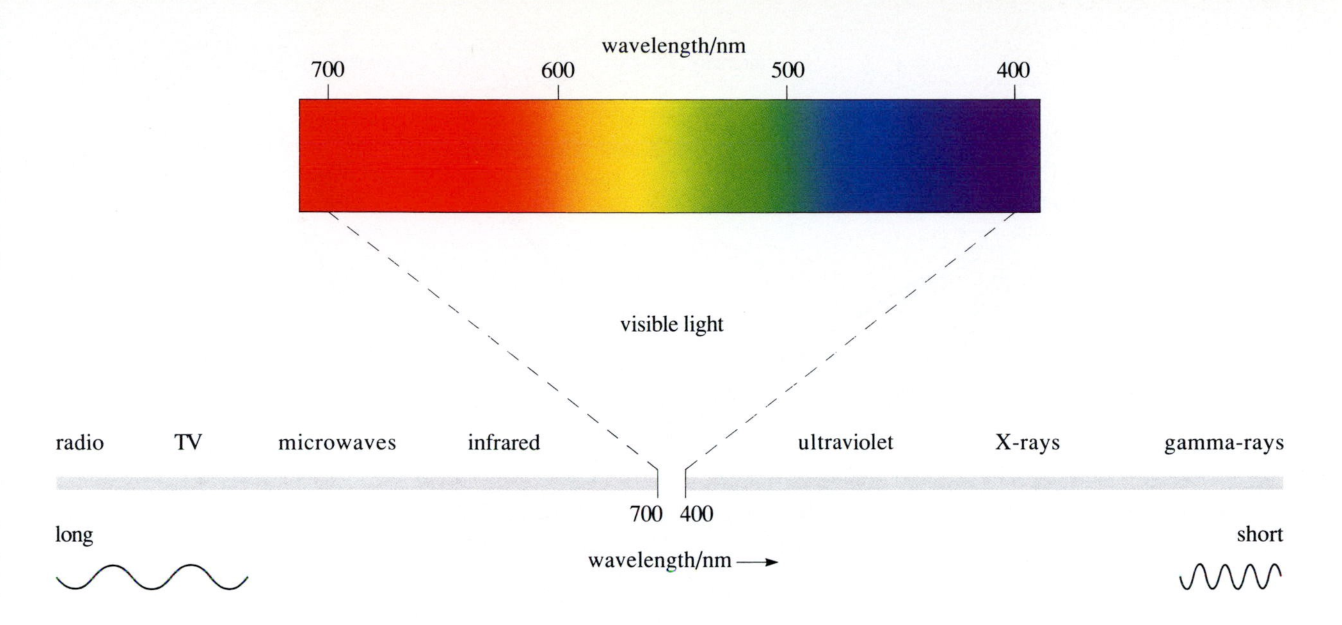

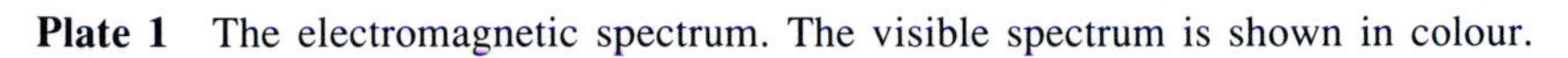

Plate 1 The electromagnetic spectrum. The visible spectrum is shown in colour.

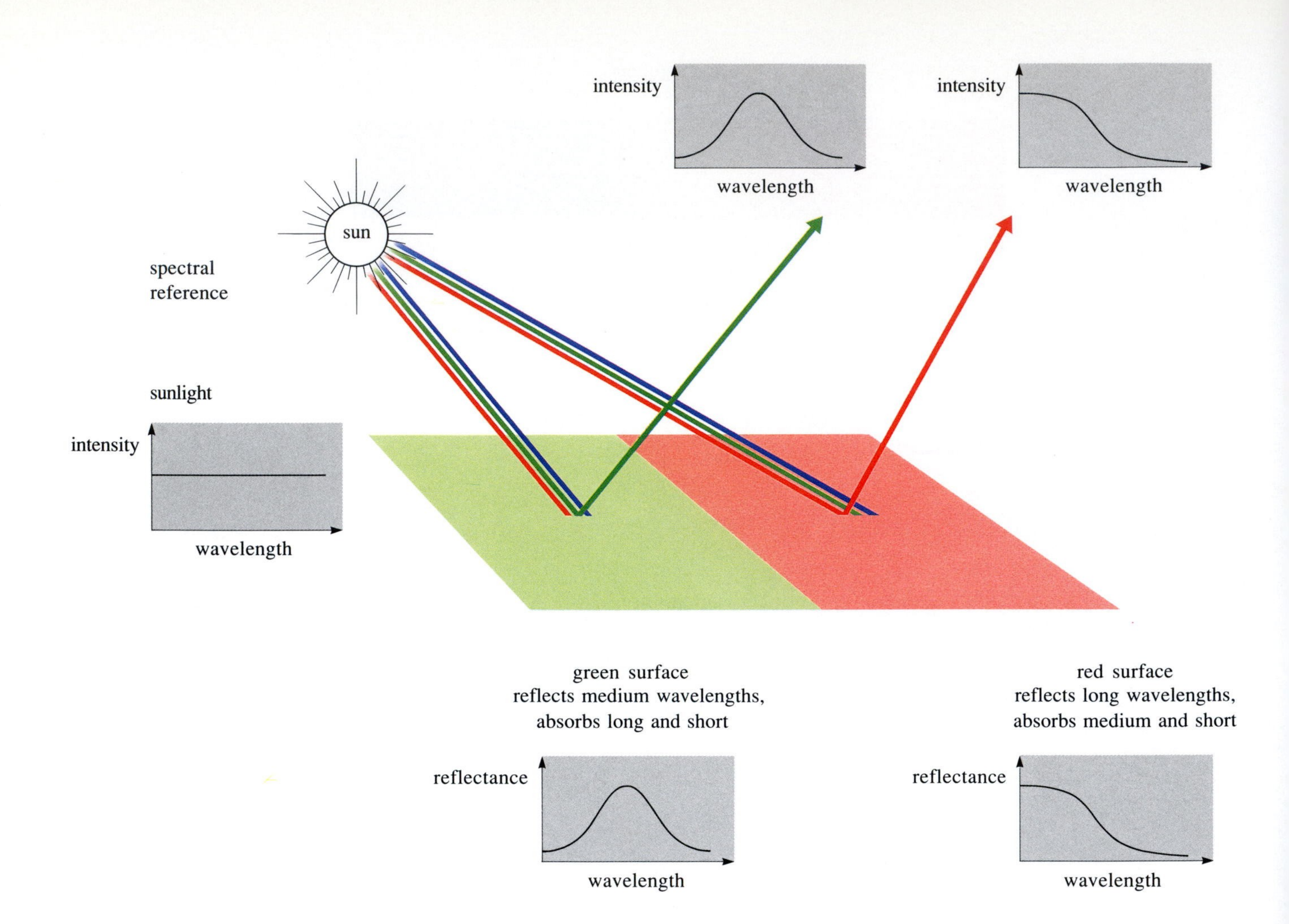

Plate 2 Spectral reflectance of different surfaces.

Plate 3 Complementary after-images. See text for instructions.

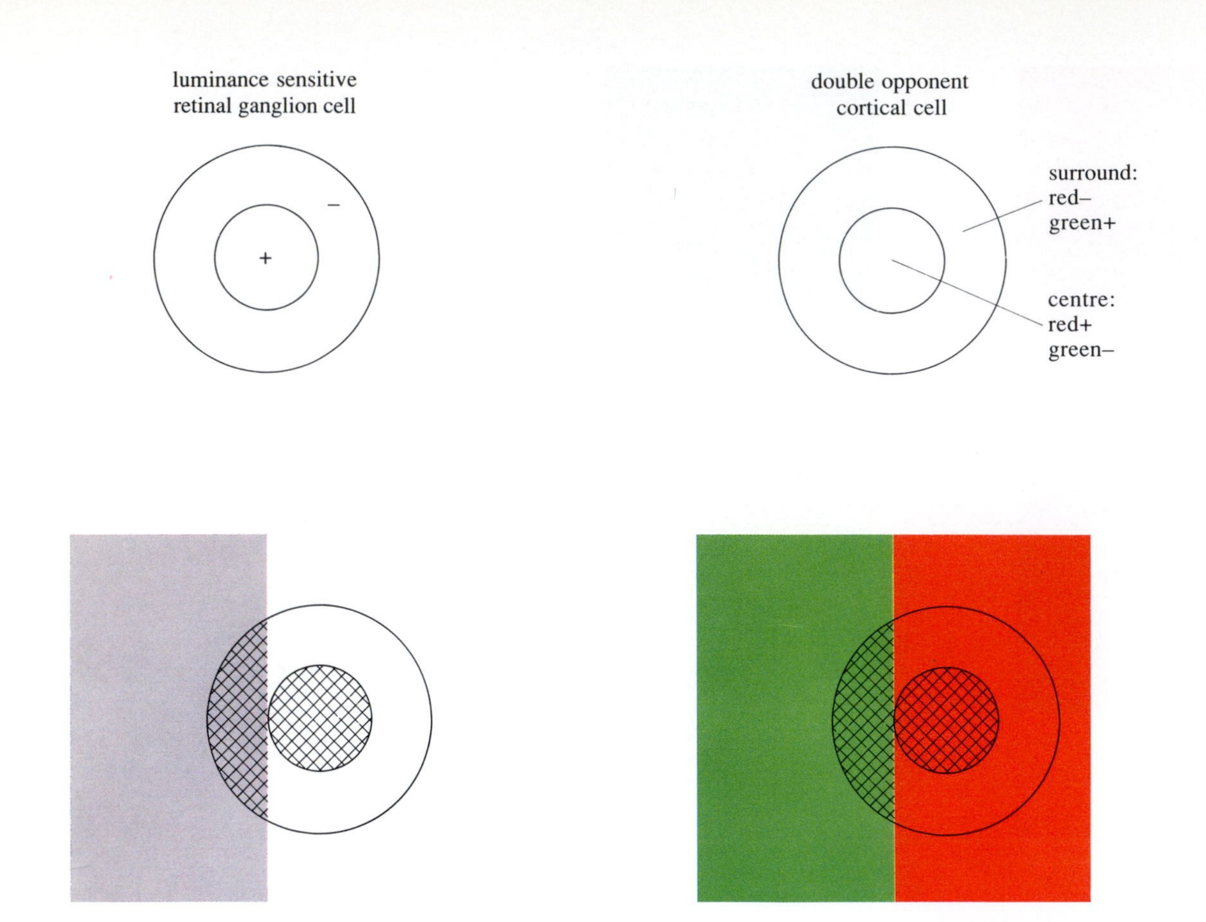

Plate 4 Double opponent receptive fields.

Plate 5 Simultaneous colour contrast.

responses are characteristically transient. Although the detailed anatomy of amacrine cells is complex and varies from species to species, many show *reciprocal synapses*, from the amacrine cell back onto itself. This arrangement would be ideal for carrying the required delayed inhibition.

Amacrine cells appear to exert different amounts of influence upon different retinal ganglion cells, thus producing the different temporal properties. Sustained cells seem to have a weak input from amacrine cells, while transient cells have a much stronger input.

Cortical processes

Rather than the specialized motion detectors of the rabbit retina, the primate retina is equipped with sustained and transient cells which together provide an efficient description emphasizing the spatial and the temporal *changes* in the image. The distinction between sustained and transient responses continues in the visual cortex, where cells of both types are found, the transient type being particularly abundant in Area 18, also called the prestriate cortex or, in primates only, V2 (see Figure 4.2, p. 78).

In addition, most cortical cells are *directionally selective*, responding better to one direction than the opposite direction of stimulus movement. Although the precise mechanisms of cortical directional selectivity are not yet known, it may arise by combining the outputs of different retinal ganglion cells with receptive fields in slightly different positions. This is very like the arrangement found in retinal mechanisms of the type shown in Figure 4.36, except that the inputs to the directionally-selective cortical cell are retinal ganglion cells rather than receptors.

□ You may have noticed a phenomenon called *apparent motion* when looking, for example, at motorway warning signs. When two lights in different positions flash on and off alternately, you see a single light which appears to move smoothly from one position to the other. How would you account for this in terms of directionally-selective cells which combine delayed inhibition from different retinal positions?

■ Such cells will respond whenever they receive inputs from the correct retinal positions in the correct temporal sequence. The stimulus does not have to move smoothly from one position to the other, it just has to appear first at one position and then the other. If two flashing lights meet this requirement, they will cause a response in the cell, which is interpreted as signalling smooth movement.

□ Why do you suppose that warning signs often use moving or flashing lights?

■ Motion and temporal change are important biological cues to potential food or danger and are therefore stimuli with a very high priority. The designers of warning signs make use of this to capture your attention in complex environments.

Cortical directionally-selective mechanisms are arranged so that there is a set of cells which can respond to all possible directions of movement at each position in the image. For the same sorts of reason that an individual orientation detector cannot signal the orientation of a stimulus by itself (Section 4.3.4 and Figure 4.26,

p. 105), however, an individual mechanism selective for a particular direction cannot signal the direction of stimulus motion by itself. The rate of response in a single mechanism could not, for example, distinguish between a dim stimulus moving in the appropriate direction and a bright stimulus moving in an inappropriate direction. In the case of direction coding there is a particularly simple solution to this general problem of ambiguity. Here, one simply needs to compare the response of mechanisms tuned to *opposite* directions of movement. Consider, for example, a 'horizontal' movement encoder which combines an excitatory input from a cell with a leftward motion preference, and an inhibitory input from a cell with a rightward movement preference. Leftward movement would result in an increased response, rightward movement in a decreased response, and a stationary stimulus would have no overall effect. Rather as in colour opponent bipolar cells, the *balance* of the response would thus signal the direction of stimulus motion. The advantage of this system is that any change in the intensity of the stimulus would affect the leftward and rightward cells equally and have no effect upon the balance of the response, which thus provides an unambiguous signal of motion direction.

There is good perceptual evidence that such a system exists in humans. If you stare for a minute or so at movement in one particular direction, you will experience a **motion after-effect**: when you shift your gaze to a stationary object, it will appear to move in the opposite direction. This is what makes you think that your train is moving backwards when it stops at a station after you have been gazing enraptured at the moving countryside. In combination with the role of visual motion in the maintenance of balance, it can even cause you to feel quite unsteady if you attempt to alight at the station.

□ How would you account for the motion after-effect in terms of the kind of balanced directional units described above?

■ Looking at one direction of movement for a prolonged period will fatigue one of the inputs (say the leftward) so that, when you shift your gaze to a stationary stimulus, the normal balance is destroyed and the other (rightward) input predominates. The resulting unbalanced response is interpreted as movement in the opposite direction to the adapting stimulus.

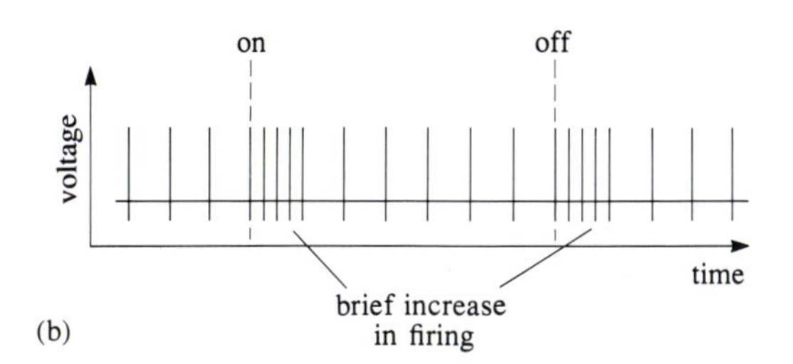

Figure 4.38 Answer to Figure 4.37b: response of transient retinal ganglion cell.

4.5.2 The analysis of retinal flow patterns

Even though the processes which encode the movement of individual points may seem complicated enough, they cannot by themselves provide an adequate description of image motion. For example, if a small region of the retinal image moves differently from its background, is it the object or the background that is moving? This is often a tricky question to answer, and the visual system sometimes gets it wrong. You may have noticed, for example, that when a cloud moves across the moon, you wrongly see the moon to be moving against an apparently stationary cloud. This phenomenon of *induced motion* suggests that, in cases of ambiguity, the visual system tends to ascribe movement to the smaller object (the one which is more likely to move in the natural world).

Despite such occasional and usually harmless lapses, the visual system is generally very good at making sense of motion. This is a very complex task. The movement of an individual point in an image may be due to the movement of your eyes, the movement of your head as you move around, the movement of an object in the world or, more than likely, to some combination of all these things. The movement of a single point then is fundamentally ambiguous and can only be understood in the overall context of movement over the whole image. Fortunately, each different potential cause of image movement has its own characteristic pattern of retinal flow. When you rotate your eyes, for example, the whole image rotates backwards at the same speed; when you walk about, the whole image expands as you approach things; and when an object in the world moves, a section of the image will move consistently and differently from its immediate surroundings. Each of these different patterns can be detected by mechanisms which simply compare the movement of individual points in neighbouring regions of the image.

Figure 4.39 shows a few examples of the types of mechanisms which could be produced by combining the outputs of several individual directionally-selective cells. Each of these mechanisms would be sensitive to a specific pattern of retinal flow. For example, the mechanism in Figure 4.39a would respond to the expansion of the image which occurs when you move towards something, that in Figure 4.39b would respond to a rotation of the head about the line of sight, and that in Figure 4.39c would respond to a horizontal rotation of the eye. Cells with very large receptive fields showing just these types of properties have been reported in the middle temporal cortex (Figure 4.2, p. 78) of the monkey, an area which receives inputs from directionally-selective units in the visual cortex. These units may function to analyse complex retinal flow patterns into simpler components which can then be interpreted separately. For example, a description of the pattern of expansion in the image might be derived by mechanisms like the one shown in Figure 4.39a.

In addition to breaking down complex patterns, cells like the ones depicted in Figure 4.39 can also provide a useful description of the resulting simpler components because they are specifically sensitive to *velocity gradients* in the image. Rather as retinal ganglion cells are sensitive to luminance gradients because they compare the luminance at neighbouring points in the image, these cells compare the velocity (speed of movement) at neighbouring points in the image. To see why this is so, look again at Figures 4.39a and b and note that the individual mechanisms on opposite sides of the circle are always selective for opposite directions of motion. As shown in Figure 4.40 (*overleaf*), this arrangement means that the mechanisms will respond only if there is a change in velocity—a velocity gradient—across the receptive field. The overall receptive

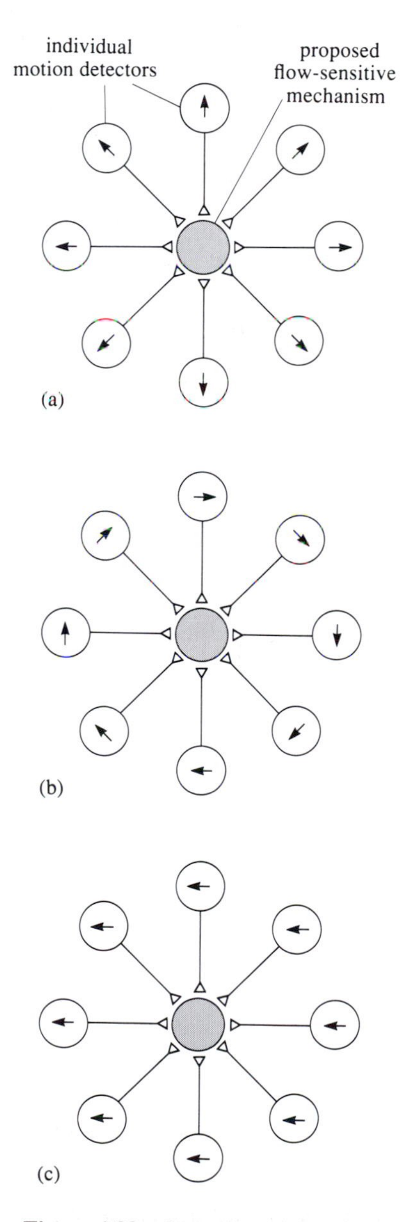

Figure 4.39 Suggested mechanisms sensitive to different types of retinal flow. See text for explanation.

fields are circular, comparing velocities at all orientations, so, again like retinal ganglion cells, these cells will respond to a velocity gradient in any direction.

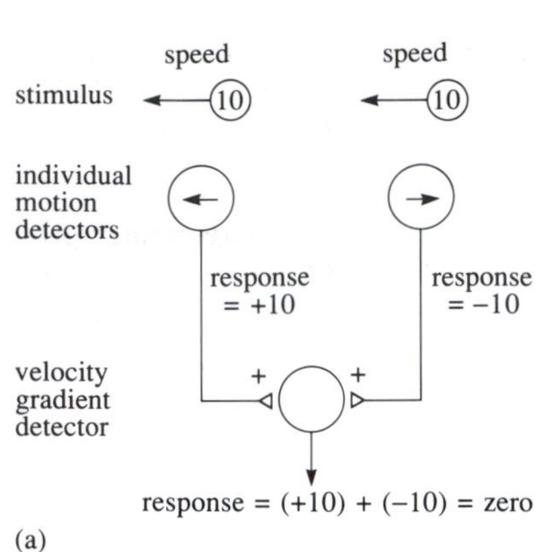

- □ From the discussions in Section 4.3.1 and in the introductory paragraphs of this section, can you suggest two possible functions of mechanisms sensitive to velocity gradients?

- ■ First, like retinal ganglion cells, they will respond at places in the image where the velocity changes abruptly, i.e. at velocity discontinuities. These will tend to occur at the boundaries of objects moving against the background. Second, they will provide a way to measure the velocity gradients produced when you move past surfaces at different orientations. Figure 4.41 shows the velocity gradients which result as you move horizontally relative to a surface which varies its slope from vertical to horizontal. The velocity gradient varies systematically, from zero when the surface is vertical to very steep when it is horizontal, and so a description of the velocity gradient provides important information about the three-dimensional layout of surfaces in the world.

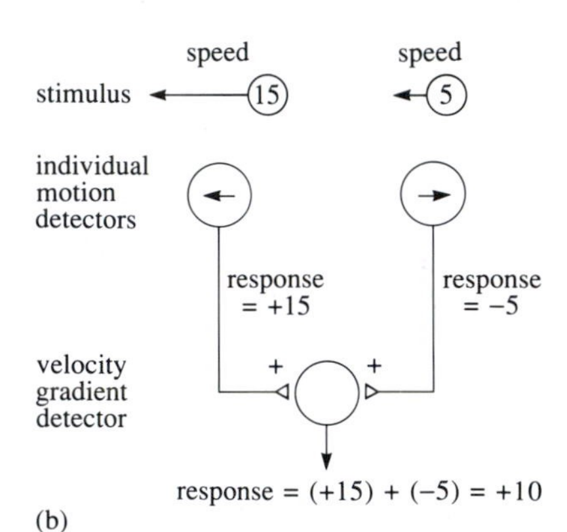

Figure 4.40 Detection of velocity gradients. In this simplified model, velocity gradient detectors add together the responses from neighbouring individual motion detectors which are selective for opposite directions of motion. Each hypothetical individual motion detector gives a positive response to motion in the preferred direction and a negative response to motion in the opposite direction, with the magnitude of the response depending upon the speed of motion (represented in the Figure by the length of the stimulus arrow). When there is no motion gradient (a), the responses of the individual motion detectors are equal and opposite, so the velocity gradient detector does not respond. When there is a velocity gradient (b), one of the individual motion detectors responds more than the other and the velocity gradient detector does respond.

Summary of Section 4.5

In certain mammals, such as primates, retinal processes make use of delayed inhibition to produce a general description of the temporal changes in the image. Early cortical processes refine this description to produce a description of the motion of individual image points. The description is further refined in the middle temporal cortex by mechanisms which combine the motions of neighbouring points and which are sensitive to the typical patterns of motion found in retinal flow patterns. The resulting description contains useful information about, for instance, the three-dimensional layout of surfaces in the world.

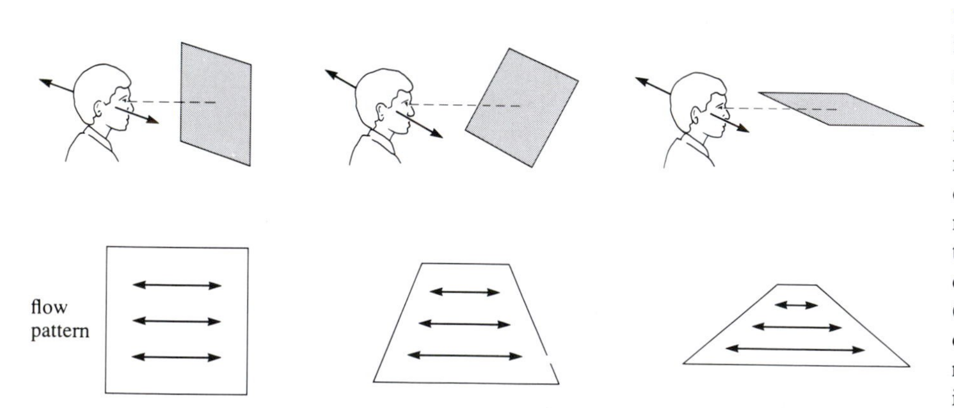

Figure 4.41 Velocity gradients and surface slant: steeply slanted surfaces produce steeper velocity gradients in the retinal image. The length of each double-headed arrow on the surface is proportional to the speed of movement.

4.6 Image description: binocular vision and stereopsis

One of the most fundamental problems of vision is that images are flat, two-dimensional things. The visual system must somehow recover from them the solid, three-dimensional structure of the world that they depict. There are several ways to do this, including monitoring the amount of lens accommodation needed to bring an object into focus, or the degree of eye convergence needed to bring an object into register upon the foveas of both eyes. This section deals in detail with **stereopsis** (literally 'solid vision')—another very important way to extract information about distance, which relies on the fact that you have two views of the world taken from slightly different positions.

Close one eye and hold a finger 25 cm or so in front of your nose. Then close the first eye and open the other. Notice that, as you do this, your finger seems to move relative to the background. This happens because the two images in your two eyes are subtly different, as shown in Figure 4.42. The differences between the images are called *disparities* and they vary systematically with the distances of objects in the world. You can confirm this to yourself by imagining what will happen to the two images of C in Figure 4.42 as it changes in distance from the eyes. As C moves away from the eyes, its image in the left eye will move to the right, past B towards A, while its image in the right eye will move to the left, past A towards B. Since disparities depend upon distance, then, turning the logic around, an analysis of disparities should provide information about the distances of the objects. Stereopsis does just this by comparing the images in the two eyes and analysing the differences between them.

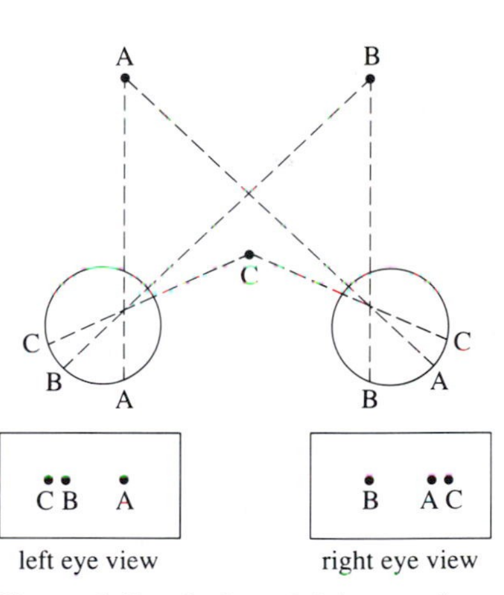

Figure 4.42 Left- and right-eye views of a simple scene illustrate the differences between the images in the two eyes.

In fact, the basis of stereopsis is almost identical to the basis of velocity gradients described in the previous section. Stereopsis makes use of the fact that you have two *simultaneous* views of the same scene taken from slightly different spatial positions, whereas velocity gradients result from a series of *successive* views of the same scene, taken from slightly different positions as you move your head.

Since stereopsis requires a comparison of the images in the two eyes, the two images have to be of more or less the same region of the world and so the two eyes have to point in more or less the same direction. By fixating some point on this page and alternately opening and closing your two eyes, you can verify that this is true of humans. Although there are regions of your visual field which only the left or right eye can see, there is a substantial area of overlap in the middle, providing binocular vision (Book 2, Section 8.8.3)—a region of the world which you can see with both your left and right eye. Only within this region are sensible comparisons, and thus stereopsis, possible.

Not all animals are so equipped. Many non-predators such as rabbits, for example, have their eyes on the sides of their heads. Consequently, they have the largest possible field of view and are able to spot danger approaching from most directions. But they have only a small area of binocular overlap and so cannot make much use of stereopsis. In contrast, many predators have their eyes on the front of their heads. They thus have a smaller field of view—they are less likely to be attacked anyway—but have a large region of binocular overlap and can use stereopsis to give them the good judgement of distance needed when attacking their prey. Predators could, in principle, extract the same information from velocity gradients if they moved their heads from side to side. But the head movement would make them more noticeable to the potential prey and so stereopsis confers an advantage. You may, however, have noticed some animals, for example owls, making characteristic sideways head movements when attempting to make crucial judgments of distance.

4.6.1 Basic anatomy

Animals that make use of stereopsis have two images of the same scene. How are the two images brought together in the brain to make comparisons between them easy? The answer, as shown in Figure 4.43, lies in the arrangement of the optic chiasma, where, in humans, half of the fibres from each eye cross over to the other side of the brain. Light from an object on the left side of the visual field projects to the right-hand side of both images. The fibres from the right-hand half of the right eye retina do not cross over at the optic chiasma and so project to the right-hand side of the brain. But the fibres from the right-hand side of the left eye retina do cross over, so they too project to the right-hand side of the brain. Using Figure 4.43 you can verify that, wherever an object is in the visual field, its two images will always project to the same side of the brain.

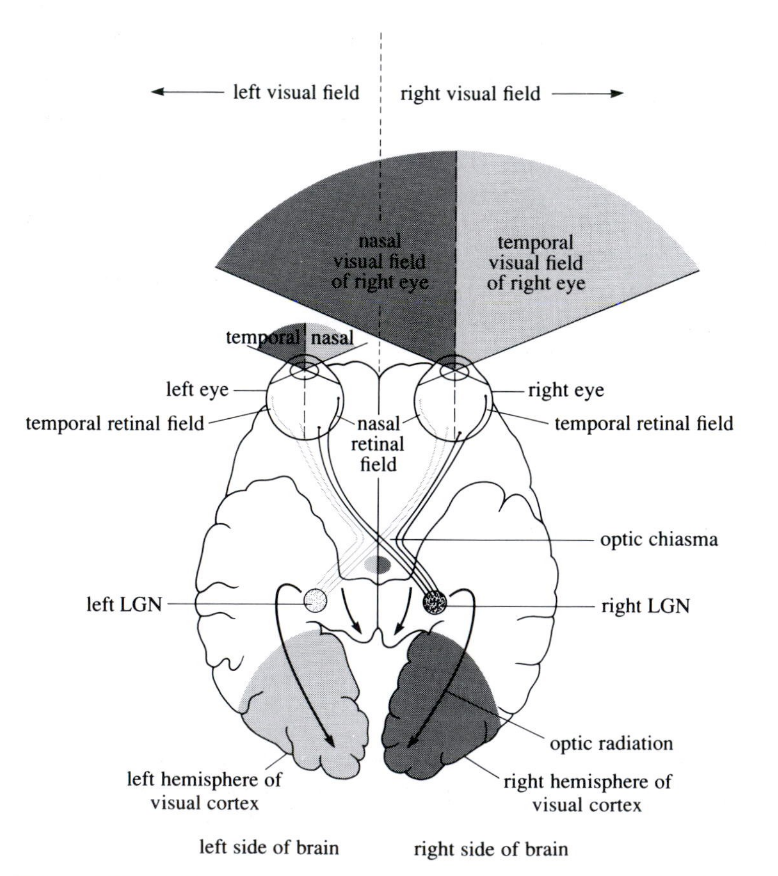

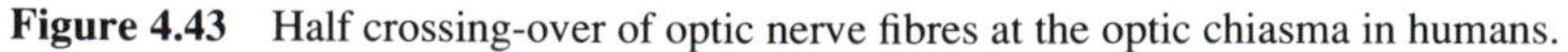

Figure 4.43 Half crossing-over of optic nerve fibres at the optic chiasma in humans.

Because of the part crossing over at the optic chiasma, each side of the LGN receives input from both eyes. However, individual cells in the LGN are all *monocularly driven*—they have a receptive field in just one of the eyes. In fact, fibres from the left and right eyes project to different layers of the LGN, as shown in Figure 4.44. Each layer of the LGN is retinotopically mapped, and the maps are registered so that cells in the same position in adjacent layers have receptive fields in corresponding positions in the two images.

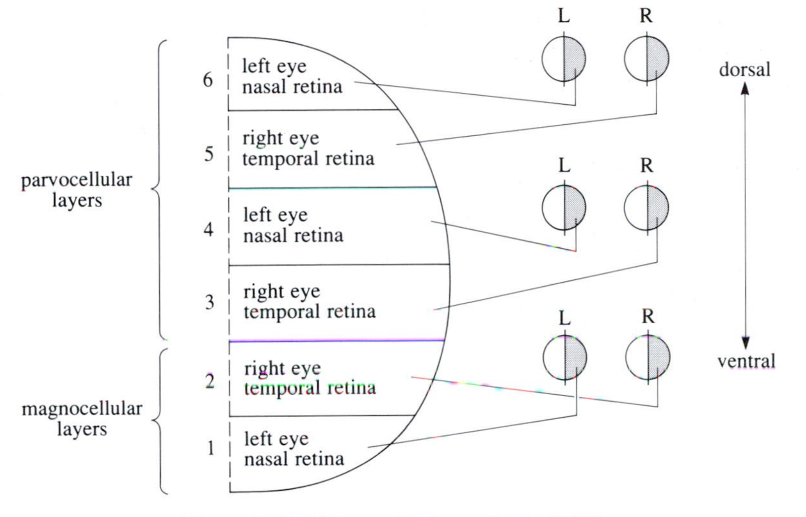

Figure 4.44 Monocular layers in the LGN.

The separation of fibres from the left and right eyes is preserved in the projection from the LGN to the visual cortex. As a microelectrode is moved across the surface of the cortex, it encounters alternating *monocular dominance stripes*, in which most of the cells have receptive fields in the same eye. As shown in Figure 4.45, this arrangement is superimposed upon the orientation columns which were described in Section 4.3.4, so that each hypercolumn is equipped with a full complement of orientation columns and with input from both of the eyes.

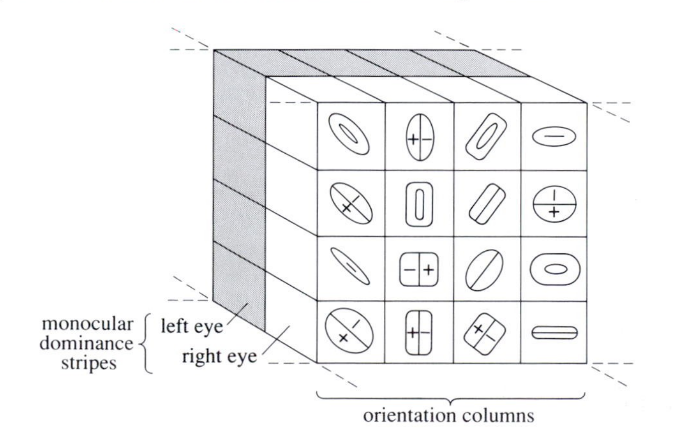

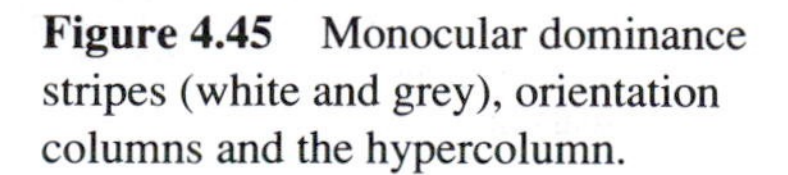

Figure 4.45 Monocular dominance stripes (white and grey), orientation columns and the hypercolumn.

The visual cortex contains both monocularly driven and *binocularly driven* cells—cells with a receptive field in both of the eyes. Indeed there is a smooth gradient of binocular influence, from cells which receive input from only one eye, through cells which have a strong input from one eye and a weaker input from the other, to cells which have a balanced input from both eyes. This gradient arises, as shown in Figure 4.46, from the arrangement of monocular stripes. Cell A, for example, lies in the middle of a left-eye stripe, and so receives its input only from this eye. Cell B, on the other hand, lies near the boundary between a left eye stripe and a right eye stripe and so receives a balanced input from both eyes. Cell C lies in a right eye stripe, but its dendrites project slightly into the adjacent left eye stripe. Its input will therefore be predominantly from the right eye, but it will also have a weaker input from the left eye.

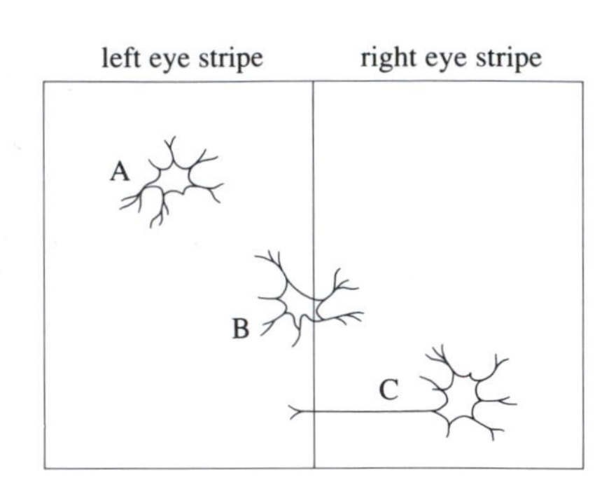

Figure 4.46 Ocular dominance stripes and monocular/binocular cells.

The important point to emerge from this anatomical arrangement is that cells with binocular input have receptive fields in roughly corresponding positions in both eyes. Moreover, recordings show that the two receptive fields have very similar properties so that, if the receptive field in the left eye is fairly large and selective for vertical orientations, so will be the receptive field in the right eye. Such cells obviously have the potential to make comparisons between the two images and so provide the required physiological basis for binocular stereopsis.

4.6.2 Disparity

What are the comparisons that actually need to be made in order to extract depth information from the two images?

□ Figure 4.47 shows the two eyes, viewed from above, fixating a point F. Considering only the left eye for the moment, what happens to the position of the image, P′, of a second point as it moves nearer to the eye from point P_1 through P_2 to P_3?

■ The answer is that nothing happens—the position of the image in one eye can signal the direction of an object, but not its distance.

□ Now consider the right eye. What happens to the position of the image in this eye as the point moves from P_1 to P_3?

■ Its image gradually changes in position from P'_1 to P'_3. As the point gets farther away its image moves in one direction, as the point gets closer its image moves in the opposite direction. It ought to be possible to exploit this change in position to extract information about the distance of the point.

To do this, the positions of the two images of an object, one in each eye, have to be compared. If the image is at point P′ in the left eye and at P'_2 in the right eye, then the object *must* be at point P_2; if it is at point P'_3 in the right eye, then the object must be at point P_3; and if it is at point P'_1 in the right eye it must be at point P_1.

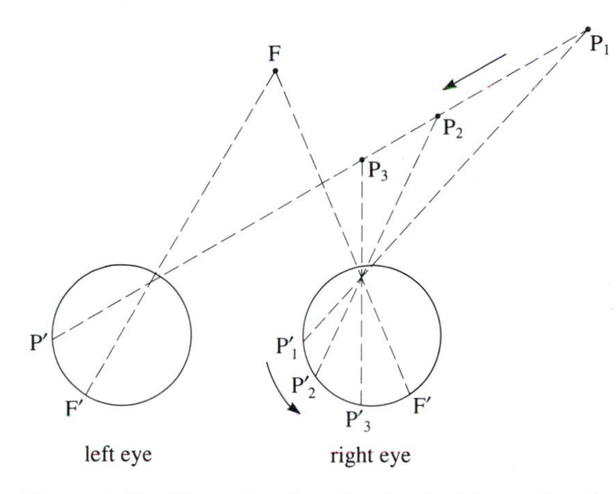

Figure 4.47 Binocular disparity. See text for explanation.

In principle, binocularly driven cells in the visual cortex are well equipped to perform just this type of comparison between the images and to measure the disparities between them. Figure 4.48 shows three such cells, each with a receptive field in the same position in the left eye, corresponding to point P′ in Figure 4.47, but with slightly offset receptive fields in the right eye, corresponding to the different positions P'_1, P'_2 and P'_3. Each cell will respond most when both its receptive fields are excited, that is, when the two images of an object fall in the appropriate positions in the two eyes. Cell B, for example, would respond most when the object is at point P_2, cell C would respond most when it is closer than this, and cell A would respond most when it is farther away.

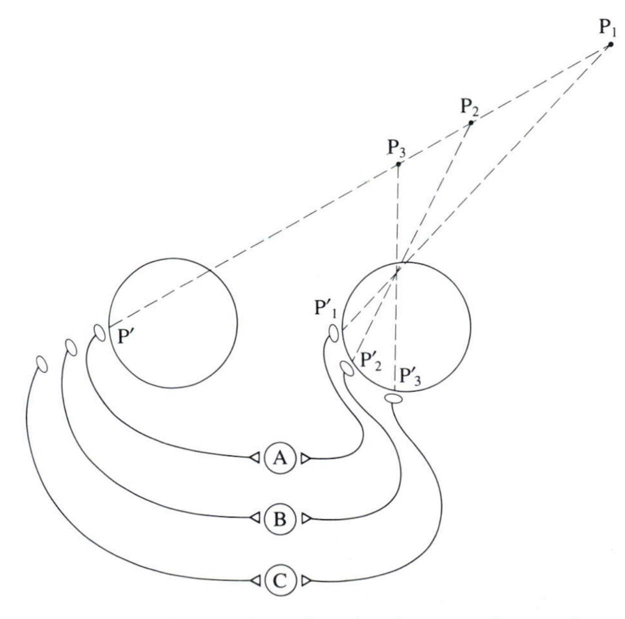

Figure 4.48 Binocularly-driven cells as disparity detectors. See text for explanation. For clarity, the three cells A–C have been drawn with widely spaced receptive fields in the right eye; the receptive fields should actually occupy roughly the same area of the retina and only be displaced by a small amount.

Summary of Section 4.6

The images in the left and right eyes are slightly different and an analysis of these differences, or disparities, can provide information about the distance of objects in the scene. The beginnings of this analysis are probably carried out in the visual cortex by binocularly-driven cells which have a receptive field in each eye and are thus equipped to make comparisons between the left- and right-eye images.

4.7 Modularity in the early stages of vision

The preceding sections dealt quite separately with luminance, colour, movement, and stereopsis. This modular approach is not just a convenient simplification—the early stages of the visual system do, in fact, seem to consist of a series of separate modules, each working in parallel to produce an independent description of one aspect of the proximal stimulus. Such a modular arrangement is what you might expect from an evolutionary viewpoint. The complex and sophisticated visual systems found in primates have all evolved from very much simpler systems. Even a very crude sensitivity to light would confer an adaptive advantage on the primitive forms which first evolved it, but how did that crude sensitivity evolve into the exquisite system found in primates?

Imagine an animal which, through a long process of small, random mutations in previous generations, has evolved a visual system which can make sense of spatial structure but which cannot discriminate between colours. Even to provide an adequate solution to the problems of encoding spatial structure, the system will have to be extremely complicated. The chances of modifying that system, by random mutation, to develop colour sensitivity are extremely slight—because any random change which confers an advantage towards colour vision will almost certainly have a much more disruptive effect on the existing processes for encoding spatial structure. It seems much more likely that different visual abilities have evolved in parallel, rather than in sequence, and that systems for dealing with, say, colour and form have maintained a functional independence. That way, random mutations in the system which deals with one sub-modality can be appropriately modified by natural selection without being confounded by almost inevitably disruptive side-effects in the systems which deal with other sub-modalities.

This section outlines anatomical and neurophysiological evidence which suggests that there are at least three independent *streams* in the early stages of vision dealing separately with form, colour and movement/stereopsis. The relevant distinctions are traced from the retina, through the LGN, to the cortex. To help you keep your bearings, Figure 4.2 (p. 78) provides a map of the main anatomical structures, while Figure 4.49 (p. 136) provides a more detailed, schematic summary.

□ In previous sections, you have already encountered the anatomical and functional bases of separate functional streams. Suggest one example from the retina, and one from V1 of the visual cortex.

- Retinal examples include distinct types of receptors (rods and cones), on- and off-bipolar and retinal ganglion cells, retinal ganglion cells with different types of colour-sensitivity, and sustained and transient retinal ganglion cells. Cortical examples include the simple, complex, hypercomplex cell classification, orientation columns, monocular dominance stripes, directionally-selective cells, and binocular cells tuned to different disparities.

4.7.1 Retinal ganglion cells

There are a number of ways of classifying retinal ganglion cells into different functional and anatomical classes—indeed the most sophisticated schemes recognize over 20 different types. One main classification, described in Section 4.5.1, divides cells into two types, sustained and transient, on the basis of their temporal response properties. It turns out that this scheme coincides rather well with one based on different functional criteria.

The responses of retinal ganglion cells to large stimuli are generally predictable from their responses to small spots of light. So, for example, if a small spot of light in one position in the receptive field produces a small increase in response, and a small spot in another position produces an equivalent decrease in response, the two spots of light presented together should produce no overall change in response. In fact, this kind of prediction is only possible for some retinal ganglion cells, which are called **X cells**. The responses of other cells, called **Y cells**, cannot be predicted from their responses to individual spots. A spot which causes an increase in response and a spot which causes a decrease in response may together, for example, produce a larger increase in response rather than the expected cancellation.

When cells are classified into X and Y types on this basis, it turns out that X cells tend to give sustained responses, while Y cells tend to give transient responses. Moreover, X cells tend to have small receptive fields and to occur predominantly in the fovea, whereas Y cells have larger receptive fields and are found predominantly in the retinal periphery. These functional differences originally led to an early attempt to identify functionally separate streams. X cells seemed ideally suited to encode spatial information (i.e. form), while Y cells seemed better suited to encode temporal information (i.e. motion). Unfortunately, although the notion of separate streams for form and motion seems correct, the basis in X and Y cells is too simple and, as is clear from the properties of cells in later stages of the visual pathway, almost certainly incorrect.

4.7.2 The LGN

You already know that cells in the LGN have concentric, antagonistic receptive fields like retinal ganglion cells (Section 4.3.4), and that the primate LGN consists of six layers, each retinotopically mapped, and each containing cells with receptive fields in just one of the eyes (Section 4.6.1). The cells in the lower layers have larger cell bodies and these are called the magnocellular layers, the cells in the upper layers have smaller cell bodies and these are called the parvocellular layers. Magno cells have large receptive fields, give transient responses, are very sensitive to luminance discontinuities and are not colour-coded. Parvo cells have small receptive fields, give sustained responses, are rather insensitive to luminance discontinuities, and are colour opponent cells. Since magno cells share some of the properties of retinal Y cells (large receptive fields, transient responses), while parvo cells show some of the properties of retinal X cells (small receptive fields, sustained responses), it was originally thought that Y cells projected to the magnocellular layers, and X cells to the parvocellular layers, so preserving the supposed separate streams for motion and form. This now seems unlikely. The chief distinction between parvo and magno cells is that the former are colour-coded whereas the latter are not. Yet both X and Y cells consist of a mixture of colour-coded and non-colour-coded cells. Moreover, X and Y cells have similar sensitivities to luminance discontinuities, but magno and parvo cells do not. Indeed, in retrospect, it seems rather strange to suggest that the proposed form system should be mediated entirely by parvo cells, which are relatively insensitive to luminance discontinuities.

Current thinking is that magno and parvo cells do indeed form the cellular basis of separate functional streams and that these two different types of cell do receive input from different types of cell in the retina. However, the retinal origins, though not yet fully understood, are almost certainly not as simple as the distinction between X and Y cells. Moreover, the distinction into just form and motion streams is also too simple, as is clear from the properties of cortical cells.

4.7.3 The visual cortex

Superimposed upon the arrangement of orientation columns and eye dominance stripes, it is possible to reveal a different kind of cortical organization by staining with cytochrome oxidase. This stain is taken up only by some cortical cells and reveals a complex and beautiful tracery of dark (stained) and pale (unstained) regions. Within each hypercolumn of V1 (Area 17), discrete groups of cells take up the stain to reveal a pattern of dark **blobs** across the surface of the cortex. Those regions of V1 which do not take up the stain are, by default, called **interblobs**. In V2 (Area 18), the pattern is equally distinct but very different. Here the stain is taken up by distinct **thin stripes** and **thick stripes**, interspersed with **pale stripes** which are relatively unstained.

How are blobs, interblobs, thick stripes, thin stripes and pale stripes related to the parvo and magno cells in the LGN and to the idea of distinct functional streams? At the moment, three functionally and anatomically distinct pathways have been identified.

1 Magno–thick stripe: a stream for motion and stereopsis?

Magno cells from the LGN project to cells in V1, which have orientationally selective receptive fields, are generally binocularly driven, and are selective for direction of motion. These cells project in turn to the middle temporal cortex where, as described in Section 4.5.2, there are many cells that are selective for different types of motion. A second pathway from the cells in V1 projects to the thick stripes in V2, where the cells respond best to moving stimuli and are frequently tuned for binocular disparity. The thick stripes in turn project to the parietal regions of the cortex. The cells in this pathway are generally not selective for colour and show no specialization for spatial form beyond basic orientation selectivity. On the other hand, most of them are selective for direction of motion and/or retinal disparity. It thus seems that the magno-thick stripe pathway may form the anatomical basis of a stream concerned primarily with motion and stereopsis. This stream is summarized in Figure 4.49a (*overleaf*).

2 Parvo–interblob: a stream for fine spatial structure?

LGN parvo cells project to cells in V1 which, despite the colour coding of the parvocellular layers, have non-colour-opponent, circular receptive fields. This pathway forms the interblob regions, where cells prefer moving stimuli, are orientation selective, and are not colour-coded. Interblobs project in turn to the pale stripes of V2, where the cells are orientation selective and tend to be end-stopped (hypercomplex) cells. The pale stripes in V2 generally project onwards in pathways leading to the temporal, rather than the parietal, cortex. Cells in this pathway are again not colour-coded, and though they 'prefer' moving stimuli, are generally not selective for the direction of motion. The projection to end-stopped cells is reminiscent of the first stages in the original hierarchical model of form perception discussed in Section 4.3.4 and this pathway is thought to be the basis of a stream concerned primarily with spatial structure. This stream is summarized in Figure 4.49b.

3 Blob–thin stripe: a stream for colour?

Cells in the blob regions of V1 have circular receptive fields with colour opponent properties. These cells project on to the thin stripes of V2, where cells are orientation selective and are also colour opponent. Thin stripes in turn project to V4 which, as described in Section 4.4.4, contains cells with fine selectivity for colour, and from there on to other regions of the temporal cortex. This pathway is thought to form the cellular basis of a separate stream concerned primarily with colour. Unfortunately, its origin in the LGN is not yet fully understood. One might expect that blobs should receive their input primarily from the parvo cells in the LGN, which are characterized by their colour opponency. But, as previously mentioned, parvo cells project to interblob cells, which are not colour-coded, so the projection is obviously a little more complex than that. Moreover, when blob regions are injected with tracer chemicals which are transported back down cell axons to the LGN, the tracers turn up, not in the parvo cells, but in a few interlaminar (between layer) cells in the LGN. It seems likely that blob cells receive their input directly from cortical, rather than LGN cells, and that this input contains elements originating in both parvocellular and magnocellular layers. This stream is summarized in Figure 4.49c.

Although this rather complex story is incomplete, and in places puzzling, the overall picture suggests a glimpse of a beautiful and intricate anatomical organization which underpins a functional separation into modules for form, colour, and motion/stereopsis. Complex as the picture seems so far, it remains just a glimpse: current classifications recognize 19 different visual areas in the primate cortex which are interconnected by over 80 known pathways.

Given this degree of complexity, it is not surprising that the anatomical details are as yet incomplete, or that different experimental techniques sometimes yield contradictory data. After all, there is tremendous variation between species, the properties of cells vary considerably with different levels of anaesthesia, and different types of stimulus tend to emphasize different cell response properties.

Indeed, the fact that a coherent story is emerging despite these complications is very encouraging in that it suggests that the overall framework is correct.

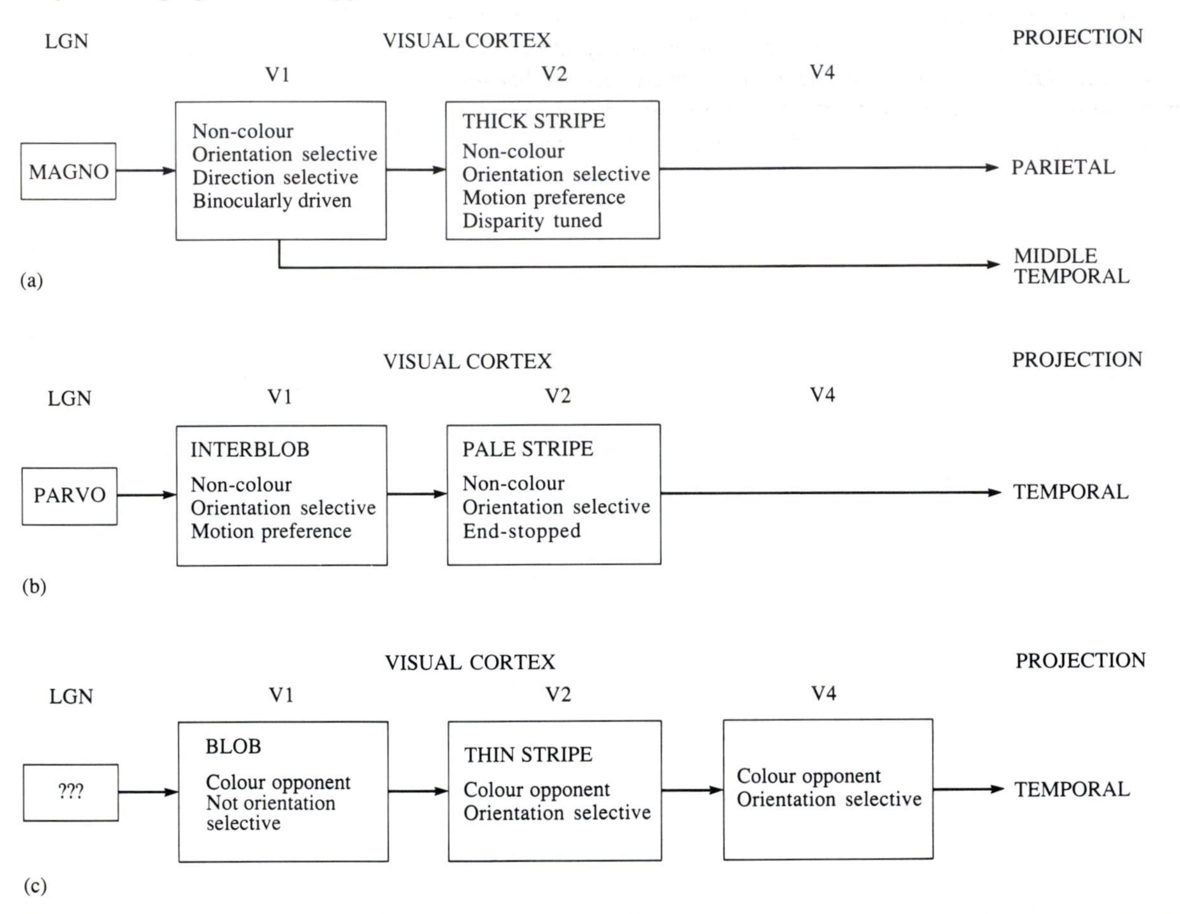

Figure 4.49 Schematic summary of anatomical and functional streams. (a) Magno-thick stripe stream. (b) Parvo-interblob stream. (c) Blob-thin stripe stream.

4.8 Later descriptive processes

The processes discussed so far are all *analytic*: they are concerned with breaking down the proximal stimulus into a number of different types of information. In this section, the emphasis shifts to *synthetic* processes, which put the information back together into a coherent whole. There are two main types of synthetic process. **Grouping processes** are concerned with building larger-scale descriptions from the rather small-scale descriptions produced in the early stages of vision. **Integrative processes** are concerned with combining the information from different visual sub-modalities.

4.8.1 Grouping processes

The first stages of vision seem to derive more or less independent descriptions of the luminance changes, colour, motion and binocular disparity present in the proximal stimulus. Each description is initially very localized: the mechanisms involved have comparatively small receptive fields and so describe only a small region of the stimulus. You can thus think of the initial description as being rather like a jigsaw, with the first cortical stages of the process providing information about individual pieces. Of course, it is actually a rather strange jigsaw because each position is represented by a stack of related pieces giving separate information about colour, motion and so forth. But the main point is that jigsaws are regarded as puzzles precisely because it is very difficult to recognize anything by looking at individual pieces. This is true even if, as in the retinotopically mapped cortex, the relative positions of the pieces are roughly known. Recognizable features of the visual world—objects and characteristic parts of objects—will tend to be represented by groups of pieces, and so each individual piece needs to be understood in its proper context. Consequently, following its initial localized stage, the descriptive process needs to group together the initial fragments into larger elements.

Figure 4.50 Examples of visual grouping: (a) Grouping by spacing. (b) Grouping by colour. (c) Grouping by shape.

The existence of grouping processes in human vision is demonstrated by Figure 4.50; you spontaneously organize each group of elements into rows and columns, according to spacing, colour, and shape. Such stimuli were much used by the Gestalt psychologists, who were intrigued by the observation that 'the whole is greater than the sum of its parts'. This catchphrase is really something of an exaggeration, but it does emphasize how the visual system imposes a structure on what is seen.

The groupings in Figure 4.50 follow simple physical rules; elements are grouped together on the basis of some shared physical similarity such as colour or form—precisely the sort of information provided by the initial localized mechanisms described in Sections 4.3 to 4.7. Again, this is rather like solving a jigsaw puzzle. If you do not know what the puzzle depicts, one strategy is to group pieces together on the basis of some shared physical similarity—colour or texture, for example. The resulting groups of meaningless pieces may then become recognizable as a part of a tree, or a face, and so forth. The assumption underpinning both jigsaw solving and visual grouping appears to be very simple: if individual fragments share a common descriptive attribute, then the chances are that they have the same physical cause, and thus belong together.

The advantages of a visual grouping process are illustrated in Figure 4.51. Look first at Figure 4.51a. A localized descriptive process might record the characteristics of each dot, in terms of colour, shape, position and so forth. But it is also possible to group these individual elements into a coherent structure—a

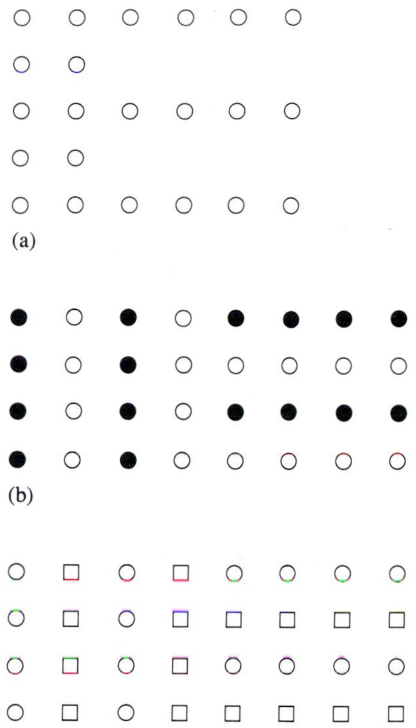

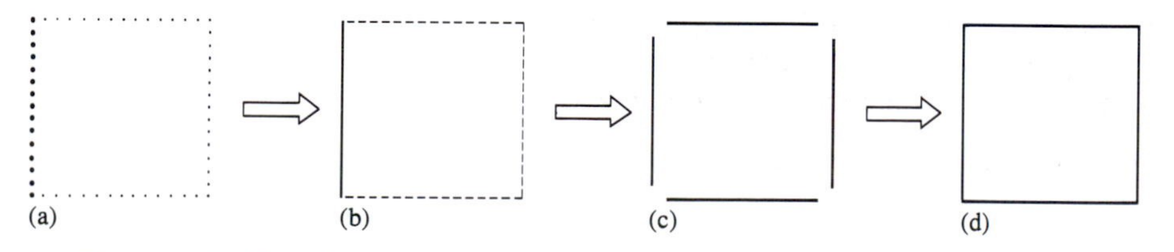

Figure 4.51 Descriptive advantages of visual grouping. See text for explanation.

line of dots with a particular orientation, as represented in Figure 4.51b. Although the orientation of this structure is *implicit* in the positions of the individual dots, the grouping process has made that information *explicit*. Figure 4.51c takes this process one stage further. Again, you could describe this figure in terms of its individual elements—4 lines, each with a particular position and orientation. But you could further organize the description into a coherent whole, as depicted in Figure 4.51d, which makes explicit the useful property of shape.

□ Can you suggest two important advantages of visual grouping?

■ First, it makes descriptions much more wieldy. Rather than a long, detailed list of individual elements, the useful information can be summarized in a coherent whole: 'A square made out of red dots' is clearly much more concise than 'Twenty red dots; the first is at position x_1, y_1; the second is at position x_2, y_2..., etc.'. Second, grouping emphasizes the properties of the stimulus, like orientation and shape, which are important in recognizing objects. Though this information is already present in a description of the individual elements, it is notoriously difficult to see the wood for the trees.

4.8.2 Integrating the description

The bulk of the evidence summarized in Section 4.7 suggests that initial localized descriptions are derived largely independently for the different sub-modalities of luminance, colour and so forth. It is possible that the grouping processes described above may also occur independently within separate functional streams. It is even clear that, under some circumstances, the information from a single stream may be sufficient for object recognition: humans can, after all, make sense of black-and-white line drawings. However, there are also strong grounds for supposing that information from different sub-modalities should, at some stage, be brought back together. The need to produce an integrated description, which draws all the different types of information together in a coherent whole, is well-exemplified by the problems of recovering the three-dimensional world from the two-dimensional image. There are many possible sources of information about the third dimension and these are generally called **depth cues**.

□ In this chapter, you have already encountered five different depth cues. Can you recall them?

■ So far, you have encountered shading, velocity gradients, accommodation, convergence of the eyes, and binocular disparity.

□ Figure 4.52 illustrates several other depth cues. You interpret this as a three-dimensional scene but you cannot be using any of the cues listed above because the stimulus is really flat, unshaded, and stationary. What kinds of cue might you be using here?

■ Possible depth cues include texture gradients (e.g. the paving stones and the flowers), height in the image (the house is further than the gate and appears above it in the image) and linear convergence (the edges of the path are really parallel but converge in the image). These are all examples of *linear perspective* (the rules for drawing solid objects on flat paper). You might also notice *occlusion* (e.g. one of the trees obscures the other and is thus seen to be in front of it) and *known size* (e.g. houses are large objects so, if they are relatively small in the image, they must be relatively distant).

Figure 4.52 Pictorial depth cues.

One final depth cue, not possible to show here, is called *aerial perspective*. Because of light scatter, distant objects appear rather blurred and bluish in colour. This is often used to convey distance in landscape paintings.

The important point here is not to remember all the individual depth cues, but to notice that there are so many of them, and that no single visual sub-modality is sufficient to specify them all. To obtain a full impression of the solidity of the world, you need to draw together information from all the visual sub-modalities.

The late David Marr, whose ideas underpinned much of vision research throughout the late 1970s and 1980s, argued that the ultimate goal of descriptive processes was to produce a single, rich, coherent description of the proximal stimulus in terms of surfaces, their properties, and their orientation in relation to the viewer. He termed this representation the *two-and-a-half-dimensional sketch* since, although it contains depth information from all the depth cues, it is not quite a full three-dimensional representation. A full three-dimensional representation would provide information about the *absolute* positions and arrangement of *all* the surfaces of each object in the scene, whereas the two-and-a-half-dimensional sketch can only provide information about the *relative* arrangement of those surfaces which are currently *visible*.

The notion of a common representation is purely theoretical—there is, as yet, no neurophysiological evidence for it. However, it does make it easier to think about vision, and it does make sense to think of a common representation where items of information from different sub-modalities can be checked against each other and used to aid the interpretation of ambiguities. Moreover, you have already encountered at least one perceptual phenomenon which demonstrates that your interpretation of depth can cause you to re-interpret information from individual visual sub-modalities.

□ Can you recall it?

■ In the Mach card demonstration (Section 4.3.3). When the apparent depth of the card reverses, your interpretation of surface reflectance changes quite dramatically.

4.9 Interpretative processes

At the beginning of this chapter, a discussion of the fundamental differences between a distal object (a cube) and its proximal stimulus (a sketch of a cube) led to the conclusion that descriptive processes are not enough, by themselves, to account for visual perception. You now know that the kind of description available to the visual system, which you can think of as a kind of two-and-a-half-dimensional sketch, is a good deal more sophisticated than a simple sketch of the cube. Since this kind of representation is clearly much more like the distal object than the original two-dimensional sketch, is the conclusion still valid? Are descriptive processes, after all, enough to explain perception? The answer, regrettably, is 'no'.

The first reason for continuing to propose that interpretative processes are also necessary is fairly straightforward. Even if, under natural conditions, you have a rich description of the proximal stimulus, you can still make sense of the world when this information is much reduced. You can, in fact, recognize the original sketch of the cube, even though your description of it must be very impoverished because there is no real depth, movement or colour to describe. In this situation the original objection, that the distal object and the proximal stimulus are fundamentally different, continues to hold.

The second reason is much more general and profound. The word 'recognize' literally means to know again (re-cognize)—and here the important word is 'again'. No matter how sophisticated your description, if you are to know something again, you must match that description with some pre-stored internal representation of the distal object. This matching process may *deal* with descriptions—of the proximal stimulus and of all known objects—but the process of matching is itself not a descriptive process: to draw a picture is one thing, but to decide that it is a likeness of something requires quite different skills.

The final stages of object recognition require such complex processes as memory and attention. Both of these topics are beyond the scope of this chapter, belonging to the domain of cognition (Book 1, Chapter 8) rather than perception. However, there are a few approaches which are particularly relevant to the perceptual scientist and which provide useful insights into interpretive processes. The first of these is the study of visual illusions. You have already encountered several of these throughout this chapter. The second approach is much more recent and involves the attempt to construct artificial vision systems using computer models.

4.9.1 Computer modelling of object recognition

Most current theories of visual object recognition propose that interpretive processes make use of specific cues present in the description of the proximal stimulus to produce an internal representation or hypothesis which can account for the data and which is, in effect, what the observer 'sees'. But what *exactly* are the cues, how *exactly* are they used to mobilize internal knowledge of the world, and how *exactly* is the internal knowledge represented? Questions of this type are often addressed by computer models, which attempt to simulate perceptual processes and which force the programmer to specify his or her ideas in exact detail. Just as experiments provide a way to break complex phenomena down into their component parts, computer models provide a way to build specific ideas up into complex, functioning systems. In other words, they complement the traditional *analytic* scientific method, by providing a method of *synthesis*—a method which seems more appropriate to the study of the mental processes involved in the later stages of perception. Computer models cannot, by themselves, show that a theory is correct but they can show whether or not a theory is actually capable of performing the task for which it was proposed.

Computer modelling of mental processes, particularly of visual perception, is itself a large subject and this short section can do no more than scratch the surface. The first thing to say is that none of the current models can even remotely approach the performance of the human visual system. There are, however, working models of many individual visual processes—models which, for example, can recover and describe the object boundaries from an image or which, given two images of a scene from adjacent viewpoints, can calculate the distances of surfaces using binocular disparity. Such models are particularly useful to the neuroscientist because they are a genuine attempt to mimic known physiological mechanisms like retinal ganglion cells. Other models are less concerned with specific neural processes and attempt instead to test more general theories of vision. Such models can, for example, interpret a sketch of a cube, like the one shown in Figure 4.1, in terms of surfaces and their three-dimensional relationships to each other. This type of model reveals that junctions between lines in the image, which correspond to corners of the object, are particularly effective cues which can be used to mobilize simple stored rules about how surfaces can be joined together in three-dimensional space.

Perhaps the most interesting kinds of model are those which go beyond simple descriptions of the stimulus and which attempt actually to recognize objects. An example of this more ambitious approach is provided by a program called WALKER, which is equipped with knowledge about the shape of the human body and about how arms and legs move during walking. Given a video of people walking through a natural setting it can pick out the people, demonstrating this by overwriting the video with matchstick figures whose movements closely match those of the actual people.

Although these achievements are considerable, they seem almost trivial in comparison with the feats that human perception can perform. Why is human object recognition so difficult to model? The answer to this brings you back to where the chapter started. Visual perception is difficult to model because the task it performs is very, very difficult—it just seems easy because humans are very, very good at it. First, vision is confronted by a huge range of data. Prior sections have described the kinds of descriptions it can form using data about luminance, colour, movement and stereopsis. No existing computer model can deal with this quantity of data, and most work only on static, black-and-white images. Although this approach reduces the amount of data which needs to be dealt with, it actually makes recognition more difficult because the possible descriptions that can be formed are inevitably very limited.

Second, the visual system has at its disposal vast amounts of internal knowledge about all kinds of objects, about their possible uses, and about the context in which they are likely to be found. Existing computer models have internal knowledge about just a few objects, and that knowledge is generally limited to physical structure. Again, this reduction in the amount of information seems to make the problem smaller and therefore easier but it may, in fact, actually make it more difficult. WALKER, for example, can track a human figure walking through a park, but is at a loss as soon as the figure sits down. This is not a problem for a human observer, who typically knows much more about people, and about parks, and about what people are likely to do in them.

In a general sense, computer models just highlight the fact that visual perception requires the condensation of the immense amount of information available in the proximal stimulus, and the matching of that information with an even more immense amount of information about the distal objects which make up the perceivable world. In particular, they force the programmer to confront the problem of how all that internal knowledge is acquired in the first place. Since it is unreasonable to suppose that babies are born with knowledge about grandmothers, Volkswagens, parks and so forth, internal knowledge must presumably be acquired through the very perceptual systems that require that knowledge in order to work properly in the first place. The only way out of this vicious circle seems to be to propose that babies' initial perceptual abilities are comparatively limited, but that they are equipped with very general rules about how to describe the proximal stimulus, how to integrate information from different senses, and how to recognize the correlations between actions and their effects upon the information available through those senses. They should then be able to learn how to perceive. Presumably the solution for computer models is rather similar: rather than attempting to equip them with a mass of knowledge, the model must be able to acquire that knowledge for itself.

In this respect, recent developments in neural networks (Parallel Distributed Processing, or PDP) are particularly exciting. These models consist of many simple processing units connected together by excitatory and inhibitory links. Apart from the fact that this is rather similar to how the brain seems to work, PDP models are attractive because they have the capacity to learn by themselves. They do this, essentially, by randomly adjusting the strengths of the connections between units: those changes that bring about an improvement in performance are further explored, those that impair performance are abandoned. This is very reminiscent of the evolutionary process—except that the time-scale is very much reduced and the programmer, rather than Nature, specifies the goals.

4.10 Concluding comments and unanswered questions

Current achievements in understanding descriptive processes, though incomplete, are really fairly impressive. It is probably fair to say that this area of study has provided some of the most convincing and detailed insights into how the brain actually works. On the other hand, turning to interpretive processes, it is easy to take a more pessimistic view. Current achievements seem so trivial in relation to the magnitude of the task and it is clear that any proper solution will require a fuller understanding not only of interpretive processes themselves, but also of how people represent their knowledge and how they acquire it in the first place. However, daunting as the problems are, they are at least fairly clearly defined, and current interdisciplinary approaches at least hold out the possibility of solutions in the foreseeable future. Moreover, the study of what may at first seem a fairly specialized ability, seeing, turns out to have much more general implications. Seeing involves the generation and manipulation of complex internal representations of the external world. This seems, in many respects, to capture the very essence of mind. Its study provides insights into areas like learning and memory. Perhaps most important, in the context of the whole course, it provides important constraints upon how to think about more general questions. You might, for example, end your study of this chapter by spending a few minutes thinking about how what you have learned bears upon each of the following issues.

What is the perceptual world like to a newborn infant? Is it, for example, like Figure 1.5 in Chapter 1 before you see the Dalmatian?

Animals, particularly mammals, seem to do a pretty good job of making sense of the world. Since this seems to require an ability to manipulate internal representations, what does this suggest about the mental capacities of these animals?

What is the relationship between an internal representation and consciousness? What are the relationships between the evolution of perception, of consciousness and, indeed, of intelligence?

Summary of Sections 4.7–4.10

Early descriptive visual processes deal with small regions of the image and with specific sub-modalities such as colour or motion. They are organized anatomically into distinct functional streams. Later stages group the small fragments into larger-scale descriptions and integrate information from the different sub-modalities.

Before objects and events can be recognized, the results of these descriptive processes must be matched with pre-stored knowledge about the world. Rather less is known about the interpretive processes that are involved in this task, but one promising way to study them is to build computer models that attempt to mimic human performance and thus provide a very rigorous test of current theories.

Objectives for Chapter 4

When you have completed this chapter you should be able to:

4.1 Define and use, or recognize definitions and applications of each of the terms printed in **bold** in the text.

4.2 Explain, with examples, the distinction between descriptive and interpretive processes.

4.3 Recall the two essential properties of a visual image.

4.4 Describe the role of the lens and of accommodation in the formation of visual images. (*Question 4.1*)

4.5 Describe the process of transduction in the retina. (*Question 4.2*)

4.6 Describe and define the receptive field of a typical retinal ganglion cell.

4.7 Describe the relationship between object boundaries and image luminance discontinuities, and the processes by which retinal ganglion cells detect and describe luminance discontinuities. (*Question 4.3*)

4.8 Describe the basic anatomy and neurophysiology of the retina, with particular reference to the processes that underpin the receptive field of retinal ganglion cells.

4.9 Account for the perceptual phenomena that occur in the Mach bands and the Craik–Cornsweet–O'Brien illusions in terms of processes occurring in the retina. (*Question 4.4*)

4.10 Describe the processes involved in dark adaptation. (*Question 4.5*)

4.11 Account for the perceptual phenomena that occur in lightness contrast and lightness constancy, in terms of processes occurring in the retina.

4.12 Describe the basic anatomy and neurophysiology of the visual cortex. (*Question 4.6*)

4.13 Outline the theory of feature detection and apply it to the coding of orientation and to the tilt after-effect.

4.14 Outline the theory of hierarchical processing and apply it to the phenomenon of face recognition. (*Question 4.7*)

4.15 Recall three advantages of colour vision.

4.16 Describe how the processes of trichromaticity and opponency occur in the retina.

4.17 Describe the role of cortical mechanisms in the phenomenon of colour constancy.

4.18 Explain why it is advantageous to be able to perceive motion.

4.19 Describe the neural basis of directional selectivity.

4.20 Account for the perceptual phenomena of apparent motion and the motion after-effect in terms of neurophysiological processes.

4.21 Describe what is meant by retinal flow patterns and retinal velocity gradients, and the cortical mechanisms that are thought to be involved in their analysis.

4.22 Describe the relationship between object distance and binocular disparity.

4.23 Describe how retinal images are projected, via the optic chiasma, to the LGN and the visual cortex.

4.24 Describe the arrangement of monocular dominance stripes and of binocularly driven cells in the visual cortex.

4.25 Recall the distinction between retinal X and Y cells, and between parvocellular and magnocellular cells in the LGN.

4.26 Describe the anatomical and neurophysiological basis of separate streams for motion/stereopsis, spatial structure, and colour.

4.27 Distinguish between grouping and integrative processes. (*Question 4.8*).

4.28 Describe depth cues.

4.29 Discuss the role of computational models of human vision. (*Question 4.9*)

Questions for Chapter 4

Question 4.1 (*Objective 4.4*)
What is the role of the lens in the human eye?

Question 4.2 (*Objective 4.5*)
How is the pigment rhodopsin involved in the process of transduction in the retina?

Question 4.3 (*Objective 4.7*)
What is the relationship between luminance discontinuities in retinal images and object boundaries in the world?

Question 4.4 (*Objective 4.9*)
What is the significance of Mach bands for theories of visual perception?

Question 4.5 (*Objective 4.10*)
What is the role of dark adaptation?

Question 4.6 (*Objective 4.12*)
What finding suggests that cortical simple cells do not provide the input to complex cells?

Question 4.7 (*Objective 4.14*)
Can the notion of hierarchical processing be applied to the process of face recognition?

Question 4.8 (*Objective 4.27*)
What sort of image features are important in determining whether or not distinct parts of the image will be grouped together?

Question 4.9 (*Objective 4.29*)
What are the advantages of computational models of visual perception?

Further reading

Bruce, V. and Green, P. R. (1997) *Visual Perception: Physiology, Psychology and Ecology*, 3rd edn, Lawrence Erlbaum Associates. An excellent, up-to-date and comprehensive text on vision.

CHAPTER 5
TOUCH AND PAIN

5.1 Introduction

Four of the traditional five senses, sight, hearing, taste and smell, have already been mentioned in this book; the fifth sense, 'touch', will be discussed in this chapter. Interestingly, pain is not included among the traditional five senses, because Aristotle did not regard pain as a sensation, but as an emotion, the opposite of pleasure. However, pain does have sensory qualities and, in this chapter, touch, pain and temperature sense will be considered together as the 'contact senses'.

In contrast to the 'distance senses' of sight, hearing and smell, described in earlier chapters of this book, the 'contact senses' arise from stimuli occurring on or near the body surface, and in some cases from within the body itself. In this context the term 'touch' will be used to refer to various tactile sensations including pressure and vibration; temperature sense relates to the awareness of changes in the temperature of the skin; pain can arise from many forms of intense stimulation of either the surface or interior of the body. The initial part of this chapter (Sections 5.2–5.3) will focus on the general aspects of the skin senses; a more comprehensive discussion of the complexities of pain will be deferred to Sections 5.4–5.9. Some aspects of sensory pathways have been described in Book 2, Chapters 8 and 9, and the present discussion will build on these foundations.

Before dealing with the biological basis of the skin senses, it is worth reconsidering some of the problems involved in communication of stimulus properties by the nervous system. Stimuli do not act directly on the brain; the brain receives information about stimuli by means of coded electrical signals that propagate along afferent nerves. The first stage in this process is the conversion or transduction of the stimulus energy into the 'language' of the nervous system—action potentials (Section 1.2). This is the role of the sensory receptor. But somehow these action potentials must encode information about various properties of the stimulus. Indeed, human sensory powers are such that they can readily feel the difference between silk and velvet or wood and metal.

□ How will the properties of action potentials influence the way in which information about the stimulus is coded?

■ All action potentials are the same, i.e. they are 'all or nothing' events, and so any coding must involve variations in the *timing* of action potentials. That is, by altering the frequency of action potentials.

□ What features of the stimulus might be coded in the signals transmitted to the brain?

- The nerve impulse discharge could code information about:
  1. What the stimulus is (*quality*).
  2. How large the stimulus is (*intensity*).
  3. The timing of the stimulus (*duration*).
  4. Where it is (*location*).

Section 5.2 will consider how information about these features may be coded and communicated along afferent nerve pathways to the brain.

5.2 The neurophysiology of sensory systems

Some of the material in this section revises topics that have been covered in earlier chapters and in Books 1 and 2, but it is included to help you to understand the conceptual basis of communication in sensory pathways.

5.2.1 Sensory coding by mechanoreceptors

The question of stimulus *quality* is determined by the type of sensory receptor that is activated by a stimulus. Receptors respond selectively to specific stimuli, and there are many different kinds of receptor, each tuned to detect their own particular *stimulus modality* (also called the *adequate stimulus*). For example, the only receptors in humans that are normally sensitive to light energy are the rods and cones of the eye; the receptors in the cochlea are sensitive to sound energy. If this were not the case, we might see thunder and hear lightning!

The skin contains a variety of receptors, each of which has its own adequate stimulus. These receptors have characteristic structures, and many are named after the person who first identified them. Some of these are shown in Figure 5.1, and their functions will be described in the following sections.

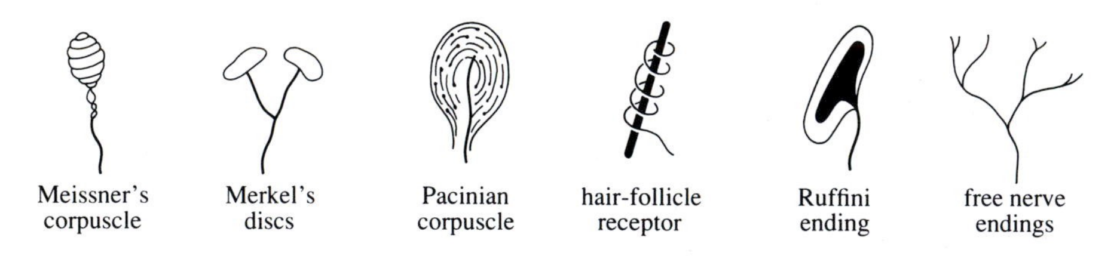

Figure 5.1 Varieties of receptors found in mammalian skin.

Many of the receptors in skin have non-neural connective tissue elements surrounding the actual ending of the afferent nerve fibres. Other receptors have no associated connective tissue elements, and are described as 'free nerve endings'.

The first stage is the transduction of the stimulus into an electrical signal. This was described in Section 1.2.1 and in Book 2, Section 2.4.1. To remind you of the events involved, consider what happens in the Pacinian corpuscle. The Pacinian corpuscle is a relatively large mechanoreceptor found in skin, joints and in some other parts of the body. Because of its size, it has been studied in detail, and its properties are well established (Figure 5.2).

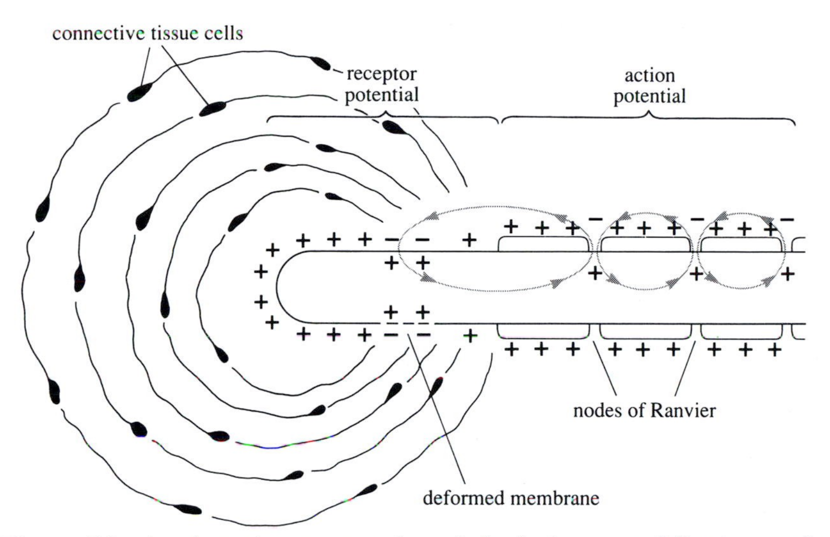

Figure 5.2 A schematic representation of the ionic events following mechanical stimulation of a Pacinian corpuscle. (Note that, for the purposes of illustration, events that in reality are sequential are shown here as occurring simultaneously.)

□ What events are involved in the transduction process?

■ The stimulus acting on the receptor distorts the ending of the sensory neuron and alters the membrane permeability by opening channels in its membrane. The result is a depolarization: the *receptor potential*.

The receptor potential is due to an increase in membrane permeability to Na^+ ions. The receptor potential influences, in turn, the ionic permeability at the first node of Ranvier of the axon. If the receptor potential is sufficiently large, its effect at this first node is to depolarize the membrane sufficiently to produce an all-or-nothing action potential. This is then propagated along the axon towards the cell body in the dorsal root ganglion and from there to the spinal cord.

□ How do receptor potentials differ from action potentials?

■ Receptor potentials are graded responses, that is, they increase in size with the intensity of the stimulus. Only when the receptor potential exceeds a certain value will action potentials be generated. Action potentials are 'all-or-nothing' events; each one is the same, regardless of the size of the stimulus that evokes it. So, action potentials also have a threshold; receptor potentials do not.

□ How might the *intensity* of the stimulus be coded?

■ Principally by varying the frequency of action potential firing.

Weak stimuli produce small receptor potentials, while stronger stimuli evoke larger receptor potentials. But action potentials are all or nothing, and so only the interval between them can be altered. Thus the intensity of the stimulus determines action potential frequency; this is called *frequency coding*.

□ From your recollection of Book 2, what property of the membrane will limit the rate at which all-or-nothing action potentials are produced?

■ You should recall that after the generation of an action potential, the membrane goes through a *refractory period* (Book 2, Sections 3.8 and 7.2.2). Early in the refractory period (absolute refractory period) it is impossible to generate another action potential. In the latter part of the refractory period (relative refractory period), another action potential can be generated, but will require a stronger depolarization. Thus, the duration of the refractory period will set an upper limit to the frequency of action potential firing.

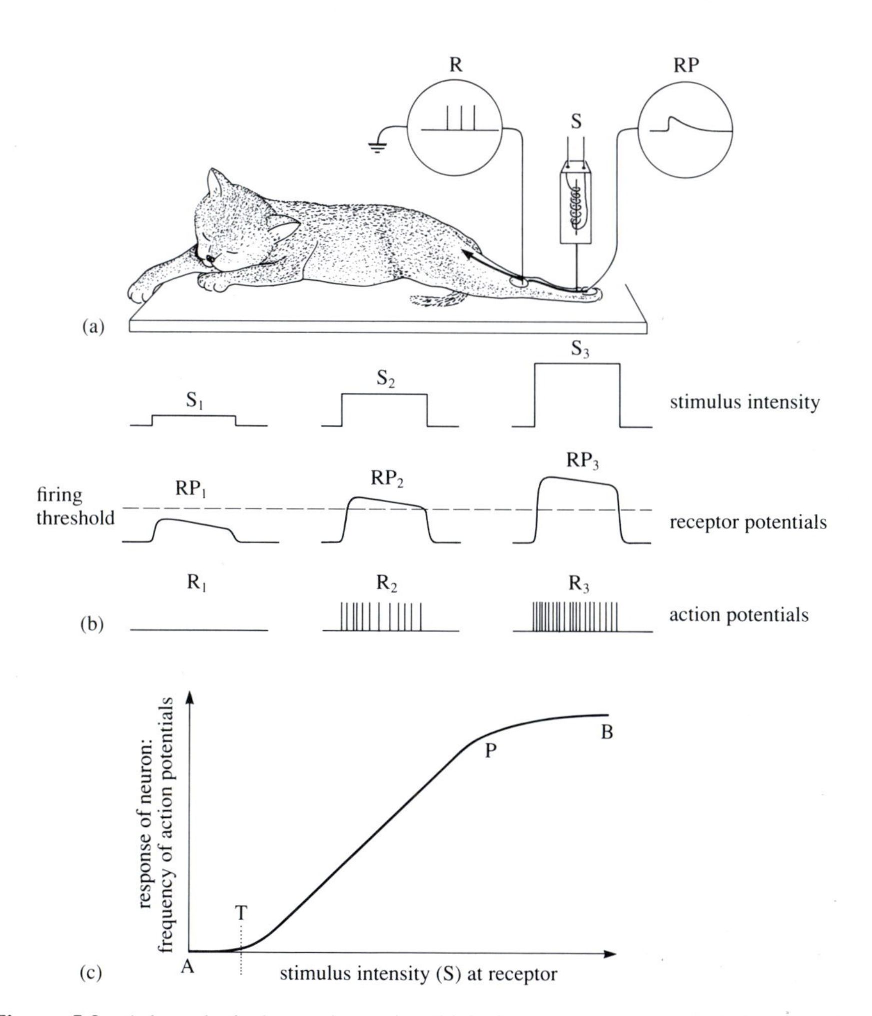

Figure 5.3 A hypothetical experiment in which the receptor potential (RP) and action potential response (R) from an afferent neuron are monitored during application of pressure stimuli (S) to the skin. See text for details.

The relationship between stimulus intensity, amplitude of receptor potential and frequency of action potentials is summarized in Figure 5.3. Figure 5.3a shows the arrangement for a hypothetical experiment upon an anaesthetized cat, in which the receptor potential and action potential responses of a sensory neuron are monitored while pressure is applied to its receptive field on the skin of a hindlimb.

Figure 5.3b shows the relationship between the intensity of stimulation, the receptor potential and the neuron's action potentials. As the intensity of stimulation is increased (S_1–S_3), you can see that the amplitude of the receptor potentials and frequency of action potentials also increase. The 'results' of this hypothetical experiment (i.e. the relationship between stimulus intensity and response of the neuron) are shown in Figure 5.3c. Here a graph has been drawn that represents the firing rate of the afferent neuron to a wide range of different stimulus intensities.

□ Look at the relationship between stimulus intensity and response of the sensory neuron. What happens between points A and T?

■ No action potentials are generated by stimuli of these intensities. This is because the receptor potential is below the threshold for action potential firing, as is the case for stimulus S_1 in Figure 5.3b.

□ What happens between points T and P?

■ With a stimulus value greater than T, action potentials are generated. The point T is called the *threshold* for the neuron. Between points T and P the graph is approximately a straight line. This indicates that the neuron firing rate between these points is roughly proportional to the intensity of the stimulus.

□ What happens between points P and B?

■ Further increases in stimulus intensity produce little increase in response.

Point P marks the start of a plateau on the graph, indicating that the linear relationship between stimulus and response has broken down. The response of the neuron is *saturated*, which means that further increases in stimulus intensity do not result in an increase in response, which remains at approximately the same level.

Individual sensory receptors do not respond over the whole range of stimulus intensities. Some have low thresholds, while others are activated only by relatively intense stimuli.

Another mechanism which contributes to the process of coding stimulus intensity is the number of receptors and afferent neurons activated. Each afferent neuron and its terminal branches is grouped in a well-defined area, which constitutes the receptive field. You should recall from Book 2, Section 9.5.2, that the dimensions of the receptive fields tend to vary depending on location on the body surface. Receptive fields of single sensory neurons are small (1–2 mm in diameter) in the fingertips, but are somewhat larger (several cm in diameter) on the forearms. Receptive fields also tend to overlap, so one stimulus will tend to activate the receptors of several afferent neurons.

□ How might the overlapping of adjacent receptive fields affect the coding of stimulus intensity?

■ A weak stimulus will excite mainly the receptors of one or two afferent neurons. Stronger stimuli will cause a greater amount of skin indentation. This increased stimulation can activate the receptive fields of other afferent neurons, which have higher thresholds for activation.

Thus the coding of stimulus intensity involves variations in (i) the frequency of action potentials in the afferent neurons, and (ii) the number of neurons that are activated.

□ How could the *duration* of a stimulus be coded?

■ The simplest way would be for the action potential discharge to continue as long as the stimulus is applied, for example, as shown in Figure 5.3b.

However, not all receptors respond in this way, as can be seen in Figure 5.4, which illustrates the action potential responses evoked by mechanical stimulation of two different receptors.

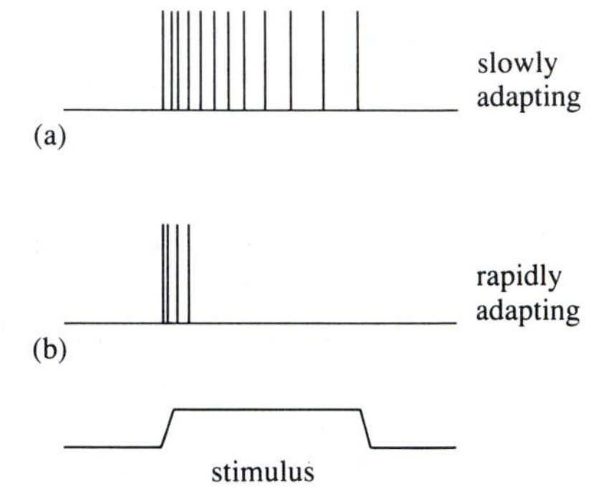

Figure 5.4 Adaptation of receptors to a stimulus that is gradually applied and then held constant. (a) Slowly adapting receptor. (b) Rapidly adapting receptor.

□ What differences are evident between the two responses?

■ The firing from the receptor in Figure 5.4a is maintained throughout the stimulus period, although the frequency declines slowly with time. However, in the receptor shown in Figure 5.4b, the firing rate subsides rapidly, and it only fires during the period when the stimulus is increasing; it does not respond to a steady stimulus.

This decline in the action potential firing rate of a receptor during a steady stimulus is called **sensory adaptation.** Different receptors display different and characteristic rates of adaptation. No receptor displays a perfectly regular discharge. Receptors such as that shown in in Figure 5.4a are *slowly adapting*. Figure 5.4b shows an example of a *rapidly adapting* receptor.

Table 5.1 Classification of mechanoreceptors according to rate of adaptation (headings above the columns) and function (at the bases of the columns).

	Adaptation to constant pressure stimulus		
	slow	rapid	very rapid
Hairless skin:	Merkel's discs	Meissner corpuscle	Pacinian corpuscle
Hairy skin:	Ruffini endings	hair-follicle receptor	Pacinian corpuscle
	intensity detector	velocity detector	vibration detector
	Classification by function		

The rates of adaptation of different mechanoreceptors present in skin are summarized in Table 5.1, but you should not learn the details of the different types of receptor. However, note that there are different kinds of mechanoreceptor present in hairy skin (e.g. the back of the hand) and non-hairy skin (e.g. the palm of the hand).

□ What function might adaptation serve?

■ It will reduce the amount of neural activity reaching the spinal cord, so that the central nervous system (CNS) is not continuously bombarded with inputs about unchanging stimuli. Can you imagine what it would be like if you were aware of your clothes all the time?

It is perhaps less important for an organism to be continually reminded of 'constant' stimuli; but it is necessary to be aware of stimuli that change. Rapidly adapting receptors (e.g. Meissner corpuscles and hair-follicle receptors) fire at the onset of a stimulus, or at its termination. They are suitable for signalling *dynamic* aspects, such as when a stimulus is changing; additionally, they can provide information about the rate of change of the stimulus. Figure 5.5 shows how the firing rate of a rapidly adapting receptor increases as the speed of application of a stimulus increases. Thus rapidly adapting receptors can function as velocity detectors (Table 5.1). Slowly adapting receptors (e.g. Ruffini endings and Merkel discs) can also serve this function, but since they generally maintain their firing during the stimulus, they can also convey information about the *static* aspects of stimulus intensity and function as intensity detectors (Table 5.1).

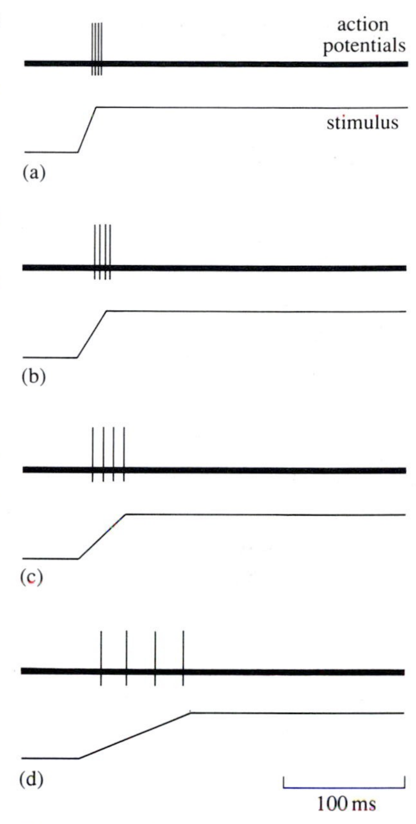

Figure 5.5 Responses from a rapidly adapting receptor to different rates of stimulus application. The rate of application is shown by the slope of the 'ramp', which is fastest in trace (a) and slowest in trace (d). The number of action potentials generated is the same in each case, but the frequency of firing is highest with the most rapidly applied stimulus (a).

The Pacinian corpuscle is essentially a rapidly adapting receptor, but it tends to respond with only one or two action potentials when a stimulus is applied or removed and is not very sensitive to the rate of application or the magnitude of the stimulus. For example, in Figures 5.6a and 5.6c the responses to two different levels of stimulation are the same. However, the Pacinian corpuscle is very sensitive to oscillating stimuli such as those shown in Figures 5.6b and 5.6d. Here, each cycle of the stimulus generates an action potential. These oscillating stimuli are felt as vibration, and the Pacinian corpuscle might be described as a vibration detector (Table 5.1)

To illustrate the rapid adaptation of the hair-follicle receptors, use a pencil tip to move a hair on your arm and hold it in the new position, without touching the skin surface. (Even better, get someone else to do this for you.) Try this several times, with different speeds and directions of movement. You will be aware of the hair's movement, but not when it is held stationary.

The final coding problem is that of stimulus *location*. You will recall from Book 2, Section 8.4.3, that different parts of the body are represented in the sensory cortex in an orderly manner. Sensory pathways from neighbouring parts of the body project to adjacent regions of the sensory cortex; this is called topographical representation. It is rather as if each part of the body surface has its own private 'telephone line' to specific parts of the brain. Not only that, but these separate lines

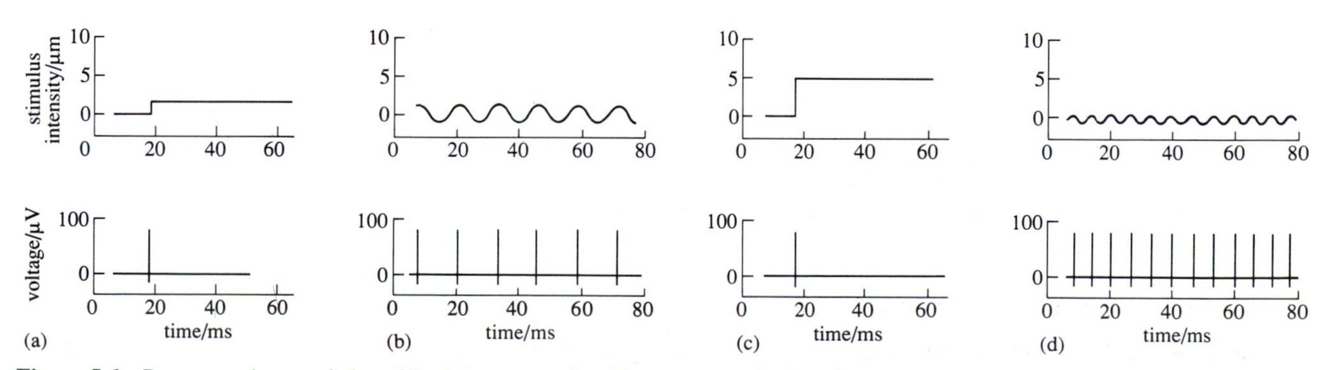

Figure 5.6 Response characteristics of Pacinian corpuscles. The upper panels show the form of the stimulus applied; the lower panels display the action potentials generated by these stimuli. In (a) and (c), single mechanical stimuli of different intensities are applied. In (b) and (d), the mechanical deformation oscillates evenly at two different frequencies.

handle only particular types, or modalities, of information. According to this view, action potentials arising at any point along a pathway will be 'felt' as originating from the peripheral receptive field. Consider what happens when you bump your 'funny bone'. As well as feeling the direct impact at the elbow, you will also feel a curious tingling on the side of the hand and some fingers. This is because the force of the bump is sufficient to generate action potentials in the axons of neurons in the ulnar nerve, which passes close to the skin at the elbow. (Of course, this is not the normal manner of generating action potentials in neurons!) Some of the afferent axons in the ulnar nerve have receptive fields on the side of the hand and on some fingers. Thus, although the action potentials were generated somewhere along the afferent nerve fibres (at the elbow), the brain *interprets* the signals as arising from the peripheral receptive fields.

5.2.2 Thermoreceptors and nociceptors

Thermoreceptors are sensitive to small changes in skin temperature. There are two types, which are distinguished on the basis of their responses to skin temperature changes in the range 30–42 °C (Figure 5.7). 'Cold' receptors increase their rate of firing in response to a fall in skin temperature over this range; below 25 °C, their firing rate decreases. 'Warm' receptors increase their rate of firing when skin temperature increases in the range 30–45 °C, but have a sudden 'cut-off' at temperatures greater than 45 °C.

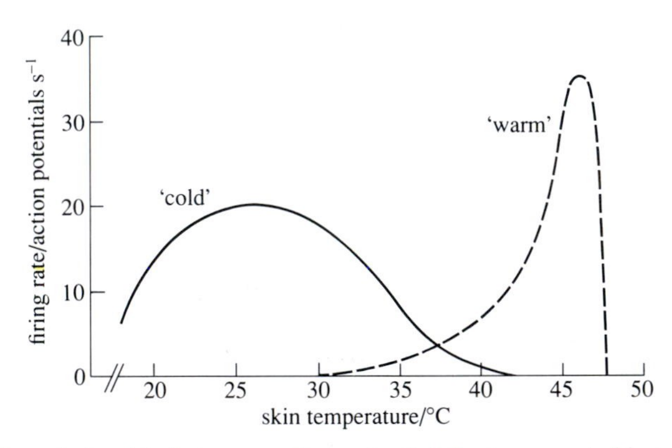

Figure 5.7 The relationship between action potential discharge and skin temperature in 'warm' (dashed line) and 'cold' (solid line) thermoreceptors.

- □ What could you infer about skin temperature if the rate of firing in a 'cold' thermoreceptor were found to be 10 action potentials s^{-1}?
- ■ The skin temperature could be either around 18 °C or 34 °C. But in order to be certain, you would also need to know the corresponding firing rate for 'warm' thermoreceptors.

Nociceptors (nociceptive receptors) respond specifically to noxious or intense stimuli. They are free nerve endings (Figure 5.1, p. 148), with high stimulus thresholds and are classified according to their responsiveness to different forms of noxious (injurious or potentially pain-producing) stimuli. There are two main types of nociceptor: mechano-nociceptors and polymodal nociceptors.

Mechano-nociceptors

In normal, i.e. undamaged, tissues these nociceptors are activated only by intense mechanical stimuli, such as heavy pressure or pinching, and respond with a slowly adapting discharge of action potentials. The receptor illustrated in Figure 5.8 is not activated by innocuous pressures, but responds only when the stimulus is damaging. These receptors do not respond to heating, cooling or the application of algogenic (pain-producing) chemicals. Although these mechano-nociceptors do not generally respond to noxious heat stimuli, there is evidence that some of them are excited by temperatures higher than 48 °C: these are referred to as mechano-thermo-nociceptors.

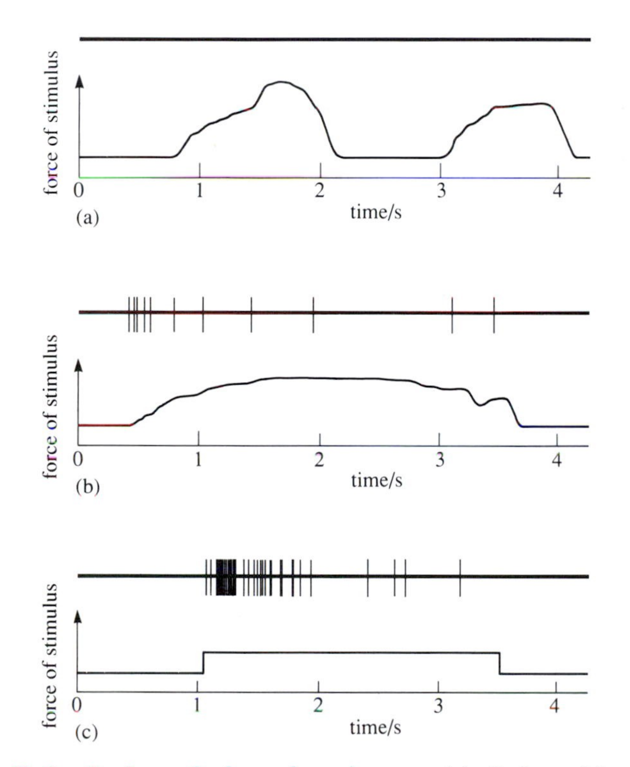

Figure 5.8 Each pair of records shows the action potentials discharged (upper trace) by a mechano-nociceptor with a myelinated axon in response to various forms of mechanical stimuli (lower trace) applied to its receptive field. (a) Pressure from a blunt probe produces no response. (b) A smaller amount of pressure applied with a sharp needle produces some action potentials. (c) A larger response is produced when the skin is crushed with forceps.

Polymodal nociceptors

This second major type of nociceptor is characterized by responses to a wide range of noxious stimuli, including strong mechanical pressure, intense heat (greater than 45 °C) or cold and various pain-producing chemicals. Figure 5.9 (*overleaf*) shows the response of a polymodal nociceptor to heat stimuli.

□ How does the response of the nociceptor shown in Figure 5.9 differ from that of a 'warm' thermoreceptor (Figure 5.7)?

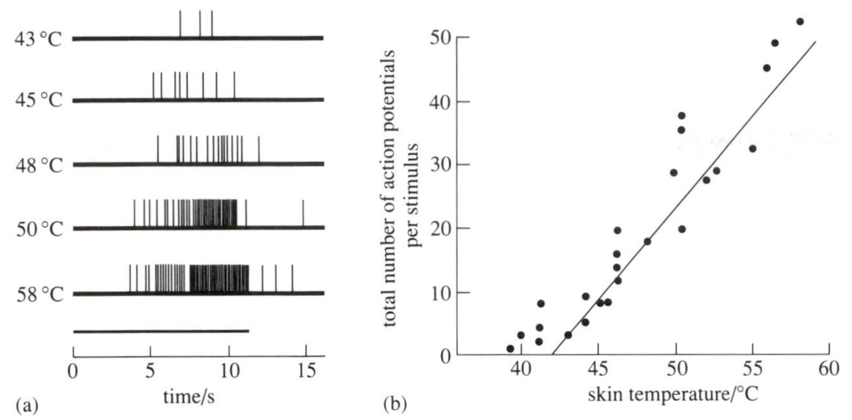

Figure 5.9 Responses of a polymodal nociceptor to heat stimuli. (a) The action potentials evoked by stimuli of different temperatures; the stimuli are applied for the period indicated by the solid bar. (b) A graph of temperature plotted against the number of action potentials evoked by various stimulus temperatures.

■ A 'warm' thermoreceptor stops firing at temperatures greater than about 45 °C. The nociceptor has a heat threshold of around 43 °C and continues firing at temperatures up to 58 °C.

The above descriptions refer to nociceptors in normal skin. However, when skin is damaged or inflamed, its sensitivity increases, and previously innocuous stimuli may cause pain. This is called **hyperalgesia**. When noxious stimuli are repeatedly applied or if tissues are damaged or inflamed, nociceptors display an increased responsiveness, known as **sensitization**. In the sensitized state, nociceptor thresholds are lowered so that the receptor responds to stimuli that formerly were ineffective and there is an increased action potential discharge to stimuli above the threshold (Figure 5.10). Sensitization of nociceptors seems to be one likely cause of hyperalgesia.

Sensitization occurs in all classes of nociceptor, and appears to be mediated by local release of chemicals in the tissues around the nociceptor endings. In damaged tissues, or where there is inflammation (e.g. due to bacterial invasion) there is local release of chemicals (*inflammatory agents*) such as histamine, serotonin (5-hydroxytryptamine), bradykinin and prostaglandins. It appears most likely that the sensitization of nociceptors is due to the release of prostaglandins and bradykinin (which itself can cause release of prostaglandins). These agents are known to increase nociceptor sensitivity. Indeed, there is now evidence that some nociceptors only respond to noxious stimuli in tissues where these chemicals have been released due to injury or inflammation.

Two of the most commonly used analgesic (pain-relieving) drugs, aspirin and paracetamol (Panadol), are known to block synthesis and release of prostaglandins.

□ Can you suggest a possible mechanism for the analgesic action of aspirin?

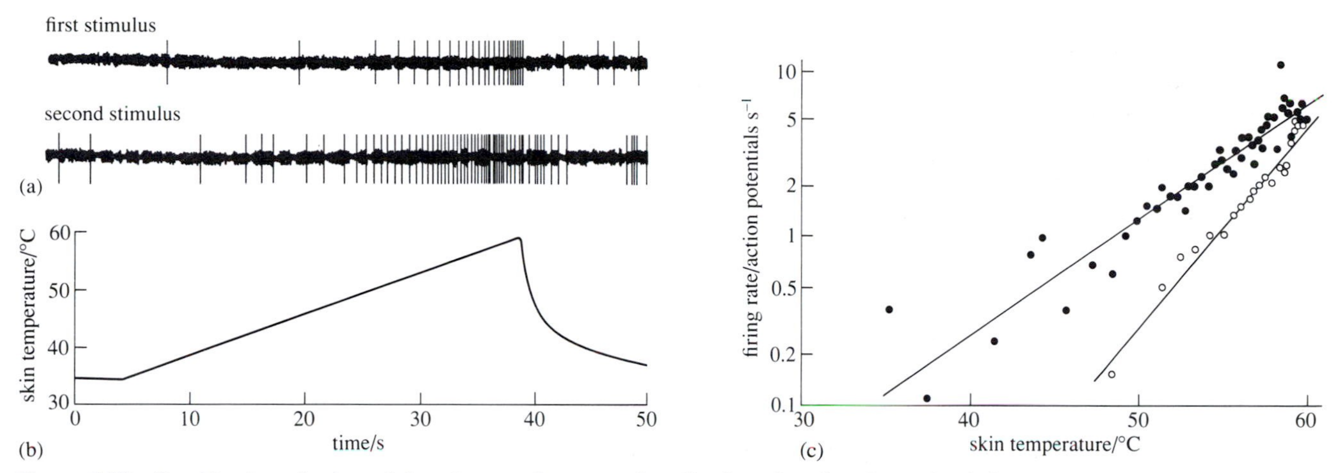

Figure 5.10 Sensitization of polymodal nociceptors by repeated application of noxious heat stimulation. (a) Nociceptor responses to the heat stimulus. The upper trace shows the action potentials generated by the first application of the stimulus. The lower trace shows the response to a second application made 4 minutes after the original stimulus. (b) The change in skin temperature during the application of the radiant heat stimulus. The skin temperature is increased progressively from an initial level of 34 °C to just over 60 °C in 35 seconds. (c) Graph of the response (the action potential frequency) plotted against skin temperature. At any given temperature, the magnitude of the response to the second stimulus (●) is greater than the response to the first stimulus (○).

- By blocking prostaglandin release, aspirin will prevent sensitization of nociceptors. Aspirin will be effective in reducing pain associated with inflammation, but it will have little effect on the sensitivity of uninflamed tissues.

5.2.3 Specificity of afferent nerves

You should recall that spinal nerves contain both afferent and efferent neurons (Book 2, Section 6.3.1). Efferent neurons convey action potentials away from the CNS, through the ventral nerve roots to effector organs such as muscle. Afferent neurons are the neurons with their cell bodies in the dorsal root ganglia. Nerve fibres have been classified according to their axon diameter and the related property of action potential *conduction velocity* (Book 2, Section 3.8). Interestingly, it turns out that this classification by size also corresponds quite well to the function of the fibres (Table 5.2). The information in the table is provided for reference only; you should not attempt to memorize the values.

You should note that the summary shown in Table 5.2 is greatly simplified, but the classification provides a convenient shorthand for naming nerve fibres. You may notice some omissions. For example, the large sensory fibres from muscle spindles (called Ia fibres) are absent. This is because they are named using a different system of classification. For the record, the Ia fibres are the largest axons present in mammals, with diameters around 15–20 μm, and conduction velocities of 90–120 $m\,s^{-1}$.

Table 5.2 Classification of mammalian nerve fibres. Groups A and B are myelinated axons, those in group C are unmyelinated.

Fibre types	Axon diameter /μm	Conduction velocity/m s^{-1}	Function
Aα	10–18	50–100	motor: skeletal muscle fibres
Aβ	6–14	36–90	mechanoreceptors (low threshold)
			proprioceptors (joints and muscles)
Aγ	2–9	10–50	motor: muscle spindle fibres
Aδ	1–6	5–36	mechano-nociceptors
			thermoreceptors (cold)
			mechanoreceptors
B	1–3	3–14	motor: pre-ganglionic autonomic
C	0.2–1	0.2–1	polymodal nociceptors
			thermoreceptors (warm and cold)
			mechanoreceptors
			motor: post-ganglionic autonomic

- □ The conduction velocity of an axon was measured as 30 m s^{-1}. What function might this neuron serve?

- ■ The answer will depend on whether the neuron is afferent or efferent. If it is an afferent fibre, it will belong to the Aδ group, and could be a mechano-nociceptor, a cold thermoreceptor or a mechanoreceptor. But if it is an efferent fibre, it will belong to the Aγ group, which is a motor neuron synapsing onto the specialized muscle fibres of a muscle spindle.

The sensory function of different types of neurons can be investigated by applying electrical stimuli to a peripheral sensory nerve. If this is done using a purely sensory nerve in human subjects, then they can report on the sensations they experience.

On electrical stimulation of a human peripheral sensory nerve, which contains only afferent neurons, low intensities of stimulation that excite only Aβ fibres (since these large fibres have the lowest electrical threshold) produce sensations that are reported as having a mechanical quality, such as 'tapping or buzzing'. Higher intensities of stimulation that, in addition, excite the Aδ fibres cause 'sharp, stinging' sensations, whereas the very strong stimulation needed to activate the C fibres (having the highest electrical threshold) results in very intense pain. Furthermore, inflating a pressure cuff around the arm results, after about 20 minutes, in the loss of touch and pressure sensations below the cuff, whereas pain sensation in the arm remains. (Pressure cuts off the blood supply and hence impairs neuronal function.) The large myelinated fibres are much more sensitive to pressure block of this kind than are the small Aδ and C fibres, and hence will cease to conduct action potentials before the smaller ones also cease to do so. (Sensations return a short while after removing the cuff.)

In recent years, techniques have been refined so that action potentials can be recorded from single afferent neurons using fine tungsten microelectrodes (Book 2, Section 3.5, Box 3.1) inserted through the skin in conscious human volunteers. Such experiments have tended to confirm the relationship between neuronal conduction velocity, and hence size, and function, and can allow comparison between a recorded action potential discharge and any accompanying sensations. This will be discussed further in Section 5.5.3.

A brief noxious stimulus, such as a pin-prick applied to the skin of the arm or leg, may elicit two sensations. First there is an immediate, sharp pain, which is brief and well localized to the site of stimulation (*first pain* or fast pain). After a short delay, a second sensation is felt. This is a more diffuse, duller feeling, which spreads beyond the stimulus site (*second pain* or slow pain). By measuring the latencies of these sensations, it can be shown that first pain involves nociceptors with myelinated, Aδ fibres, while nociceptors with slowly conducting, unmyelinated C fibres are responsible for second pain. In the 1930s, the eminent British physiologist Sir Thomas Lewis carried out a simple experiment to demonstrate that first and second pains were distinct and that the latter are mediated by unmyelinated C fibres. Lewis applied a noxious stimulus to a subject's foot. The subject was instructed to press a switch when he felt the onset of second pain. The latency between the application of the stimulus and reaction to the onset of second pain was measured. This reaction time on its own does not tell us much, as the neural pathway involved is long and complex. However, Lewis used a simple, but effective way of eliminating the unknown variables by means of the 'method of differences'. He repeated the experiment, this time applying the stimulus to the subject's thigh. This produced a faster reaction time, because the total conduction pathway is shorter. The difference between the two reaction times is due to the time necessary for impulse transmission in the afferent pathway between the foot and thigh: all other components were unaltered. Typical results are shown in Table 5.3.

Table 5.3 The mean reaction times for second pain produced by a noxious stimulus applied to two different sites on a subject's leg.

Stimulus site	Reaction time (mean)/s
Toe	1.82
Thigh	0.75

The difference between reaction times, 1.07 s, is the time required for afferent signals to travel the distance between the two stimulation sites; this distance was 75 cm (0.75 m).

□ Using this information, calculate the conduction velocity of action potentials along the nerve between the two stimulation sites.

■ The conduction velocity is 0.75 m in 1.07 s, or 0.7 $m s^{-1}$, which is well within the range for C fibres, and too slow for action potential conduction in myelinated nerve fibres (Table 5.2).

□ Why should there be two sets of nerves responding to noxious stimuli?

■ The more rapidly transmitted signals in Aδ fibres warn of the onset of a noxious stimulus; the more slowly conducted signals in C fibres signify that tissue damage has occurred. The Aδ input triggers reflex withdrawal reactions, while the C fibre discharge prompts 'guarding' and immobility of the injured part.

5.2.4 Pathways in the CNS

The central axons of the sensory neurons, which have their cell bodies in the dorsal root ganglia, enter the spinal cord via the dorsal roots. Once in the spinal cord, they synapse with other neurons, which relay the signals to the brain via several ascending pathways. There are many such ascending systems, and the following account will deal only with their general features.

Some pathways are *specific*; in these, the information coded by the receptors is preserved and accurately transmitted at successive synapses. Inputs in these pathways are likely to form the basis for sensory perception. However, there are also *non-specific* pathways, where there is considerable convergence of inputs from different modalities, so that much of the detailed information coded at the receptors is lost. The non-specific pathways probably serve to alert the animal to events, rather than providing detailed sensory information.

The basic structure of the spinal cord was described in Book 2, Section 8.2. You should also recall from Book 2, Section 8.8.1, that differences in the microscopic structure of cells in the brain have been used to classify the cortex into many individual areas (Brodmann's areas). On a similar basis of neuronal structure, the Swedish neuroanatomist Bror Rexed has classified the spinal cord grey matter into ten zones, known as Rexed's laminae (Figure 5.11a). As with Brodmann's areas in the brain, Rexed's laminae can be used (like postal codes) to identify different 'addresses' in the spinal cord grey matter. The dorsal horn comprises laminae I–VI; lamina I is nearest to the surface and lamina VI is the deepest.

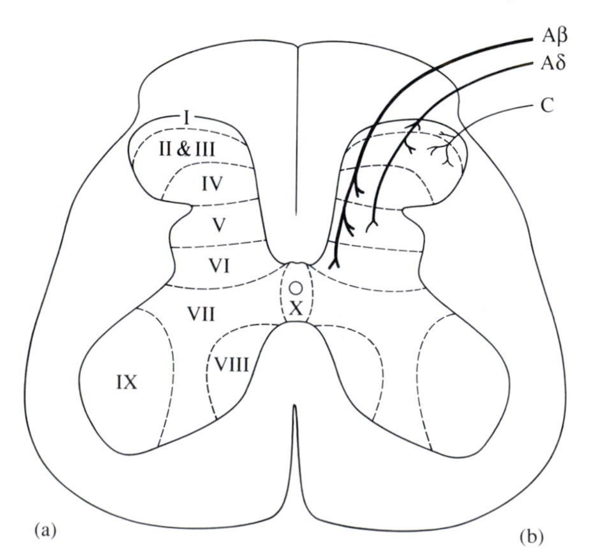

Figure 5.11 (a) Diagram of the spinal cord showing the layout of Rexed's laminae in the grey matter. The dorsal horn comprises laminae I–VI. (b) The projections of the different types of primary afferent nerve fibres in the dorsal horn. Small diameter fibres project mainly to the superficial laminae; large fibres make connections with cells in deeper laminae.

Much has been learnt about the anatomy and physiology of the dorsal horn by using techniques such as the HRP method of labelling neurons (Book 2, Section 8.7.1, Box 8.1) to trace the spread of afferent neuronal terminals and microelectrode recording to investigate the properties of the dorsal horn cells. The picture that emerges is a complicated one, and the following account is simplified to summarize the key features.

It is a fairly good rule of thumb that the larger the diameter of an afferent fibre, the deeper it penetrates into the dorsal horn. C-fibre afferents terminate in the superficial laminae I and II. Aδ fibres have a similar distribution, but a few penetrate to lamina V. Large mechanosensitive (Aβ) fibres project mainly to laminae IV–VI, but some terminate in the superficial laminae as well (Figure 5.11b).

Most peripheral sensory neurons display a high degree of specificity to particular stimuli, but only about 30% of the neurons in the dorsal horn are similarly modality specific. The remaining 70% of dorsal horn cells respond to a variety of stimuli (i.e. they are *multimodal* neurons)

□ Can you suggest an explanation for this?

■ Since dorsal horn neurons receive their inputs from the modality specific afferent neurons, it follows that the multimodal dorsal horn neurons receive inputs from several different types of afferent neurons.

Multimodal neurons are found in all laminae of the dorsal horn and, as already mentioned, these are the most abundant type of neurons in the dorsal horn. However, there is some localization of modality-specific neurons in certain laminae.

Lamina I contains neurons that respond only to noxious stimuli (nociceptive-specific neurons).

Laminae II and III form the *substantia gelatinosa*, which contains small interneurons that can influence the activity of other neurons in the dorsal horn. Their function in the modulation of nociceptive inputs will be discussed further in Sections 5.7 and 5.8.

Laminae IV and VI contain neurons that respond to light mechanical stimulation of skin and inputs from proprioceptors in joints and muscle.

Lamina V contains multimodal neurons, which receive inputs from mechanoreceptors, thermoreceptors and nociceptors. In addition, some of these cells also receive inputs from receptors in internal organs (*enteroceptors*).

The location of some of the main ascending projection pathways within the white matter of the spinal cord is shown in Figure 5.12. The projection pathways for the various sensory neurons entering the spinal cord and from cells in the dorsal horn are shown in Figure 5.13 (*overleaf*).

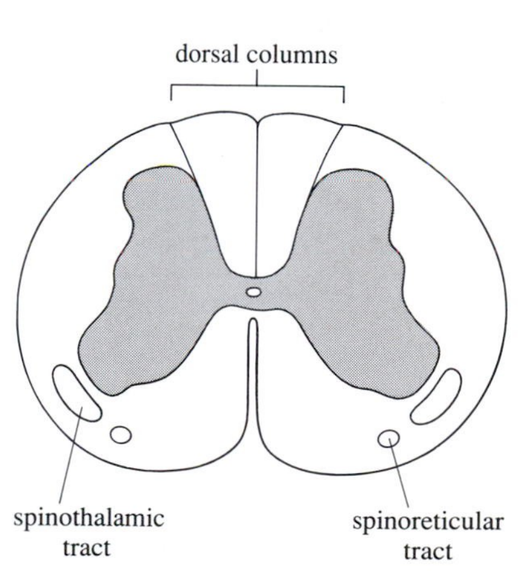

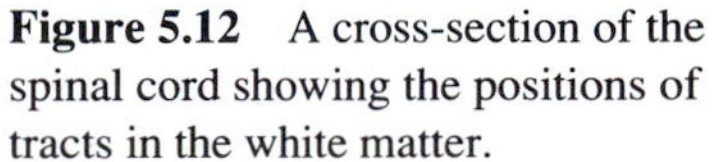
Figure 5.12 A cross-section of the spinal cord showing the positions of tracts in the white matter.

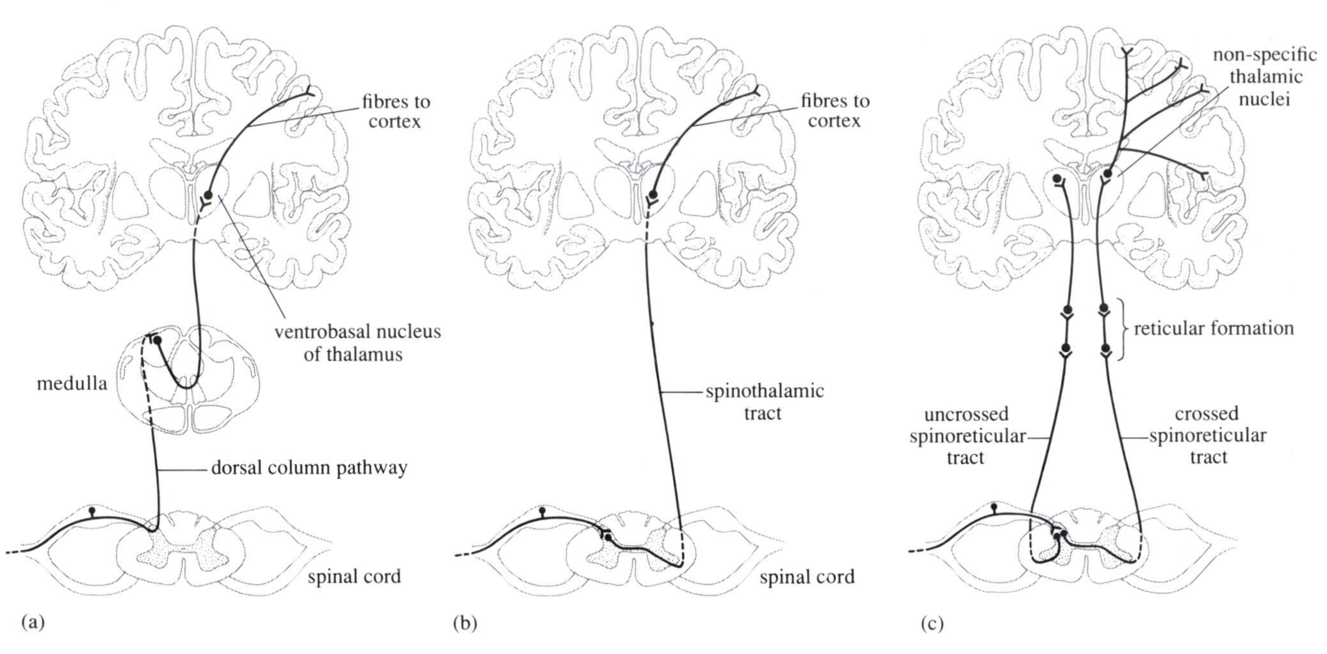

Figure 5.13 Ascending pathways in the spinal cord. (a) The dorsal column (DC). (b) The spinothalamic tract (STT). (c) The spinoreticular tract (SRT).

Two specific pathways are shown, the dorsal column (DC) pathway (Figure 5.13a) and the spinothalamic tracts (STT, Figure 5.13b), and one non-specific pathway, the spinoreticular tract (SRT, Figure 5.13c). Neurons in each of these pathways synapse in the thalamus, an important 'relay-station' in most sensory pathways. From here, neurons project to parts of the forebrain.

The DC and STT pathways have several features in common.

1 The synapse between the sensory neuron and the next neuron in the pathway occurs in grey matter on the ipsilateral side of the CNS.

2 The axon of the second neuron in the pathway *crosses the midline* and synapses with a neuron in the contralateral thalamus.

3 The axons of the neurons in the thalamus project to the sensory cortex.

Look at Figures 5.13a and b, and confirm these points for the DC and spinothalamic tract systems.

□ What *differences* do you see between the DC and STT pathways?

■ The main difference between these pathways is the site of the first synapse. In the spinothalamic tract, the first synapse is in the dorsal horn of the spinal cord near the level of entry of the spinal nerve. In the DC system, the axon of the first neuron in the pathway extends to the top of the spinal cord and terminates in the dorsal column nucleus, which is located at the lower (caudal) end of the medulla of the brain stem.

What are the functions of these pathways? Electrical recordings from cells in the pathways can reveal how they respond to various stimuli, but other methods, such as lesioning, have to be used to investigate the roles of the pathways.

The inputs to the DC system are from low-threshold mechanoreceptors and proprioceptors. Neurons in the dorsal column nuclei project to the ventrobasal nucleus of the contralateral thalamus; the thalamic neurons project in an orderly manner to the sensory cortex. The neurons at each stage in this pathway have small receptive fields and are highly modality-specific: there is thus little convergence between adjacent channels. Lesions to the DC system result in a reduction of sense of vibration and proprioception; there is also impairment of touch localization and spatial discrimination of stimuli.

The spinothalamic tracts contain the axons of modality-specific dorsal horn neurons, activated by inputs from either mechanosensitive, nociceptive or thermosensitive afferent axons. These modality-specific neurons also project to the specific neurons in the ventrobasal nucleus of the contralateral thalamus, which in turn project to the sensory cortex. However, some STT neurons are multimodal and can be excited by a range of stimuli, including inputs from internal organs (discussed in Section 5.2.5). Some multimodal neurons project to non-specific thalamic nuclei, and others make connections with cells of the brain stem reticular formation. Lesions to the spinothalamic tracts produce a variety of sensory deficits, including reduction in pain and temperature sense. Touch sense is affected to a very limited degree.

The third pathway, the spinoreticular tract (SRT), is shown in Figure 5.13c. This is even less specific. Notice that projections from the dorsal horn are bilateral (they project on both sides of the spinal cord), and the SRT contains more neurons and synapses than either the DC or the STT pathways. There is a lot more convergence, so that the neurons are multimodal and can have very large receptive fields. The SRT projects to cells in the reticular formation and from there to the non-specific thalamic nuclei. The forebrain projections of the SRT are very diffuse, and discrete lesioning of this pathway is virtually impossible. It is likely that the SRT pathway provides non-specific information that is used primarily for alerting the organism to the presence of stimuli.

5.2.5 Referred pain

One peculiar aspect of pain is that irritation or disease of an internal organ or deep structure can cause pain to be felt as if originating from some other, usually superficial, site. This is known as **referred pain**, and there are many examples.

□ Can you think of any examples of referred pain?

■ Pain associated with the heart (e.g. as in angina) is often referred to the left shoulder and arm; pain arising from a kidney stone may be referred to the groin; pain due to appendicitis is classically referred to the midline of the abdomen near the umbilicus (belly button).

These patterns of referral may appear haphazard, but they reflect the embryonic origins of the structures. For example, the heart develops in the neck of the embryo and receives its innervation (nerve supply) from the cervical segments of the spinal cord. During development, the heart migrates to its mature position in the chest, taking its nerve supply with it. But these nerves retain their embryonic links with the cervical segments of the spinal cord, which also supply the neck, shoulder and part of the arm. *Visceral* (internal) structures are innervated by afferent neurons, which terminate in laminae I or V in the dorsal horn. But there are no 'viscero-specific' cells; thus cells that receive inputs from visceral structures also receive inputs from sites on the skin (Figure 5.14).

When the dorsal horn cell illustrated in Figure 5.14 is activated, the brain cannot distinguish between inputs from the skin and inputs from the internal organ; any sensations are interpreted as originating from the more frequently stimulated site on the skin, which is where the pain is felt.

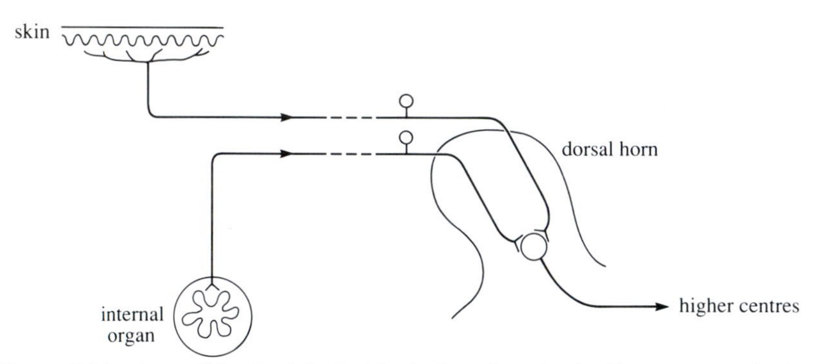

Figure 5.14 A possible physiological basis for referred pain. Some neurons in the dorsal horn receive inputs from both skin and internal structures. Although each afferent pathway can be activated independently, they converge on cells in the spinal cord and share a common projection to the brain. Neurons in the brain are, however, unable to distinguish between activity arising in one or other of the peripheral pathways, and any activity is assumed to originate from the more commonly stimulated site on the skin.

Summary of Section 5.2

Knowledge of the properties of receptors, afferent nerve fibres and central pathways gives a picture of how sensory information can be communicated within the nervous system. Thus, information about an object touching the skin will be coded by different receptors: stimulus quality is coded by the specific type of receptors activated; stimulus intensity is coded by the frequency of action potentials and the number of active neurons; the duration of the stimulus is coded by the duration of the action potential discharge in slowly adapting receptors and information about the rate of change of the stimulus can be signalled by rapidly adapting receptors. The location of a stimulus is signified by the region of sensory cortex activated by the afferent input.

Different types of mechanoreceptor perform separate functions as intensity detectors, velocity detectors or vibration detectors. The two classes of thermoreceptors provide precise information about skin temperature. Nociceptors are activated only by intense stimulation, and can be sensitized by the presence of inflammation, which causes a reduction in threshold and increased firing.

The action potentials which code different aspects of the stimulus are transmitted via separate ascending pathways and are also separately represented in the cortex.

Three ascending pathways that are concerned with this are the dorsal column (DC) pathway, the spinothalamic tract (STT) and the spinoreticular tract (SRT). The last pathway seems to give non-specific information about the occurrence of stimuli, in contrast to the very specific information about their qualities and location provided by the STT and DC pathways.

5.3 The value of touch

Touch can provide animals with information about objects in their surroundings. As mentioned in Chapter 8 of Book 2, humans can use their hands to feel the area ahead when moving across strange territory in the dark. The information gained in this way is poor compared with that provided by sight. But this is not so for all animals. Rats, mice, raccoons and sheep have all been classified as 'feelers' on the basis that touch is their primary sense, in contrast to humans who use vision as their primary sense and are therefore 'beholders'. In 'feelers' the somatosensory pathways are emphasized. The ventrobasal nucleus of the thalamus is relatively larger than the other thalamic sensory nuclei and likewise the somatosensory cortex is the largest sensory area of the cortex. In contrast, the lateral geniculate nucleus of the thalamus is prominent in 'beholders', while the medial geniculate nucleus is largest in 'listeners' (Figure 5.15a). Not all animals appear to use touch in the same way. This can be inferred not only from observing their behaviour, but also from the organization of their brains. For example, mapping the somatosensory cortex to produce an 'animalculus' (cf. the homunculus described in Book 2) shows the raccoon to be a forepaw feeler, the rat a whisker feeler and the sheep a tongue and lips feeler (Figure 5.15b).

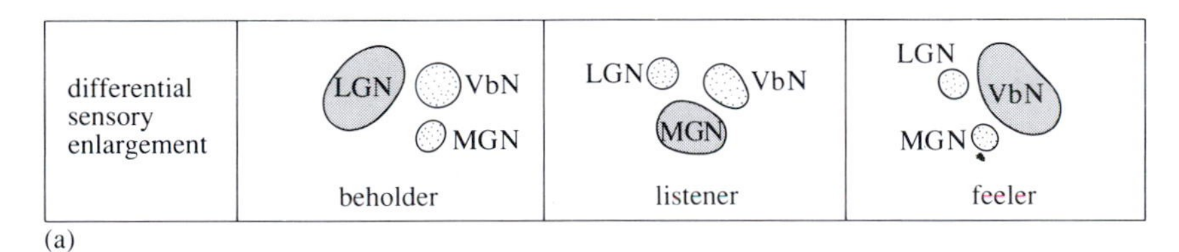

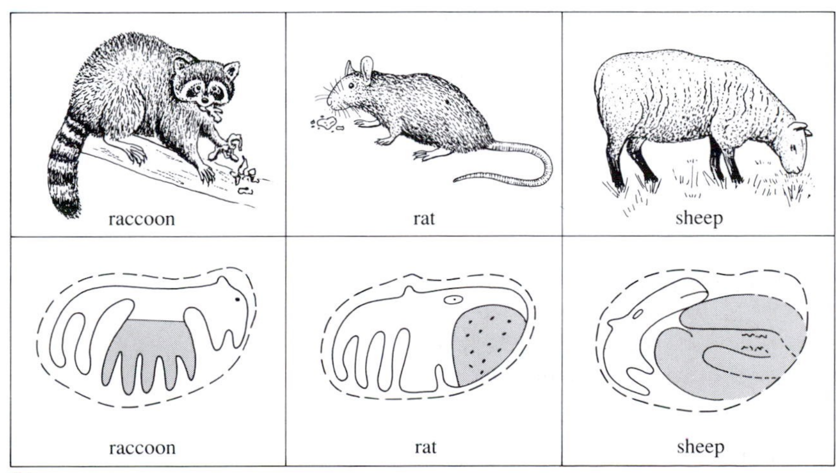

Figure 5.15 (a) The relative sizes of thalamic nuclei in a beholder, a listener and a feeler. (MGN, medial geniculate nucleus; LGN, lateral geniculate nucleus; VbN, ventrobasal nucleus.) (b) The sensory cortices in a raccoon, rat and sheep, showing the 'animalculus' for each of them.

The brain map, or homunculus, is not a real picture of the body that is 'painted' on the surface of the brain; it is a neurological image, arising from the pattern of projections of the sensory nerve pathways. Nor are the brain maps for various animals faithful representations of the relative sizes of different parts of their bodies. Some areas of the body have a much larger representation in the sensory cortex than they ought to have on the basis of their anatomical size alone.

These disparities are analogous to the differences between geographical maps and electoral maps. Figure 5.16a is a geographical map of Great Britain, showing the relative proportions of the counties. Compare this with Figure 5.16b, which has been redrawn to represent the parliamentary proportions, based on population density. Here, the more densely populated areas, such as the cities, receive a larger representation in Parliament than might be justified by actual geographical size. Although there is considerable variation in the geographical sizes of different

Figure 5.16 Maps of Great Britain. (a) The actual geographical size of counties (pre-1974). (b) Electoral map depicting population density of the counties.

constituencies, each should contain roughly similar numbers of electors. Thus the parliamentary representation of the country is based on the number of people in an area, rather than on its geographical size.

- □ Returning to the brain, how might the analogy of electoral maps help to explain why certain parts of the body have a disproportionately large representation in the brain map: the sensory homunculus? (You may wish to refer to Figure 8.10b of Book 2, Section 8.4.3.)

- ■ Electoral maps are based on population density, that is, the number of people per square kilometre of land. Brain maps reflect the innervation density, that is, the numbers of sensory neurons per square centimetre of body surface.

Those parts of the body, such as the fingers and lips, that in humans have the largest representation relative to size in the sensory cortex, have more sensory neurons per unit area of surface, than say, the legs or trunk. Furthermore, in those regions where there are more sensory neurons per unit area, the receptive fields also tend to be smaller (see Section 5.2.1)

These differences in innervation have functional implications. The regions of the body that have the greatest density of neurons and the greatest cortical representation are those parts of the body that have the best powers of discrimination. Although humans are primarily 'beholders', their tactile senses are not insignificant. For example, most people can put their hand into a pocket or purse and bring out a particular coin without having to look at it. This ability to recognize the shape of an object by touch alone is called *stereognosis*. The use of Braille 'letters' by the unsighted is also made possible by the dense innervation of the fingers and the resultant high degree of spatial discrimination. To convey similar meaning, Braille characters would have to be many times larger, if people had to use the skin on their forearms instead of fingertips to read them.

5.3.1 Social aspects of touch

Touching a conspecific can also provide the toucher with information about the individual touched but, for many animals, touching conspecifics has other important roles. Animals that live in social groups can be observed to touch in particular ways. When rejoining a group, some primates have greeting ceremonies that involve touching (see Figure 5.17).

Figure 5.17 (a) An adult female chimpanzee approaches and greets a male. She holds out her hand and he responds by holding out his. (b) A juvenile female chimpanzee greets a juvenile male with a kiss.

Allogrooming (where one individual grooms another) probably has a hygienic function (the removal of dirt, dead skin and parasites) and inaccessible parts of the body are groomed preferentially. However, more allogrooming is performed than is necessary for body surface maintenance and different individuals groom and are groomed by different amounts. Typically, adult males are groomees more often than groomers and the reverse is true for adult females. Allogrooming is thought to be important in the maintenance of social relationships and it has also been cited as an example of altruistic behaviour (Book 1, Section 10.4). Unrelated vervet monkeys (*Cercopithecus aethiops*) that have allogroomed recently are more likely to respond to each other's calls for assistance.

- □ What kind of altruism is this?

■ This is an example of reciprocal altruism (Book 1, Section 10.4.2).

Other studies have shown a rise in the level of a particular class of endorphins (β-endorphin) in grooming talapoin monkeys (*Miopithecus talapoin*).

□ Recall from Book 2, Section 4.5.6, the function of β-endorphin.

■ It is an endogenous opioid. In rats it produces analgesia and paralysis. It is also known to be produced in response to stress. (This is discussed further in Section 5.8.4.)

□ What effect would you expect it to have on the monkeys?

■ You would not expect them to be reduced to a state of paralysis but it might have a relaxing effect.

People indulge in various kinds of body contact such as stroking, caressing, massaging, hugging and cuddling. Such behaviour patterns have a role in maintaining social cohesion. They are also performed selectively; they are more likely to occur between some individuals than others. There are unwritten rules about who may touch whom and where. These rules vary slightly between people of different cultures, suggesting that they have to be learned. Figure 5.18 gives one example.

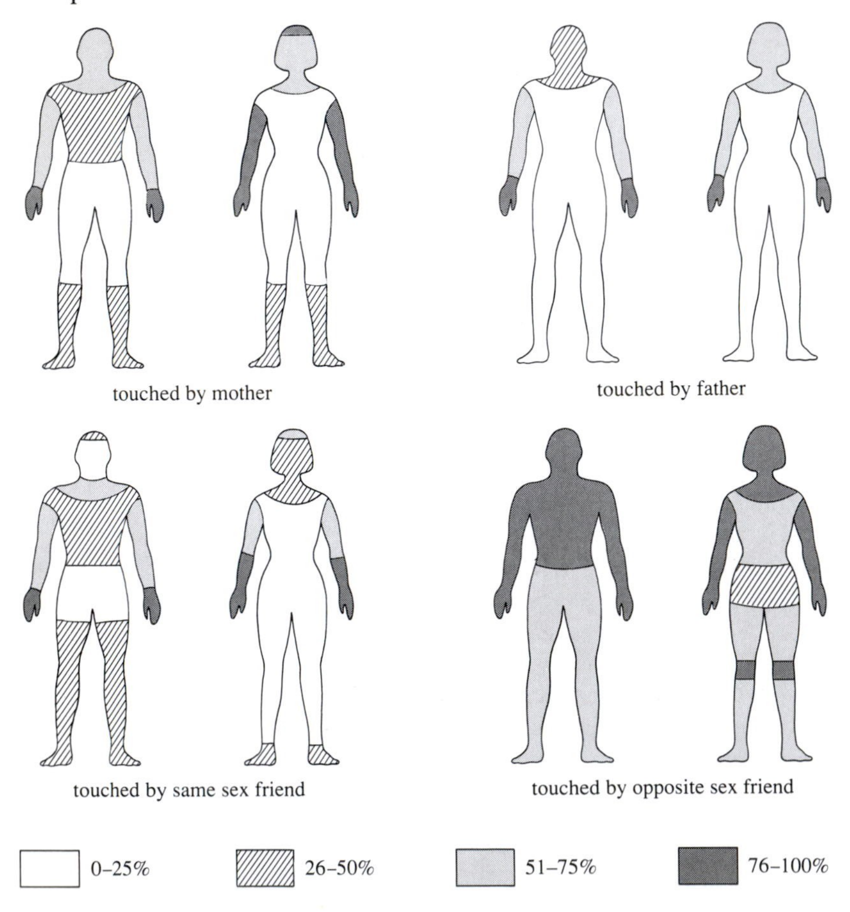

Figure 5.18 The 'taboo' zones of American college graduates. These body zones differ in touchability both by sex and by relationship to the toucher, the percentage indicating how many graduates reported being touched in a particular body zone by different kinds of 'touchers'.

It is not known whether touching raises levels of β-endorphin in humans but the health of elderly people living alone is improved if they have a pet to stroke, and recovery after heart surgery is better in pet-owners than non-pet-owners. There is more about stress and about the body's ability to fight infection (the immune system) later, in Books 5 and 6 respectively, and in Chapter 4 of Book 4 the role of touch in normal development will be considered.

Finally, hugging can be used as a positive reinforcer but there is evidence that even minimal touching can be reinforcing. A survey of people's perception of a library's facilities was carried out by interview as they left the library. The librarians had been instructed not to smile at anyone but half of the visitors had their hand contacted lightly by the librarian's hand as the book was returned to them after being checked out. The group that had had their hands touched were more likely to rate highly the library's facilities and the quality of service and were also more likely to think that the librarian had smiled at them as they left.

In conclusion, the behavioural evidence as well as knowledge of receptors and pathways show that the processing of a proximal stimulus such as touch is no more straightforward than the processing of distal information such as that provided by light (Chapter 4).

5.4 Pain

The remaining sections of this chapter will be devoted to a more detailed consideration of the biology of pain.

5.4.1 What is pain?

All animals (and some plants) react to noxious or potentially damaging stimuli. Humans certainly display this ability, except for a few rare individuals, and the reaction is usually associated with an experience they call 'pain'. This is known, because they are able to say so. It is reasonable to suppose that many animals experience similar feelings, although they are unable to communicate this in quite the same way, and what they feel can only be inferred from observations of behaviour. Consequently, throughout the remaining sections, emphasis will be placed on pain in humans.

Most people have some personal experience of pain and while the actual individual experiences may differ widely, most would agree that pain is unpleasant and would associate it with some form of injury. The word 'pain' derives from the Latin word *poena* (meaning punishment or penalty), which may reflect an early belief that pain was some sort of divine retribution as a consequence of a misdeed. Pain ranges from the minor irritation of a small cut or scratch, to the torment associated with severe injuries or disease; on a different plane, the effect of a screeching discord on the hearing of a sensitive musician might well be to cause pain.

☐ Pause for a moment, and try to form a definition for pain, based on your own experiences. Write this down in some convenient place—you will be asked to return to this at the end of this section.

■ Individual definitions are likely to differ considerably, but your definition will probably include some, or all of the following components of pain:

(a) *Sensory component* Pain can be described in terms of how it feels, e.g. aching, sharp, drawing, etc. or perhaps in terms of the cause, e.g. burning, stabbing. In this aspect, pain is much like any other sensation, with the intensity of the feeling tending to increase with the magnitude of the stimulus.

(b) *Affective or emotional component* Some sensations, like seeing the colour of a carpet, normally arouse little or no emotional reaction, but pain is rarely perceived with indifference. There is always some emotional element, which, for most people, is unpleasant or disturbing. In other words, pain hurts.

(c) *Autonomic component* Pain is often accompanied by changes in certain internal, involuntary body processes, e.g. heart rate, blood pressure, sweating, etc. The breathing pattern may also change.

(d) *Motor component* When an area of the body is exposed to a noxious, or damaging stimulus, there is usually an immediate withdrawal of the injured part from the source of the injury. This reaction is a reflex, before there is any conscious awareness of any sensation, and it is often accompanied by some form of vocalization.

□ Does pain always include all of these components?

■ No. These components may occur in isolation or in various combinations and proportions.

Two possible relationships are shown in Figure 5.19. Since the components can occur independently, the linear (in series) arrangement shown in Figure 5.19a is unlikely. Also, it is by no means certain that the components occur in the order shown, i.e one component does not appear to 'depend' on the others; it seems better, therefore, to think of the elements as being arranged 'in parallel', as in Figure 5.19b.

Before concluding this section, it is necessary to explain a few more terms that will be used to characterize features of pain such as its origin (or location) and its time course (duration or frequency).

Pain associated with skin or the body surface is referred to as *superficial*; pain from deeper body structures, such as muscles, joints or bones, is called *deep pain*. The former has a sharp, distinct quality and is easy to locate; the latter is duller, more persistent and is less well located. Both superficial and deep pains are associated with *somatic* structures (such as skin and muscles). Pain from *visceral* structures (internal organs, such as the heart, kidneys, intestines, etc.) is poorly localized, and follows a much longer time course of build-up and decline than is often the case with somatic pains. Another feature of visceral pains is that they may be referred to other parts of the body (see Section 5.2.5). Figure 5.20 illustrates the relative time course and intensity of several different types of pain.

The terms acute and chronic refer to the duration of the pain. *Acute* pains have a sudden onset but subside once the cause is removed and healing established. (An example would be acute appendicitis, due to inflammation of the appendix, part of the intestines.) *Chronic* pains are more stubborn and persist long after healing is apparently complete. There is no clear dividing line, but in general, pain is not usually classified by doctors as 'chronic' until it has lasted more than about six months.

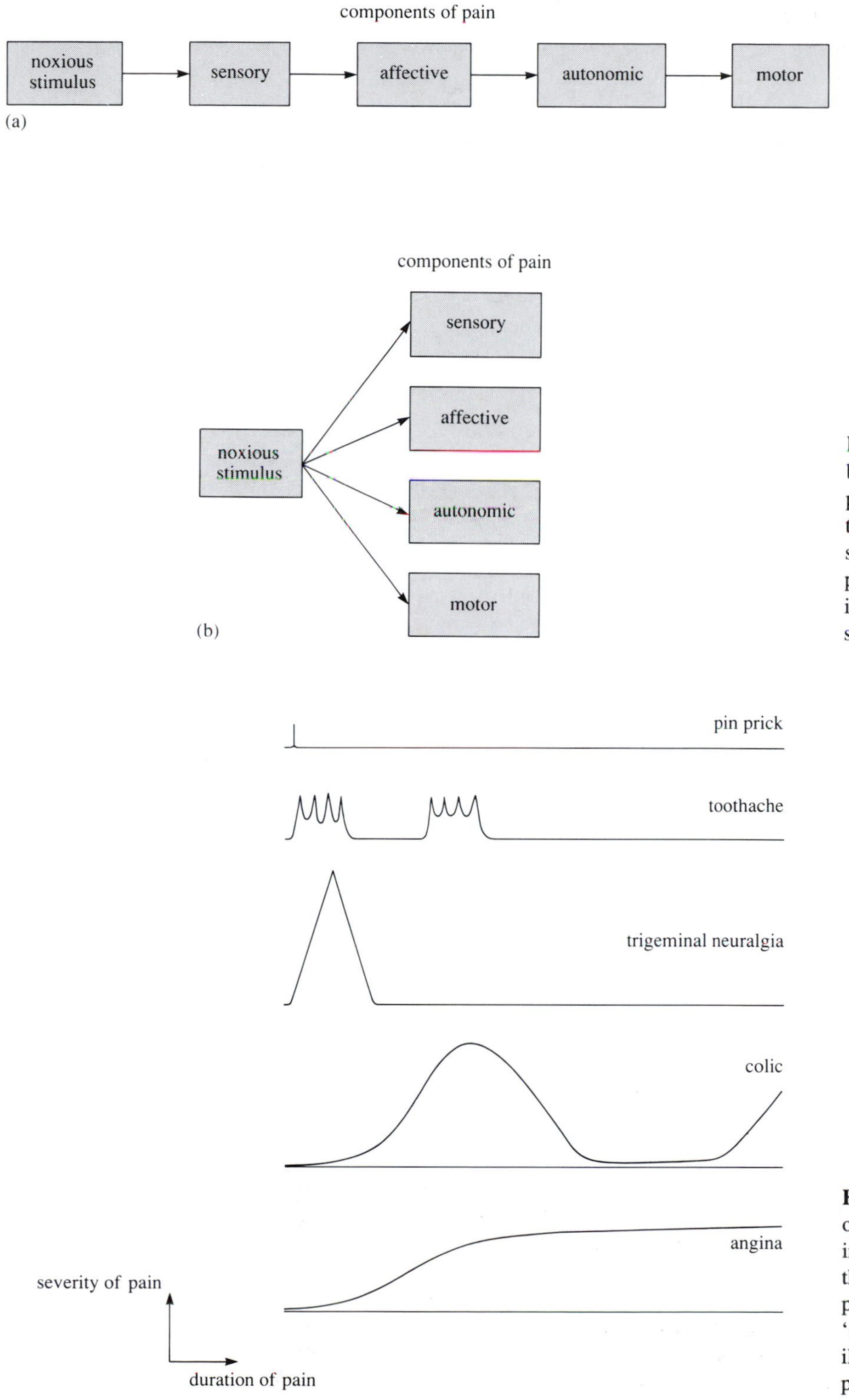

Figure 5.19 Possible relationships between the components of pain produced by a noxious stimulus. In (a) the components are represented 'in series', with one following on from the previous one; in (b) each component is independent of the others, which are shown 'in parallel'.

Figure 5.20 Schematic representation of the relative time courses and intensities of different types of pain. The three uppermost examples depict somatic pains, where the pain is generally 'sharp'. The two lower examples illustrate the typically slower and more prolonged pattern of visceral pains.

5.4.2 The value of pain

Sensations provide animals with information about their environment, allowing them to take appropriate action such as finding food or avoiding potential danger. Pain serves to warn the animal about threats to the body. Like an alarm system, it spurs the animal into avoidance, for example to minimize damage by pulling away from a source of intense heat, or resting an injured limb.

□ What is the value of pain?

■ To alert the organism to tissue injury before serious damage occurs. This has undoubted survival value, serving to increase fitness (Book 1, Section 4.3.4).

In the specific context of pain:

1. The initial experience of pain prompts avoidance, to prevent further damage.
2. Pain due to an injury or illness limits activity, forcing the animal to rest the injured part, thus allowing repair to take place and preventing aggravation of the injury.
3. The experience modifies behaviour, with the result that an animal learns to avoid similar dangers in the future. The 'bright, noisy water' experiment, which was described in Book 1, Section 6.4.1, is another example of an animal learning to avoid danger.

Thus, like a fire alarm, the warning bell of pain has definite survival value. But would life not be so much better if animals were spared the misery and anguish of pain? The survival value of pain becomes clearer when one considers what happens to those few people who are born without the ability to feel pain. Such individuals are described as *congenitally insensitive* to pain, and lack the ability to feel pain because of some developmental abnormality in the structure or function of their nervous system.

Individuals who are congenitally insensitive to pain may sustain severe burns and cuts or fracture limbs without feeling pain, and the failure to feel pain arising from illness, such as acute appendicitis (normally very painful and debilitating), can be fatal if the underlying cause remains undetected. Only with great difficulty and effort do such individuals learn to avoid damaging stimuli and to recognize other signs of illness, such as loss of appetite. Examination of the nervous system of individuals who are congenitally insensitive to pain often fails to reveal any abnormality. So far as can be detected, the wiring of the 'alarm' circuit is intact, but the system is somehow not 'switched on'.

There are many instances of congenital insensitivity to pain, but that of a Canadian, Miss C, is one of the best documented, possibly because her father was a physician. As a child, Miss C bit off the tip of her tongue, and suffered severe burns while kneeling on a radiator. On clinical testing, she felt no pain when noxious stimuli (such as strong electric currents, pinching, prolonged exposure to very hot water or ice) were applied. Also, she displayed none of the usual involuntary reactions to these stimuli, such as altered heart rate or blood pressure. Normal protective reflexes, such as blinking and coughing, were absent. Miss C displayed pathological changes in many joints, which were attributed to the lack of protection afforded by sensations of pain. Because she did not feel discomfort, she failed to shift her weight when standing, or to turn over in bed, or to avoid certain postures. Persistent injury and damage to the joints results in dead or dying tissue (known as Charcot joint). Blood supply to the region is reduced and the area becomes a perfect culture medium for bacteria. This happened to Miss C and she died from osteomyelitis at the age of 29 years. Only during her last months did she complain of some pain and discomfort, which was alleviated by analgesic drugs. Careful examination of her nervous system failed to reveal any abnormality; receptors in her skin and joints and her CNS all appeared structurally normal.

□ Would it be true to say that all pains are beneficial?

■ No. Most acute pains can be seen to have some protective value, but chronic pains (such as those in rheumatoid arthritis—a painful condition of joints—or the severe pain associated with the final stages of cancer) have no obvious biological value. If anything, these are detrimental to survival. Unlike acute pains, which usually have an identifiable cause, the sufferer cannot give any meaning to chronic pain, and it is little wonder that in the past such pains were often regarded as a form of punishment. The problem of chronic pain is discussed further in later sections of this chapter.

Pain also has diagnostic value. Pain is a prominent symptom of many conditions, and often the description of symptoms, or the pattern of signs elicited by examination can be so characteristic as to clinch a diagnosis.

5.4.3 Pain mechanisms

From the accounts of sensory pathways given in Section 5.2, it seems logical to think of pain being served by some sort of 'pain pathway'. That was the view of René Descartes, who thought of pain in terms of a 'fire alarm' system, which is depicted in Figure 5.21. This idea formed the basis of neural specificity theories that emerged during the 19th century and which became enshrined in medical texts in the 20th century. Figure 5.22 (*overleaf*) is a more modern version of Descartes' concept, and you might recognize the components of the pathway from Section 5.2. This type of diagram is still found in many textbooks, with reference to features such as 'pain receptors', 'pain fibres' and 'pain pathways'.

□ What is wrong with the terms 'pain receptor' or 'pain fibres'?

■ Pain is not a stimulus: it is a feeling. To say that a receptor responds to noxious stimulation is acceptable. But to call it a 'pain receptor' implies that activation of the receptor always elicits pain and pain alone. Similar objections can be applied to the terms 'pain fibres' or 'pain pathways'. The convention now is to use the term *nociceptor* for sensory endings that respond specifically to tissue damage.

The notion of specific neural pathways can be compelling, and can be applied to many sensory systems, such as vision. There is no reason why it should not also apply to pain and, indeed, many aspects of pain can be explained on the basis of a specific pathway. However, there are some aspects of pain that are not easily explained in terms of specific neural pathways. The amount of pain perceived often bears little predictable relationship to the magnitude of the stimulus. Also, cutting the spinothalamic tracts, supposed to mediate pain, does not always abolish it. Because the specificity theory cannot account for all pain phenomena, alternatives

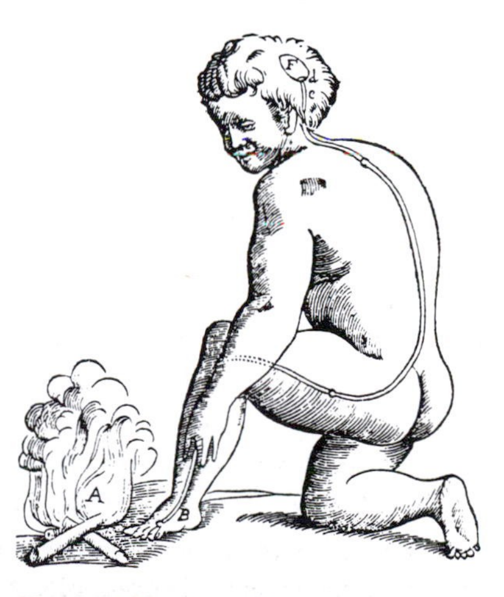

Figure 5.21 René Descartes' concept of the mechanism of pain, which provided the basis for present-day specificity theories by postulating the existence of a direct connection between the periphery of the body and 'centres' in the brain.

have been proposed. For example, pain can be produced by intense stimulation of almost any type and it is possible that pain could be signalled not by the type of nerve activated, but by the nature or pattern of the transmitted action potentials (*pattern theory*). However, neither the specificity theory nor the pattern theory adequately accounts for all aspects of pain. The following section considers some phenomena which must be explained by any satisfactory theory of pain.

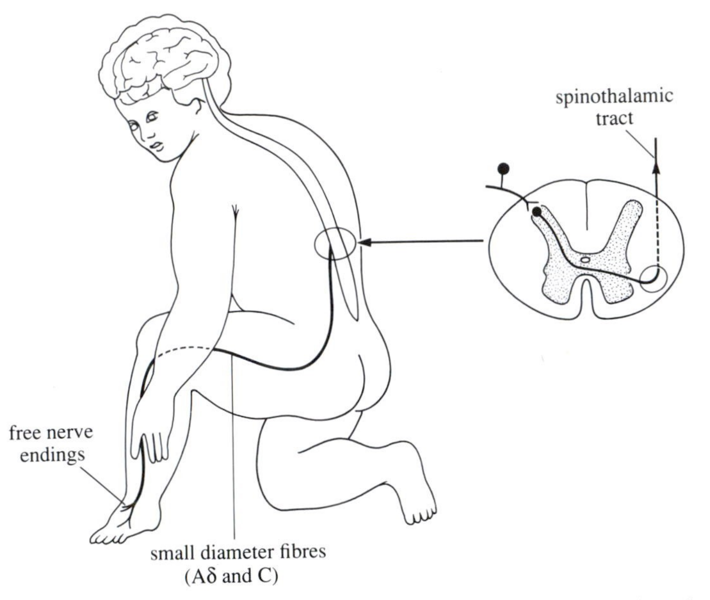

Figure 5.22 A modern interpretation of Descartes' diagram, showing the essential elements of a specific neural pathway mediating pain. The receptors are free nerve endings, from which action potentials pass along small-diameter (Aδ and C) nerve fibres to the spinal cord. From the dorsal horn, neurons project via the spinothalamic tracts to particular regions of the brain.

5.4.4 Some puzzles of pain

Some of the complexities of pain have now been established, and it is clear that things are not as simple as Descartes' fire-alarm system based upon a specific neural pathway. However, before this concept of pain is rejected, the fire-alarm analogy provides a useful way to illustrate some additional puzzles of pain.

(a) There are occasions when one might expect pain to occur, but the alarm bell does not ring.

It has been frequently observed that the amount of pain experienced bears little relation to the extent of injuries inflicted. About 60% of soldiers severely wounded in battle, and around 20% of people undergoing major surgery report little or no pain during the period following the injury or surgery. The differences between the prevalence of pain in military and civilian casualties are themselves of interest, but it is clear that the absence of pain is not simply due to some overall depression of perception, as these individuals are quick to complain about a clumsy injection. Some individuals are able to push skewers and needles into their flesh or through their cheeks, without apparently feeling pain; others can lie on a bed of nails. The problem is to explain why pain may be absent in such circumstances.

(b) Pain can occur in the absence of damage: the alarm bell rings when there is no fire.

The skin of the face, mouth and anterior part of the scalp is innervated by the trigeminal nerve, the 5th cranial nerve, which is connected to the brain stem. In *trigeminal neuralgia* (look back at Figure 5.20), severe bursts of pain are felt in one of the areas served by the trigeminal nerve—often the cheek. Pains can occur spontaneously, in the absence of any stimulation, or may be triggered by a gentle touch or light breeze. During an attack, there is spasm of the facial muscles, contorting the face—hence the alternative name, *tic dolloureux*. There is usually no history of injury or prior infection nor abnormality of the trigeminal nerve.

Many individuals who have lost a limb through amputation experience 'phantom' sensations of the missing limb. At first, the phantom limb may 'feel' normal, but later the phantom becomes less distinct and may fade completely. However, some people experience *phantom limb pain*. The pains can be severe, as if the limb were held for long periods in an uncomfortable position with the muscles contracted continuously. Again, the cause of the pain is unknown.

Tabes dorsalis occurs in the terminal stages of syphilis and is characterized by degeneration of the dorsal horn and dorsal spinal nerve roots. Repeated applications of a warm stimulus to the skin of a tabes sufferer, at first felt only as warmth, become increasingly hot until the person cries out as if burned. This phenomenon is not evident in people not suffering from tabes. Similarly, a single pin prick, normally felt as a brief pain, produces a prolonged diffuse pain in tabes sufferers.

(c) Surgical interruption of 'central pathways' often fails to relieve pain permanently—the alarm continues to ring even after all the 'wires' are apparently cut.

In people suffering from chronic pains which fail to respond to drug therapy, clinicians may resort to surgical interruption of the pathways supposed to transmit the action potentials that mediate pain—essentially cutting the 'wires' leading to the 'fire-alarm bell'. Numerous operations have been described for cutting peripheral nerves, dorsal roots, spinothalamic tracts, etc. (these are discussed in Section 5.9.2) but, although these may provide some relief, the pains frequently return, driving the sufferers to seek further, more drastic operations.

Clearly, pain is not the inevitable consequence of exciting certain nerves. It is highly variable. It may vary, not only from person to person, but also from one time to another in the same individual. Whatever neural mechanisms are involved, it appears that they are not fixed or 'hard wired'.

5.4.5 Conclusion to Section 5.4

This section has revealed something of the complexity of pain, and at this point it might be useful to re-examine your own definition of pain, which you wrote down earlier. Do this now, and amend your definition if you wish.

You might compare your definition with one produced by a committee of experts of the International Association for the Study of Pain:

> Pain is an unpleasant sensory and emotional experience associated with actual or potential tissue damage, or described in terms of such damage.

This definition goes a long way to encapsulating the essence of pain, although it remains somewhat inadequate. The word 'unpleasant' surely does not adequately describe a feeling that can cause such misery and anguish and may drive sufferers to undergo crippling operations or even to take their own lives.

Clearly, any satisfactory explanation of the nature and mechanisms of pain must include all aspects of pain. The specificity and pattern theories have their merits, but each fails to provide a comprehensive explanation. An alternative account will be considered in Sections 5.7 and 5.8.

5.5 Measurement of pain

5.5.1 Can sensations be measured?

Measurement of biological variables is an important aspect of clinical and experimental science; in the words of the physicist, Lord Kelvin:

> When you cannot measure it, when you cannot express it in numbers—you have scarcely, in your thoughts, advanced to the stage of science, whatever the matter may be.

For example, by measuring a patient's temperature or blood pressure, a doctor can assess health or monitor the progress of treatment.

□ Can you suggest how measurements of pain might be useful?

■ A measure of pain may be useful to gauge its severity at a given time. Also, it could be used to assess the effects of treatment.

□ Recall the components of pain (Section 5.4.1). Which of these could be measured?

■ Pain has sensory, emotional, autonomic and motor components. The first two represent internal, private feelings and cannot be observed directly. The autonomic and motor components are less subjective and can be measured externally. It is, however, the sensory and emotional components that are likely to be more important to the sufferer.

Since it is not yet possible to look into a person's brain to measure pain, or any other sensation, the only way to find out about the 'feelings' is to ask them to describe the pain. Some regard this subjective information as scientifically invalid, possibly because there is no independent means of checking it. (If one has reason to doubt a temperature measurement, one can always repeat the measurement using a different thermometer; but this is not possible with subjective reports.) In rejecting the evidence of the only first-hand witness, however, valuable information is lost. Although it is true that some people may not be able to give accurate information (e.g. infants, or those in coma) or may deliberately lie about their condition, the great majority of people have no motive for being untruthful about their pain.

Because of the difficulties in verifying subjective assessments of pain, this section will first consider more objective methods of investigating sensations and sensory capabilities. These include measuring sensations in terms of the stimuli that produce them, and measuring some of the associated physiological events, such as electrical activity in nerves, or involuntary behaviour.

5.5.2 Threshold measurements

Threshold measurements allow sensations or sensory abilities to be expressed in terms of the stimuli that evoke them. The *threshold* may be defined as the least amount of stimulus required to produce a sensation or response. Alternatively, one could measure *discrimination*, the smallest change in stimulus intensity that can be detected.

In relation to pain, there are several different thresholds that can be distinguished (Figure 5.23):

(a) **Sensory detection threshold**: the lowest stimulus intensity at which a sensation, of any sort, can be detected.

(b) **Pain perception threshold**: the lowest stimulus intensity required for the stimulus to produce a sensation that is distinctly, but just, painful.

(c) **Pain tolerance**: the stimulus intensity at which the subject is forced to withdraw, or to request stimulation to cease.

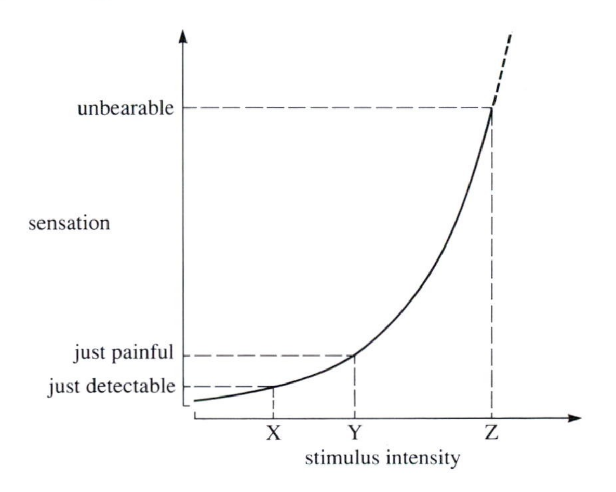

Figure 5.23 The relationship between the intensity of a noxious stimulus and the sensation evoked. X is the detection threshold: stimuli below this level of intensity produce no sensation. Y is the pain perception threshold: the intensity at which the stimulus becomes just noticeably painful. Z is the tolerance level: this is the maximum stimulus intensity which the subject can endure. The interval between X and Y is the 'pre-pain' range (see text).

The concept of pain threshold seems simple enough: a determination is made of the minimum level of stimulus required to produce pain. Allowing for the expected individual variation in any set of measurements, the detection thresholds for a given stimulus are fairly uniform. However, pain perception thresholds may display greater inter-subject variations. For example, detection thresholds for electrical stimuli in a group of Nepalese porters were no different from the thresholds among a group of Western climbers, but the porters required significantly higher currents in order to produce sensations that they regarded as 'painful'.

□ Can we infer that the Nepalese porters are less 'sensitive' than the Western climbers?

■ No. Sensitivity as measured by detection thresholds was similar in the two groups. However, the pain perception thresholds were higher for the Nepalese porters.

The differences in pain perception thresholds tend to reflect differences in the subjective criteria used by individuals to classify sensations, and it would appear that the porters are perhaps more stoical in their attitude. These variations are most marked when the stimulus produces non-painful sensations ('pre-pain') over a wide range of intensities (the 'pre-pain' range corresponds to the region X–Y in Figure 5.23). In other words, a given stimulus may evoke a sensation described by one person as 'warm' or 'hot', but a different person might regard the sensation produced by the same stimulus as 'painful'; the pain threshold will then appear lower in the second person compared with the first. Likewise, the same individual may perceive the same stimulus differently on different occasions. Even greater individual variations have been observed with pain tolerance levels. Experiments that seek to determine detection and perception thresholds do not mimic naturally occurring pains, and are thus less useful models for it. On the other hand, tolerance to a persistent stimulus such as immersing a hand in ice-cold water or exercising an arm to which the blood supply has been cut off, appear to provide more realistic models for natural pains than those afforded by brief, discrete stimuli. Other ways in which pain is investigated in the laboratory will be considered in Section 5.5.5.

5.5.3 Physiological correlates of pain

Here, ways to measure some of the physiological events that accompany pain will briefly be examined. For example, it is possible to record changes in involuntary activities, such as heart rate, that are triggered by painful stimuli. Also, techniques have been developed to record the electrical activity in peripheral nerves using microelectrodes inserted through the skin. When carried out on conscious subjects it is possible to demonstrate correlations between action potential discharge in single nerve fibres and simultaneous reports of pain in response to painful stimuli (Figure 5.24).

□ Is there evidence of a causal relationship between the activity in the single nerve fibre and the level of pain shown in Figure 5.24?

■ In some cases, the two events are closely related, but even here a causal relationship cannot be inferred from this evidence. The stimulus is likely to excite other nerve fibres as well, and their activity could be more influential.

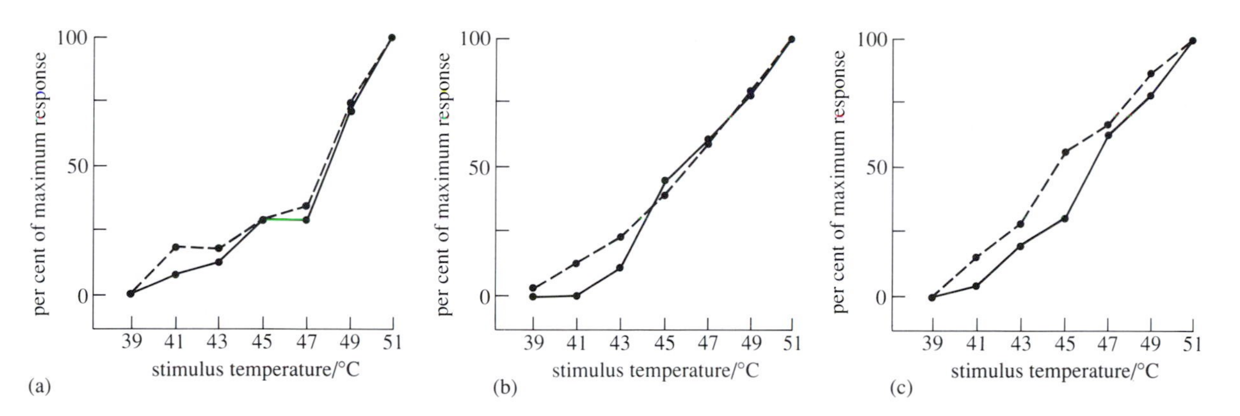

Figure 5.24 Graphs showing the relationship between stimulus intensity and response (pain or action potential firing in single nerve fibres) in three different subjects (a), (b) and (c). In the experiments illustrated, noxious heat stimuli were applied to the skin of human subjects. The firing rate in single unmyelinated nerve fibres (solid lines) was recorded along with the subjects' assessment of the amount of pain evoked by each stimulus intensity (dashed lines). The magnitudes of these responses have been expressed as percentages of the maximum nerve fibre firing frequency or amount of pain in each subject.

Even though one can measure such neural activity, the relationship between this and subjective sensations can be a tenuous one. Indeed, neural activity and withdrawal reflexes can be evoked by noxious stimuli in people with completely severed spinal cords, who are unable to experience accompanying sensations. Also, measures such as these are applicable only to phenomena evoked by stimuli, and they are less useful in assessing the spontaneous and continuous feelings typical of most forms of clinical pain. Therefore we must now consider ways of recording individuals' subjective reports.

5.5.4 Subjective measurement of pain

The simplest way of finding out about pain is to ask the person who is experiencing the pain. The first stage in such a subjective evaluation might be to record the actual words used by the person to describe their pain. This could provide a qualitative expression, but the adequacy of this depends upon the person's vocabulary and ability (or capability) to communicate:

> The merest schoolgirl, when she falls in love, has Shakespeare and Keats to speak for her; ... but let a sufferer try to describe a pain in his head to a doctor and language at once runs dry. (Virginia Woolf: 'On being ill'.)

However, there are ways to help the person to express pain in a quantitative manner. The **visual analogue scale (VAS)** is a simple, direct way of assessing pain severity (Figure 5.25, *overleaf*). The sufferer is presented with a line—usually 10 cm long—and told that one end represents no pain and the other represents the worst pain imaginable. The person is invited to make a mark on the line at the point that represents the present level of pain. The distance from the zero-end can provide a numerical assessment of the pain. The VAS can be used clinically to monitor the effects of analgesic drugs, as shown in Figure 5.26 (*overleaf*). The VAS offers a quick and simple method of scoring pain. It makes few demands on the sufferer and can be used frequently, but has the disadvantage of scoring pain in a single dimension, and the score can be dominated by the single most prominent feature of the pain.

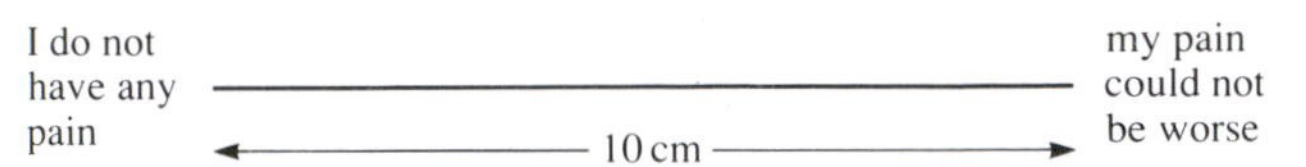

Figure 5.25 The visual analogue scale (VAS) for assessing pain.

The **McGill Pain Questionnaire (MPQ)** devised by Ronald Melzack at McGill University in Canada provides a means of assessing separately the sensory, affective and other aspects of pain (Figure 5.27). The MPQ consists of sets of words, each expressing different levels of one aspect of pain. Within each set, the words are ranked and, on the basis of prior studies, have been assigned a numerical 'severity value' unknown to the sufferer. There are five anchoring or reference levels of pain (1 = mild, 2 = discomforting, 3 = distressing, 4 = horrible, 5 = excruciating), which are used to provide a single assessment of the pain: the Present Pain Index (PPI). The different sets of words in the questionnaire reflect different aspects of pain: sensory, affective and evaluative, and the sufferer can score various subcategories of the pain, e.g. sensory (S), affective (A), evaluative (E) and miscellaneous (M).

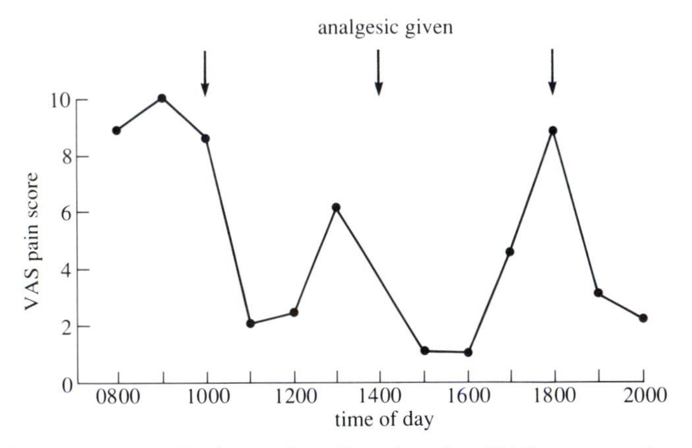

Figure 5.26 The pain profile for a pain sufferer based on VAS scores made over a 12-hour period. Each point represents a single assessment. At the times arrowed, the person received a dose of an analgesic drug after which the pain score is seen to decrease.

The sufferer goes through the questionnaire, selecting the word from each section that best describes the current pain. From these answers, a total pain score, the Pain Rating Index (PRI) can be calculated. It has been discovered that the questionnaire can characterize different types of pain according to a distinctive constellation of words selected. For example, it reveals that in women, labour pains are on average the most intense of those investigated, outscoring the pains associated with cancer, arthritis and many neuralgias. Although the MPQ provides a comprehensive analysis of pain, using it is a lengthy and demanding procedure, and not ideal for repeated use on debilitated sufferers.

5.5.5 Experimental pain

Although the information gained from observations on actual sufferers is clinically more 'realistic' and relevant, the experimental variables are less easily controlled and the results are more difficult to interpret. For this reason, there have been numerous attempts to investigate pain in the more readily controlled environment of the laboratory, where experimental pain may be studied either in animals or in human subjects.

McGill-Melzack Pain Questionnaire

Patient's name Date Time am/pm
Analgesic(s) Dosage Time given am/pm
Dosage Time given am/pm

Analgesic time difference (hours): +4 +1 +2 +3
PRI: S (1–10) A (11–15) E (16) M(S) (17–19) M(AE) (20) M(T) (17–20) PRI(T) (1–20)

1 *Flickering, Quivering, Pulsing, Throbbing, Beating, Pounding*
2 *Jumping, Flashing, Shooting*
3 *Pricking, Boring, Drilling, Stabbing, Lancinating*
4 *Sharp, Cutting, Lacerating*
5 *Pinching, Pressing, Gnawing, Cramping, Crushing*
6 *Tugging, Pulling, Wrenching*
7 *Hot, Burning, Scalding, Searing*
8 *Tingling, Itchy, Smarting, Stinging*
9 *Dull, Sore, Hurting, Aching, Heavy*
10 *Tender, Taut, Rasping, Splitting*
11 *Tiring, Exhausting*
12 *Sickening, Suffocating*
13 *Fearful, Frightful, Terrifying*
14 *Punishing, Gruelling, Cruel, Vicious, Killing*
15 *Wretched, Blinding*
16 *Annoying, Troublesome, Miserable, Intense, Unbearable*
17 *Spreading, Radiating, Penetrating, Piercing*
18 *Tight, Numb, Drawing, Squeezing, Tearing*
19 *Cool, Cold, Freezing*
20 *Nagging, Nauseating, Agonizing, Dreadful, Torturing*

PPI
0 No pain, 1 Mild, 2 Discomforting, 3 Distressing, 4 Horrible, 5 Excruciating

PPI Comments:

Constant, Periodic, Brief

Accompanying symptoms: *Nausea, Headache, Dizziness, Drowsiness, Constipation, Diarrhoea*
Comments:

Sleep: *Good, Fitful, Can't sleep*
Comments:

Food intake: *Good, Some, Little, None*
Comments:

Activity: *Good, Some, Little, None*
Comments:

Figure 5.27 A version of the McGill Pain Questionnaire. From each of the applicable sections, the sufferer chooses the word that best describes the current feelings and sensations evoked by the pain. The questionnaire allows people to assess separately various sub-classes of pain, e.g. sensory (S; numbers 1–10), affective (A; numbers 11–16), evaluative (E; numbers 17–19) and miscellaneous (M; number 20) categories. From the individual responses, a Pain Rating Index (PRI) can be calculated, based on the numerical values previously ascribed to each word. The Present Pain Index (PPI) is a single assessment of overall pain severity.

Experimental pain in animals

Much insight about basic physiological processes has been gained from animal experiments. Of course, the whole issue of animal experimentation is a contentious one, and the moral and legal aspects of this are considered separately in the audiotape, *The Use of Animals in Research*, associated with this course. Animal experiments offer certain advantages, not least in providing a greater degree of control over experimental variables and a wider range of possible operative procedures than is possible with human subjects. However, unlike humans, animals cannot communicate their feelings (although it is possible to some extent to use conditioning methods to gauge sensory thresholds). It is therefore necessary to investigate more objective aspects of behaviour, such as withdrawal responses, or recordings of neural activity evoked by noxious stimuli. One such test—the **tail flick test**—involves measuring the latency of the withdrawal response when a rat's tail is placed on a plate, heated by a light bulb. This test has been used extensively to assess the effects of analgesic drugs or procedures such as stimulation-produced analgesia (which will be discussed in Section 5.8.1).

□ Although such methods are 'objective', they can have only a limited application to human pain. Why?

■ 1 The tail flick reflex obviously cannot occur in humans, and although there may be some comparable reactions, reflexes present in animals are not always evident in humans.

2 Such involuntary spinal reflex responses say nothing about the higher brain events implicated in pain. Indeed highly coordinated spinal reflex responses can occur, without any associated pain, in anaesthetized animals and in humans in whom the spinal cord has been completely severed through injury.

Experimental pain in humans

It is difficult, if not impossible, to mimic chronic pain states in the laboratory, although a number of studies have investigated acute pains elicited in human volunteers by various forms of acute stimulation. Intense electrical stimuli are frequently used, as they are readily controlled, reproducible and can be quantified. However, electrical stimuli tend to excite nerve axons, rather than the sensory receptors normally activated by pain-producing stimuli.

Tests which do appear to provide reasonably accurate models of pain are the **cold pressor test** (in which the subject's hand is immersed in ice-cold water) and the **tourniquet test** (which involves arresting the blood supply to a limb which is then exercised). In each case, the stimulus is persistent, and the pain builds up until it becomes unbearable or intolerable. These tests can be used to assess pain tolerance, and you will see later how they are useful for evaluating the effectiveness of analgesic drugs and other forms of pain-relieving treatment.

5.6 The psychology of pain

All animals have the ability to react to noxious stimuli, and it seems reasonable to suppose that this ability is inherited. The exceptions are those rare individuals who are congenitally insensitive to pain (Section 5.4.2).

However, the nature and intensity of the pain reaction and accompanying sensory or emotional components do differ widely among different individuals; even in the same person, pain may be felt differently at different times. The extent to which such variations in pain are due to inherited differences or to environmental factors operating after birth, are considered next.

5.6.1 The influence of development and learning

During the early weeks and years of life, young animals are exposed to a rich variety of experiences. They may be exposed to varying degrees of injury sustained in play, exploration or fighting with other animals. Through these episodes, young animals learn about different environmental stimuli, and this experience is carried into adulthood. The importance of such experience of noxious stimuli was demonstrated in experiments by Melzack and Scott. (These were discussed in Book 1, Section 5.3.2.) They raised dogs in isolation, protected from the customary bumps and scrapes of puppyhood. When these animals reached maturity, they displayed abnormal behaviour towards noxious stimuli. They would repeatedly sniff at a lighted match presented to them and, although they reflexly withdrew from the stimulus, they did not exhibit any affective reactions (cowering, whimpering). Also, they tended to approach and sniff the match repeatedly. In contrast, siblings reared in a 'normal' environment were quick to learn the potential danger of lighted matches, and could rarely be induced to approach a match a second time. Heredity does have some bearing, as different breeds of dog display different degrees of disturbance of behaviour, but the experiments do suggest that early experience does play some role in pain perception.

Even established reactions to noxious stimuli, such as withdrawal responses to strong electric shocks, can be modified by circumstances or context. Amongst his many experiments (described in Book 1, Section 6.2), Pavlov conditioned dogs by presenting food immediately following the application of a strong electrical stimulus to a particular paw in each animal. Eventually, the dogs reacted to the shocks by salivating, wagging their tail and approaching the food bowl. However, if similar shocks were applied to other paws, not used in the conditioning stages, the dogs would react with violent withdrawal. Thus, even apparently 'unchangeable' reflexes can be altered by appropriate conditioning. This suggests that in some way the afferent signal evoked by a stimulus is analysed before the sensory and behavioural responses are expressed. These two examples illustrate how even the most fundamental components of pain can be modified by the experiences of early development and learning through conditioning.

The evidence (Section 5.5.2) that *detection* thresholds are similar among different individuals of one species or even between related species (such as humans and apes), suggests that the basic neural mechanisms are similar, and that external factors have little effect. Differences in *pain* thresholds and pain tolerance observed between different ethnic groups appear to reflect different attitudes towards pain.

□ It has been reported that American women of Italian descent are less tolerant of intense electric shocks than American women of Northern European or Jewish origin. How might these differences in pain tolerance be related to developmental differences?

■ The differences in pain tolerances of the two sets of women could arise from (a) differences in genetic or inherited factors, or (b) different environmental factors present during childhood.

There is little evidence that these differences in pain tolerance are genetic in origin, and they are perhaps more likely to arise from factors operating after birth. Individuals of Northern European or Jewish origin tend to have a more pragmatic attitude towards pain; those of Italian stock are more openly expressive of their feelings. These individual attitudes could be shaped early in life through interaction with family groups. By her immediate response (which acts as reinforcement) to her child's accident, a mother can shape subsequent behaviour. Some societies cultivate a stoical attitude ('big boys don't cry'), while others exhibit greater emotionality.

5.6.2 Pain and personality

Pain is a symptom in a high proportion of people with psychiatric illness or emotional disorders. But is there any relationship between pain and personality? This question has been studied in numerous investigations, and the relationship does not appear to be a simple one. It is important to appreciate that, although personality can affect pain perception, pain experience can influence personality.

This is illustrated by an investigation by M. R. Bond and I. B. Pearson (1969) into the incidence of pain in women with cancer of the cervix and its relation to their personalities. The study involved 52 women of similar ages with cancers of comparable durations and clinical state. The women were divided into three groups on the basis of pain experience and treatment administered. Group I reported no pain; Group II had pain, but did not request nor receive analgesics; the women in Group III also had pain, for which drugs had been administered (Table 5.4).

Table 5.4 Personality traits of three groups of cancer sufferers.

Characteristic	Group I (no pain)	Group II (pain: no analgesics)	Group III (pain: analgesics given)
Extroversion	high	low	high
Neuroticism	low	high	high

The women's personalities were assessed by the Eysenck Personality Inventory, which measures the two dimensions, stability–neuroticism and introversion–extroversion. (In this context, *neuroticism* is regarded as emotional instability and *extroversion* refers to an uninhibited, outgoing personality.) Overall, the women's mean scores did not differ significantly from those of a control group; however, differences were apparent between the different groups. The pain-free group (Group I) and those in pain who received drugs (Group III) tended to be extroverts. Those in pain who did not request drugs (Group II) tended to be introverts. The pain-free group had significantly lower neuroticism scores compared with both pain groups, among whom the scores for neuroticism were somewhat higher in the introverts.

□ What can be deduced from these findings in relation to the complaints of pain, and in the treatment received?

■ The women without pain (Group I) had low scores for neuroticism, whereas patients with pain (Groups II and III) displayed greater emotionality. Neuroticism was highest among the more introverted women who did not request nor receive drugs (Group II), and who tended to suffer in silence.

□ Can we relate the pain experience and behaviour of these women to pre-existing personality traits?

■ No. There is no information about the woman's personality before the onset of illness, so it cannot be assumed that these scores reflect basic personalities.

Although there appears to be a link between pain and neuroticism, a causal relationship cannot be inferred. It is quite possible that the high neuroticism scores are the result of pain, rather than the cause. Indeed, Bond has reported that, following successful surgical treatment for cancer pain, mean neuroticism scores are reduced post-operatively, although there are differences in the changes for individuals. It is likely that individual personality can be altered by illness, although the extent of such changes may be influenced by individual differences.

The nature of any personality changes resulting from illness or pain are also dependent on the nature of the clinical condition. Although one must be cautious in making generalizations, it is nevertheless apparent that acute pains tend to be associated with anxiety, whereas people with chronic pain are more likely to suffer depression. An individual with an acute chest pain may experience greater pain if anxiety is heightened by the news that a friend or relative died of a heart attack; in contrast pain due to indigestion is less sinister, and will afford less anxiety. Among sufferers of chronic pain, the depression may arise from feelings of helplessness or hopelessness, or a sense of loss or even guilt and inadequacy. Drugs that reduce anxiety, such as the minor tranquillizers, the benzodiazepines (e.g. diazepam), can successfully reduce the pain associated with anxiety. Similarly, antidepressants are of value in treating severe depression where pain is a prominent symptom, or even depression arising from chronic pain. (This is discussed further in Section 5.9.1.)

5.7 Modulation of afferent signals by 'gate control'

Until the mid-1960s, the predominant teaching was that pain was a distinct modality served by a specific set of neurons—much like any other sensation. However, in 1965, the established thinking was challenged by the publication of a new theory for pain—**gate control theory**—by Patrick Wall and Ronald Melzack. The publication of gate control theory was a landmark in the study of pain. It was a controversial hypothesis, which generated considerable, often heated, debate. Much of the controversy centred on the details of the proposed 'gate'. In this respect, it is important to distinguish between the overall concept (which is generally accepted) and the precise mechanisms of operation (which remain uncertain). As a result, the theory has undergone several modifications, and it is likely to evolve further in the future. The following discussion will concentrate on the concepts of gate control rather than the fine detail.

Gate control theory was an attempt to formulate a comprehensive theory for pain based on the known anatomical and physiological properties of nociceptive neurons, but which also allowed scope for modulation of the signals within the CNS. Basically, the theory proposed that a neural mechanism in the dorsal horn of the spinal cord acted like a 'gate', which regulated the effects of afferent action potentials on neurons in the CNS. The concept is shown diagrammatically in Figure 5.28. When the gate is closed (Figure 5.28a), signals in the small-diameter nociceptive afferent axons cannot excite the dorsal horn neurons (called T cells but do not confuse these T cells, in which 'T' stands for Transmission, with the T lymphocytes mentioned in Books 2 and 5) that project axons to the brain, and no

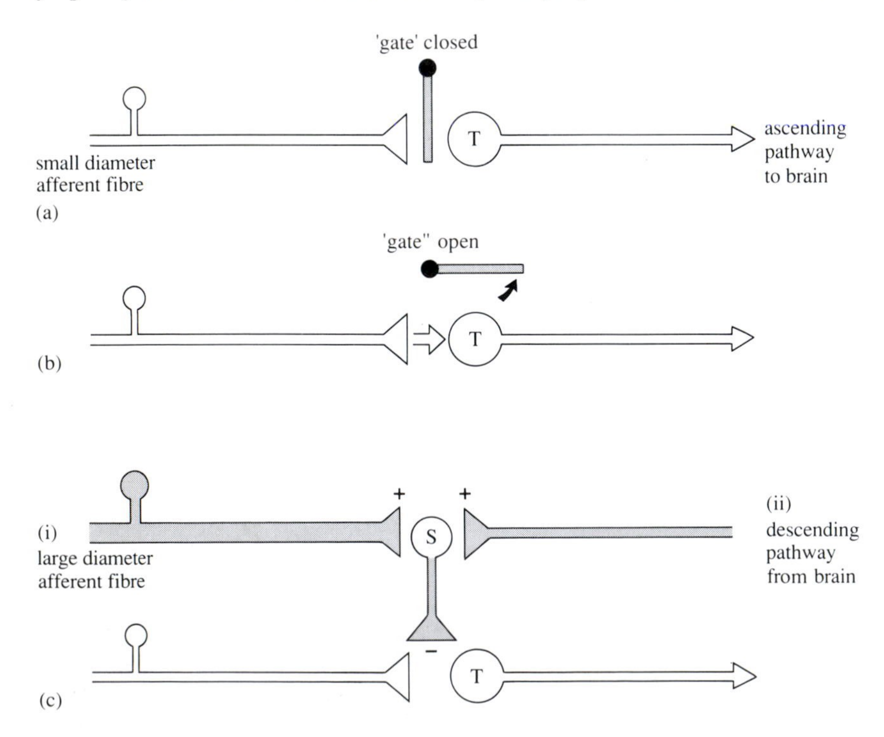

Figure 5.28 The concept of gate control. The diagrams represent a synapse between a small-diameter nociceptor afferent terminal and a transmission (T) cell in the dorsal horn. Transmission at this synapse is regulated by a hypothetical 'gate'. (a) When the gate is in the 'closed' position, there is no transmission at the synapse. (b) When the gate is 'open', activity in the afferent neuron is able to excite the dorsal horn cell. (c) Small interneurons in the substantia gelatinosa (S) (see Section 5.2.4) act as the gate, and prevent activation of the T cell. The action of the gate is determined by (i) inputs in other afferent neurons, and (ii) descending signals from the brain.

pain is felt. However, when the gate is open (Figure 5.28b), nociceptive afferent neurons are able to excite the T cells and pain may result. The axons of the dorsal horn T cells project to the brain by the spinothalamic and spinoreticular tracts. Melzack and Wall (1965) proposed that small inhibitory interneurons in the substantia gelatinosa acted as the gate, by regulating the amount of excitation reaching the T cells (Figure 5.28c). (The mechanism of the inhibition exerted by the substantia gelatinosa neurons may be either presynaptic or postsynaptic, and is considered in more detail in Section 5.8.4.) The setting of the gate (i.e. whether it is open or closed) is controlled by two factors: (i) the pattern of activity in other afferent inputs to the spinal cord, and (ii) by descending pathways from the brain stem (Figure 5.28c). The influence of afferent inputs on the activity of T cells is shown in Figure 5.29 (*overleaf*). The mechanisms of gate function and the way in which the descending pathways might influence the gate will be considered in Section 5.8.

Figure 5.29a (*overleaf*) shows the effect of noxious heat stimulation on the receptive field of a multimodal neuron in lamina V of the dorsal horn. When the heat stimulus is applied, the firing rate of the cell increases from less than 10 action potentials s^{-1} to roughly 30–40 action potentials s^{-1}. This is repeated in Figure 5.29b, but for part of the time the heat is applied, large afferent (Aβ) fibres are also stimulated electrically (shown in the figure by the bar marked 'nerve').

☐ Look at Figure 5.29b. What effect does stimulation of large fibres have on the heat-evoked discharge of the dorsal horn cell?

■ Stimulation of large fibres suppresses the heat-evoked response of the dorsal horn neuron.

Of course, there is no indication of the precise mechanism of the inhibition, but it demonstrates one important principle of gate control: selective activation of large-diameter afferent axons has the effect of closing the gate. (It was suggested at one time that activity in small-diameter afferent axons might actually keep the gate open, but there is no definite evidence to support this idea.) The setting of the gate thus seems to be determined by the relative amount of activity at any given moment in large- and small-diameter afferent axons.

☐ From your knowledge of gate control theory, how might the proposed gating mechanism be applied in the relief of pain?

■ Activity in large-diameter afferent axons closes the gate. Thus, selective stimulation of large-diameter axons should reduce pain.

Indeed most people will do this without thinking about it! You will almost certainly have done something like rubbing a knee after you have bumped into a chair. In a similar way, animals tend to lick their wounds. This mechanical stimulation will activate large-diameter afferent axons, and any pain relief may be due to closing the gate. It has been demonstrated on numerous occasions that selective activation of large-diameter afferent axons by gentle mechanical stimuli or brief periods of weak electrical stimulation will cause pain relief lasting many hours. From this has evolved the principle of Transcutaneous Electrical Nerve Stimulation (TENS), which has become an important method of pain control (discussed further in Section 5.9.4).

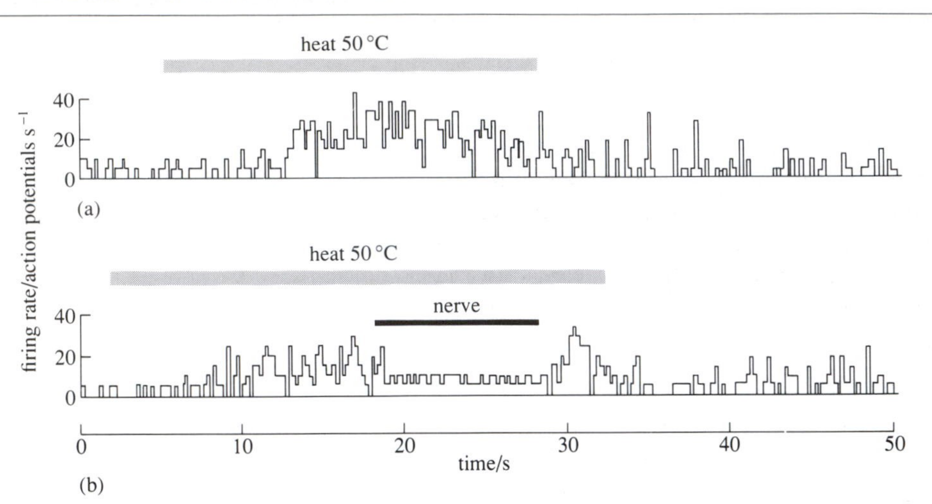

Figure 5.29 Responses of multimodal dorsal horn cells to noxious stimuli can be modified by inputs in other afferent fibres and by stimulation of spinal pathways. (a) The pattern of action potential discharge generated by the application of a noxious heat stimulus, which is applied during the period marked by the solid line labelled 'heat 50 °C'. (b) The response to heating is suppressed by the activation of large-diameter afferent nerve fibres by electrical stimulation, indicated by the black line labelled 'nerve' above the graph.

5.8 Endogenous analgesic systems

Previous sections of this chapter have described the mechanisms by which noxious stimuli may cause pain. It is evident that pain is a highly variable phenomenon, which can be affected by many factors, such as personality. Nociceptive signals in afferent neurons can be filtered in the spinal cord by a gate control mechanism. The following sections will address the problem of why pain does not always occur in situations when it might be expected to arise. Specifically, the mechanisms that may enable the brain to activate the gate control system to inhibit pain will be considered next.

5.8.1 Behavioural evidence

It is clear that pain is an extremely variable phenomenon and it is logical to suppose that the brain can somehow influence what sensations are perceived—'mind over matter', as it were.

In the late 1960s and early 1970s it was demonstrated in rats that electrical stimulation of certain regions in the brain stem with implanted electrodes abolished aversive responses to a variety of noxious stimuli. This phenomenon was called **stimulation produced analgesia** (SPA), and was investigated in detail by David Mayer and his group at the University of California at Los Angeles (Mayer *et al.*, 1971).

SPA could be produced by electrodes placed in specific sites such as (a) the periventricular grey matter (PVGM) in the walls of the third ventricle, (b) the periaqueductal grey matter (PAGM) surrounding the cerebral aqueduct (the channel connecting the third ventricle to the fourth ventricle) in the midbrain, and (c) sites in the medulla, such as the raphe nuclei (RN) (Figure 5.30).

In their experiments, Mayer and his colleagues found that the effects of SPA were confined to specific areas of skin which were usually restricted to one half or one quarter of the body. Stimuli applied outside these areas evoked normal aversive reactions. The analgesia occurred during the stimulation period and lasted for a few minutes after stimulation ceased. Other sensory and motor activities were undiminished, and stimulation did not disrupt other normal behaviour patterns, such as feeding. How then does SPA operate?

SPA could have resulted from a temporary block of action potential transmission in ascending nociceptive pathways passing close to the stimulating electrodes. However, permanent lesions at SPA sites did not produce any obvious reduction in pain sensitivity. SPA seemed to be due to the activation of particular neural systems, rather than the blocking of impulse transmission.

It was possible that SPA was due to the apparently 'rewarding' properties associated with stimulation of particular brain regions. It is known that rats will press a bar in a Skinner box (Book 1, Section 6.3.4) for long periods of time simply to receive brief electrical stimuli delivered through electrodes implanted in certain areas of the brain. This is called self-stimulation, and is described in Book 2, Section 10.5, and Book 4. The rewarding effects of the SPA were assessed by using it as reinforcement for bar pressing. It was found that many of the sites that produced analgesia also yielded high rates of bar pressing for self-stimulation. However, some sites which were associated with high self-stimulation rates produced no analgesia. Furthermore, lowering the levels of serotonin (Book 2, Section 4.5.5) in the CNS reduces the effects of SPA, but it does not affect the self-stimulation motivational effect evoked from the same site. Thus, the effects of SPA cannot be ascribed completely to activation of a 'reward' system.

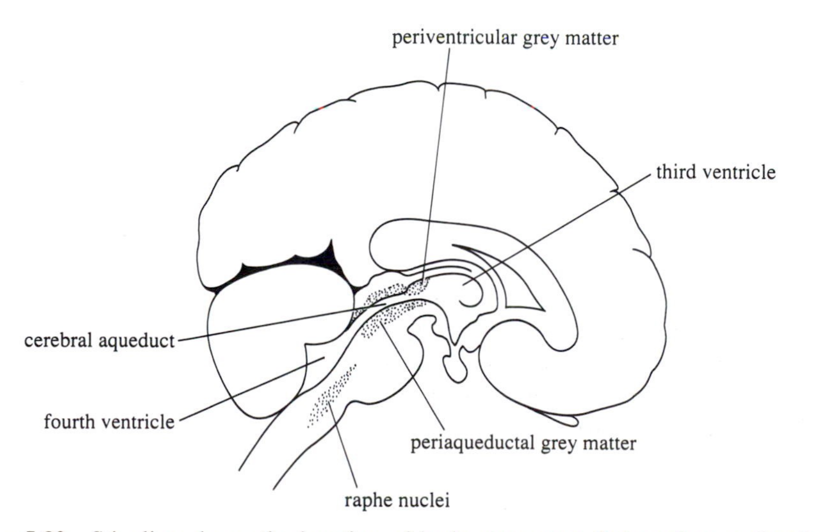

Figure 5.30 Stippling shows the location of brain stem areas that produce analgesic effects when stimulated (SPA).

Other sites which produced analgesia, but no self-stimulation, afforded an opportunity to test whether the analgesic effect alone could reinforce the operant of bar pressing. Animals were given access to two levers. Pressing one lever (the active lever) caused a stimulus pulse to be delivered; the other (inactive) lever did nothing. In the absence of any noxious stimuli, it was found that the animals tended to press the inactive lever more often than the one delivering a stimulus to the brain.

□ Can you think of an explanation for this?

■ The brain stimulation alone could have been slightly aversive.

However, during 1-minute test periods, when strong inescapable electric shocks were delivered to the rats' tails, the rats vigorously pressed the 'active' lever, which controlled brain stimulation. Thus, in the presence of noxious stimulation, the normally non-rewarding brain stimulation can reinforce an operant.

□ Is this brain stimulation acting as a positive or negative reinforcer?

■ Such brain stimulation alone is not rewarding and does not normally reinforce lever pressing; thus it is not a positive reinforcer. Lever pressing does not remove the noxious stimulation of the tail, but it does appear to diminish its unpleasant effect, and so is acting as negative reinforcement.

Another possible explanation was that SPA was due to activation of a neural system capable of modulating nociceptive information, possibly by triggering the gate control mechanism. Before attempting to explain how stimulation of brain stem areas such as the periaqueductal grey matter and raphe nuclei might influence transmission at synapses in the spinal cord, it is helpful to consider several additional lines of evidence, which are described in the next two sections.

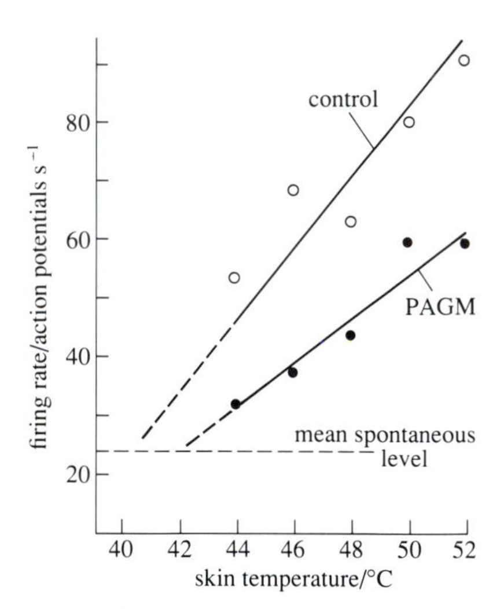

Figure 5.31 Descending inhibition of spinal dorsal horn neurons. The mean responses (action potentials s^{-1}) evoked by noxious heat stimuli are plotted against the skin temperature. The normal (control) responses are shown by the open dots. When the thermal stimuli were applied during electrical stimulation of the PAGM (black dots), the responses were reduced compared with the controls. The horizontal dotted line shows the cell's background firing rate

5.8.2 Anatomical and physiological evidence

One of the features of the SPA described in the previous section was the suppression of spinal reflex responses to noxious stimuli (e.g. the tail-flick response). Since these reflexes do not involve synapses at levels higher than the spinal cord, it is likely therefore that nociceptive responses of spinal neurons are inhibited by descending pathways originating from brain stem regions such as the PAGM and the raphe nuclei. The following lines of evidence support this.

The responses of neurons in the dorsal horn to noxious stimulation are suppressed by electrical stimulation of SPA sites in the brain stem. Figure 5.31 shows the firing rate of a dorsal horn neuron in response to different levels of heat stimulation. When the PAGM is stimulated during the application of heat stimuli, the numbers of action potentials produced at any given temperature are reduced compared with the controls. Also, it has been found that after the spinal cord is cut through completely, the responses of dorsal horn cells to noxious stimuli are enhanced. Their background firing rate is also increased, suggesting that these cells normally receive continuous inhibition from structures at a higher level in the CNS. Other experiments indicate that the fibres activated by stimulation of the PAGM or raphe nuclei pass in a common pathway descending in the dorsolateral funiculus (DLF) ('funiculus' means pathway) and that neurons in this pathway utilize serotonin as a transmitter.

From studies of neuronal connections, it is likely that neurons in the PAGM have excitatory connections with cells in the raphe nuclei, which project via the DLF to the dorsal horn, where they inhibit neurons normally excited by inputs in nociceptive afferent fibres. The possible nature and mechanism of this inhibition will be considered in Section 5.8.4.

Other studies have revealed the existence of several distinct inhibitory systems descending in the DLF from various regions of the brain stem, but the details are beyond the scope of this chapter.

The analgesic effects of brain stimulation are remarkably similar to the effects of the narcotic analgesic, morphine, and Figure 5.32 shows how the effect of SPA can be mimicked by morphine.

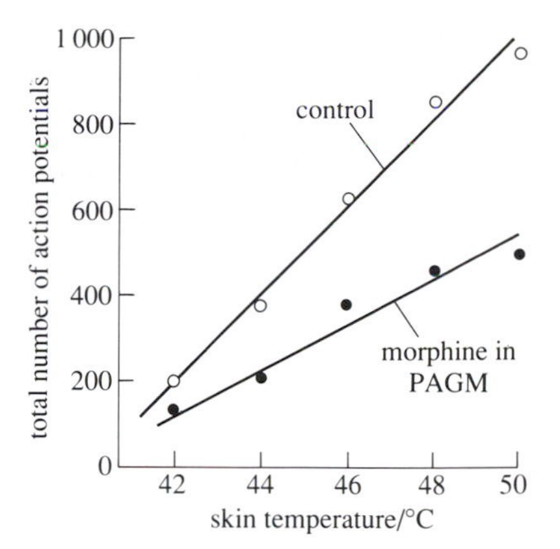

Figure 5.32 Mean responses (total number of action potentials) of dorsal horn neurons to noxious heat stimuli applied before (open dots—controls) and after the injection of morphine (black dots) into the PAGM. The effect of morphine is to reduce the firing rate at any one temperature compared with the controls.

□ Compare the effects shown in Figures 5.31 and 5.32. What similarities are apparent?

■ Morphine (Figure 5.32) suppresses the rate of firing of the dorsal horn cells in a manner very similar to that seen with PAGM stimulation (Figure 5.31). It is possible that the two effects are due to a common mechanism.

Section 5.8.3 will now look more closely at the mode of action of drugs such as morphine.

5.8.3 Neurochemical evidence: opiate receptors and opioid peptides

The euphoric and analgesic effects of opium, a mixture of alkaloids extracted from the seed cases of the poppy (*Papaver somniferum*) have been known for centuries. The component responsible for the powerful analgesic effects of opium—morphine—was isolated early in the 19th century and is called an **opiate**. But how does morphine work? It was reasonable to suppose that it acted on the brain, but for a long time there was no direct way of confirming this.

In 1973, Candace Pert and Sol Snyder of the Johns Hopkins Medical School in Baltimore successfully used radioactively labelled ligands to demonstrate the presence of specific **opiate receptors** in brain tissue. (Ligands are compounds, not necessarily neurotransmitters, that bind specifically to particular membrane receptors.) There is now evidence for several different kinds of opiate receptors, which differ in their distribution in the nervous system and in their affinities for different ligands. These various receptors appear to be involved in a wide range of functions such as feeding, regulation of fluid balance, control of hormone release, locomotor behaviour and thermoregulation, as well as pain.

□ Why should the CNS contain receptors that bind substances like morphine, which do not occur naturally in animals?

■ The most logical interpretation is that animal brains also contain endogenous (naturally occurring) neurotransmitters that normally act on these opiate receptors.

Shortly after the identification of opiate receptors in the CNS, the predicted, naturally occurring, ligands were discovered. In 1975, a group led by John Hughes and Hans Kosterlitz in Aberdeen described two endogenous **opioids** as small peptide molecules, which are called Met-enkephalin and Leu-enkephalin. Since the initial discovery of enkephalins, several other endogenous opioid peptides such as endorphins and dynorphins have been identified. Attempts have been made to correlate these ligands with the various opiate receptors, but so far it has not been possible to assign specific roles to these various ligands and receptors.

The term 'opiate' was originally used to describe drugs extracted from opium, including morphine, and chemical derivatives of morphine. However, with the synthesis of totally new drugs (and the discovery of endogenous substances) with morphine-like effects but different chemical structures, the term 'opioid' was introduced. Opioids include all drugs, either natural or synthetic, with morphine-like actions. Thus, morphine is an opiate and an opioid, whereas enkephalins are opioids.

5.8.4 Endogenous analgesia: an overview

The preceding sections have described a system of neurons in the brain stem and spinal cord, involving opioid peptides and opiate receptors, which can suppress the responsiveness of dorsal horn cells to noxious stimuli. How does such a system work, and how might this serve to enhance fitness?

□ Earlier in this chapter, the possible value of pain was considered. What are the benefits of pain?

■ Pain serves as a warning of tissue damage. This experience can alter present and future behaviour.

But pain can be distracting and could actually handicap the animal at times of danger. However, sometimes physical injury and tissue damage can occur but no pain arises. Is it possible that under certain circumstances an endogenous system such as that involved in SPA could be activated to suppress pain?

Normally, animals do not have electrodes conveniently implanted in their brains so that they can 'switch off' pain. There is, however, anecdotal evidence that something equivalent can occur naturally. The following account, made over a century ago, lends support to the idea that an *endogenous analgesic system* can be activated at times of life-threatening stress.

During a lion-hunting expedition in Africa, the Scottish missionary and explorer, David Livingstone, was mauled by a lion. The story starts at the point after Livingstone has shot and wounded a lion, and is busy reloading his gun:

> ...I heard a shout. Starting, and looking half round, I saw the lion just in the act of springing upon me. I was upon a little height; he caught my shoulder as he sprang, and we both came to the ground below together. Growling horribly close to my ear, he shook me as a terrier does a rat. The shock produced a stupor similar to that which seems to be felt by a mouse after the first shake of the cat. It caused a sort of dreaminess in which there was no sense of pain nor feeling of terror, though quite

> conscious of all that was happening. It was like what patients partially under the influence of chloroform describe, who see all the operation, but feel not the knife. This singular condition was not the result of any mental process. The shake annihilated fear, and allowed no sense of horror in looking round at the beast. This peculiar state is probably produced in all animals killed by the carnivora; and if so, is a merciful provision by our benevolent Creator for lessening the pain of death. (David Livingstone, *Missionary Travels and Researches in South Africa*, 1857)

(Livingstone was rescued by companions, and survived to recount his experience. Interestingly, Livingstone's left humerus (upper arm bone) is preserved at the Royal College of Physicians and Surgeons in Glasgow, and bears scars consistent with the described attack.)

Perhaps the state described by Livingstone could have been produced by the activation of an endogenous analgesic system. The possible physiological basis of such a system will now be presented. The main elements are summarized in Figure 5.33, which should be referred to when reading the following account. You should not attempt to learn the details in this diagram, which is included solely to illustrate the following discussion. Note that some of the connections that will be referred to in Figure 5.34 (*overleaf*) have been omitted for clarity.

Starting with Figure 5.33a, it can be seen than neurons in some forebrain regions (such as the hypothalamus and frontal cortex) send excitatory projections to cells in the midbrain periaqueductal grey matter (PAGM). These forebrain regions,

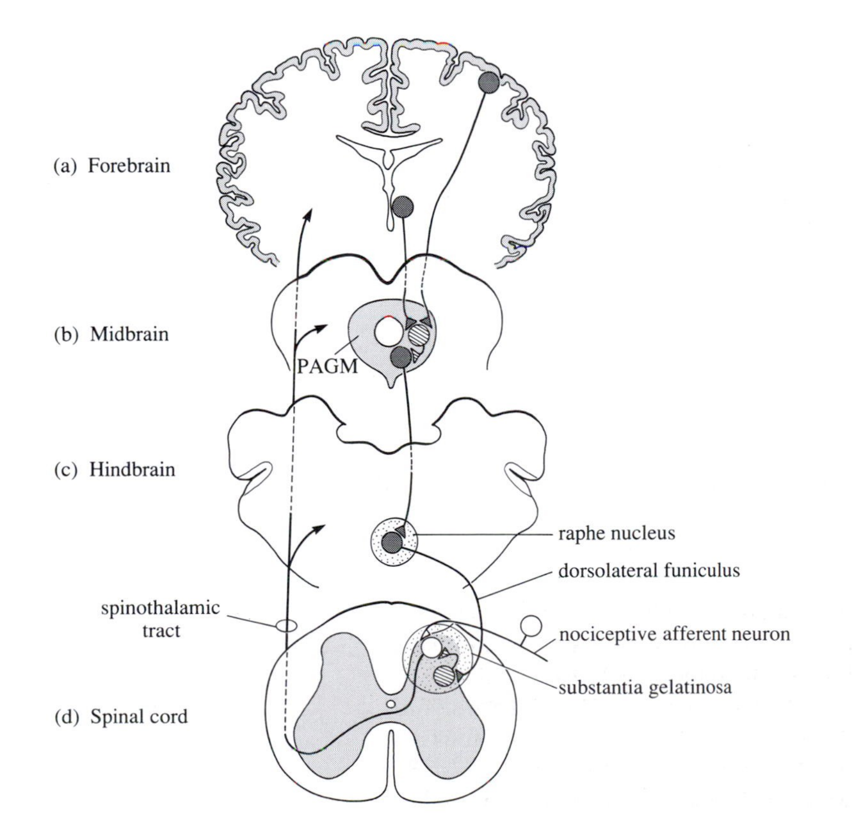

Figure 5.33 A summary of the principal elements in an endogenous analgesic system (stippled) in the central nervous system. See text for explanation. (a) Neurons in certain forebrain areas project to (b) the midbrain periaqueductal grey matter (PAGM). (c) Neurons in the PAGM in turn project to the raphe nuclei in the medulla of the hindbrain. Pathways descend from the medulla to the dorsal horn in the dorsolateral funiculus (DLF). (d) The pathway from the raphe nucleus to the dorsal horn of the spinal cord employs the transmitter serotonin and terminates on interneurons in the substantia gelatinosa (circled). Neurons containing opioid peptides are depicted by cross-hatching. This system receives connections at various levels from ascending fibres in the nociceptive pathway (shown by the arrows on the left-hand side of the diagram).

which contain opiate receptors and opioid peptides, also receive ascending projections from nociceptive pathways, such as the spinothalamic tract (shown by the arrows on the left-hand side of the diagram), and which could provide an excitatory 'triggering' input.

The PAGM (Figure 5.33b) also contains opiate receptors and opioid peptides. Neurons in the PAGM in turn make excitatory connections with neurons in the raphe nuclei in the medulla (Figure 5.33c). (This part of the medulla contains many other neurons, employing a wide range of neurotransmitters, and receives inputs from a variety of sources including the ascending axons from nociceptor neurons in the dorsal horn. However, these are not shown in the diagram.) Axons of the neurons of the raphe nuclei project to the substantia gelatinosa of the dorsal horn of the spinal cord via the dorsolateral funiculus; these descending neurons release the transmitter serotonin from their terminals in the dorsal horn of the spinal cord. The raphe nuclei do not contain the same density of opioids as the PAGM, but some opiate receptors are present.

The dorsal horn (Figure 5.33d) contains opiate receptors and cells containing opioid peptides such as enkephalins and dynorphins; these opioid peptides appear to be concentrated in areas such as the substantia gelatinosa (circled in the figure). In addition, branches of many other sensory neurons project to the substantia gelatinosa, although, in order to simplify the explanation, these connections are not shown in the figure. These will be considered again (see Figure 5.34). The enkephalin-containing interneurons in the substantia gelatinosa appear to be the best candidates for mediating gate control inhibition, initiated by activity in either large-diameter afferent fibres or descending pathways from the brain.

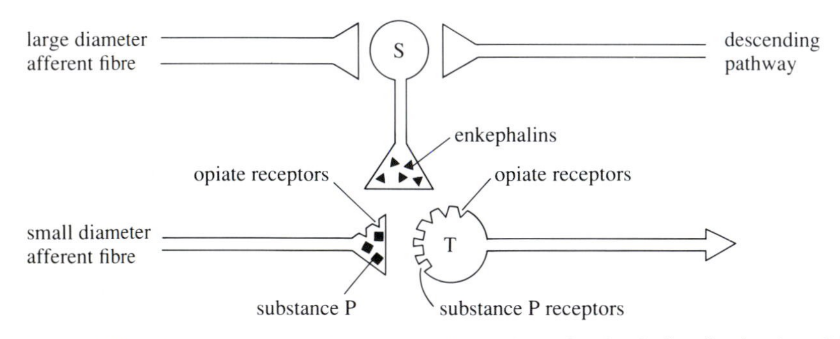

Figure 5.34 One possible mechanism for the action of enkephalins in the dorsal horn. Enkephalins released from spinal interneurons in the substantia gelatinosa (S) inhibit the release of substance P from nociceptor axon terminals. Opiate receptors are present on the axon terminals, suggesting that at least part of the action is due to presynaptic inhibition. However, opiate receptors may also be present on the membranes of the T cells, so that some of the effects may be postsynaptic in nature.

One possible mechanism for gate control is shown in more detail in Figure 5.34, which is based on the principle shown earlier in Figure 5.28 and 'fleshes out' the simplified pathway outlined in Figure 5.33. Almost 50% of the opiate receptor sites in the dorsal horn are present on the terminals of sensory C fibres, many of which release the peptide substance P (Book 2, Section 4.5.6). It has been shown that enkephalins will block release of substance P from sensory axon terminals, and it is possible that some of the endogenous control operates through presynaptic inhibition to suppress release of substance P (Figure 5.34). However, although some of the inhibition is presynaptic, some of it is postsynaptic and exerted directly on the next neuron in the pathway (T cell).

Recent evidence indicates that polymodal C fibres do more than merely relay messages about noxious stimuli. The C-nociceptor neurons contain a veritable chemical cocktail of various peptide molecules. One of these—substance P—is released as a conventional excitatory transmitter, but evidence is accumulating that substance P and other peptides can exert slower, longer lasting neuromodulatory effects that alter the excitability of cells in the vicinity. Sustained C fibre activity as a result of continued stimulation or altered activity (as may occur following nerve damage) increases the excitability of neurons in the dorsal horn (opening the gate). These effects persist when action potential discharge is blocked by local anaesthetics, or if the C fibres are damaged by crushing. However, the effects are abolished when C fibres are destroyed by selectively acting neurotoxins or when their axons are cut. It would appear that these slower, neuromodulator effects are mediated by transport of peptides along the axons from the peripheral endings, and are released locally in the dorsal horn. Significantly, narcotic drugs, such as morphine, are very effective in abolishing these peptide-mediated excitability changes. The effects of transmitters like substance P are not confined to the synapses in the CNS. There is now evidence that substance P is also released from the *peripheral* end of sensory neurons in response to stimulation, and that it may modify the excitability of peripheral neuronal terminals as well as affecting the local blood flow. (This is discussed in Section 5.9.1.)

It must be emphasized that this account is greatly simplified and is incomplete. However, it does appear to provide some tangible basis for phenomena like those described by David Livingstone. At times of stress, the body prepares for 'fight or flight' through a neuroendocrine response involving the action of the sympathetic nervous system and secretions from the adrenal gland. (This will be taken up in Book 5.) As shown in Figure 5.33, the parts of the brain responsible for this (limbic system and hypothalamus) are linked to the endogenous analgesic system, and may coordinate the overall response. This system can be 'primed' by signals evoked by noxious stimuli. The initial signals, travelling via rapidly conducting routes, alert the brain, and, depending on circumstances, the brain can trigger the endogenous analgesic system, which exerts descending inhibition on nociceptive traffic in the spinal cord. From a survival standpoint, it would seem beneficial to suppress the potentially distracting and immobilizing aspects of pain, at least until immediate danger has passed. Consistent with this, there is now good evidence that β-endorphin levels in the cerebrospinal fluid are raised during exercise.

5.9 The treatment of pain

In this final section, the various approaches to pain control and some of the problems that may be encountered in the management of pain will be considered. In addition to examining the traditional ways to treat pain, the impact of new knowledge and the role of 'unconventional' methods will be discussed. There are three main categories of treatment for pain, namely pharmacological, surgical and psychological methods.

5.9.1 The pharmacological treatment of pain

Drugs are widely used for treating pain, particularly in the Western world, and the main classes used are outlined in Table 5.5.

Table 5.5 Classes of drugs used in the treatment of pain.

Type	Example	Site and mechanism of action
Analgesic (mild)	aspirin, paracetamol	peripheral action, blocking synthesis of prostaglandins in damaged tissues
Analgesic (narcotic)	morphine, pethidine	central action on opiate receptors in the CNS
Local anaesthetic	lignocaine	prevents action potential propagation in axons by blocking sodium channels
Tranquillizer	benzodiazepines, phenothiazines	alteration of CNS neurotransmitter functions
Antidepressant	tricyclic compounds, monoamine oxidase (MAO)-inhibitors	alteration of CNS neurotransmitter functions
Anticonvulsant	barbiturates, carbamazepine	alteration of CNS neurotransmitter functions

Analgesics

Analgesics selectively abolish pain without diminishing other sensations. The mild analgesics, such as aspirin and paracetamol, act peripherally, preventing the synthesis of prostaglandins that sensitize nociceptors in damaged or inflamed tissues (Section 5.2.2). They have no effect on pain thresholds in normal tissue, and are most successful in dealing with mild pains, associated with inflammatory changes. The mechanism of action of these mild analgesics is summarized in Figure 5.35.

Injury can generate activity in nociceptive afferent neurons which excite spinal cord pathways, but the nociceptors are also affected by chemicals released at the site of injury. Chemicals released from the nerve terminals (substance P) and the damaged tissues (bradykinin, serotonin and prostaglandins) can affect the state of local blood vessels and also sensitize the nociceptors. Activity in sympathetic nerves, which release noradrenalin, causes constriction of blood vessels in the area and can increase nociceptor sensitivity. Drugs such as aspirin ('blocker 1' in Figure 5.35) prevent prostaglandin release, thus reducing sensitization. Local anaesthetics ('blocker 2') prevent the propagation of action potentials in afferent and efferent nerves. Specific sympathetic blocking drugs ('blocker 3') can reduce activity in sympathetic nerves, and can be effective in treating pains where there is excessive activity in sympathetic nerves. The precise role of the sympathetic nervous system is not clear, but in some cases, pain can be associated with a reduced blood flow and coldness in the area. In such cases, pain can be alleviated by using sympathetic blockers, which prevent release of noradrenalin.

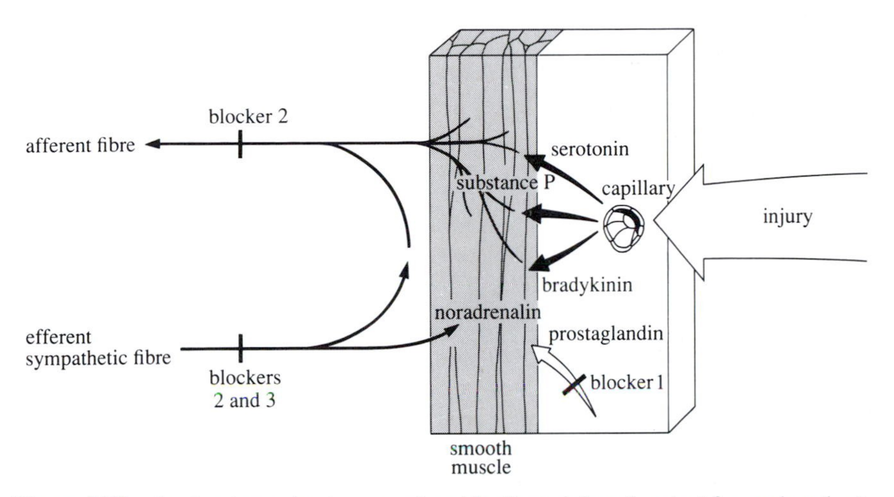

Figure 5.35 Analgesic mechanisms produced by tissue injury (see text for explanation).

Narcotic analgesics such as morphine and pethidine are used in the treatment of severe pains. Their principal effect is to suppress affective, autonomic and motor components of pain. These agents act centrally on the various opiate receptors in the CNS. (Contrast this action with the peripheral action of the mild analgesics, aspirin and paracetamol.) The narcotics act by mimicking the endogenous opioids (e.g. enkephalins and endorphins). Important sites of action are likely to be receptors in the midbrain (PAGM) and dorsal horn. However, the involvement of the numerous opiate receptors in the forebrain cannot be discounted.

Prolonged and repeated administration of narcotics can have two consequences: *dependence* and *tolerance*.

Dependence, or addiction, is a state in which an individual who has taken narcotics over a period of time experiences 'withdrawal symptoms' when the drug is no longer taken. When tolerance to a drug develops, the initial dose loses its effectiveness, and the dose has to be increased to maintain a given effect.

Although dependence and tolerance are undoubtedly major problems among drug abusers, these phenomena are infrequently encountered when narcotics are administered for pain control. Of the many thousands of Israeli soldiers wounded in the Yom Kippur war of 1973 who received injections of narcotics for pain relief, not a single case of addiction was reported. Cancer sufferers can take steady doses of narcotics for months or even years, with no sign of tolerance. Also, in such cases which are later successfully treated by surgery, narcotic intake can be tapered off over a few days with no evidence of withdrawal symptoms.

Local anaesthetics

Local anaesthetics are widely used for pain control during minor surgery, such as dental treatment. These drugs, such as lignocaine, prevent action potential propagation by blocking voltage-sensitive Na^+ channels in axon membranes. The main effect will be to reduce activity in afferent nerves, but possible effects on efferent (sympathetic) fibres cannot be discounted.

□ What is the difference between an anaesthetic and an analgesic?

■ By blocking action potential transmission, anaesthetics can abolish all sensations. An analgesic acts selectively to abolish pain, but other sensory modalities are unaffected.

Psychotropic drugs

Psychotropic drugs are the third main group of drugs used to treat pain. These drugs include tranquillizers and antidepressants, which mainly affect synaptic transmission in the CNS. They may be used alone or in association with analgesics to combat pain. They are particularly effective where there is a strong emotional component or when pain arises with no obvious physical cause and when it may be related to psychological factors. Both minor (e.g. diazepines) and major (e.g. phenothiazines) tranquillizers reduce pain-related anxiety; antidepressants (e.g. tricyclic compounds or monoamine oxidase (MAO) inhibitors) appear to have some analgesic effect that is independent of their main action. Anticonvulsants (e.g. barbiturates or carbamezepine) have proved to be very effective in many people in preventing the pain of trigeminal neuralgia (Section 5.4.4).

The placebo effect

In the course of many clinical investigations, it has been found that pain can often be relieved by giving a sufferer something like a sugar pill or a salt solution in place of an analgesic drug. An innocuous substance, such as sugar or salt, given in this way instead of a drug, is called a *placebo*; a word which literally means 'I will please'. The analgesia produced by a placebo is called a *placebo effect*. In the strict sense of the word, a placebo is a harmless substance given to someone, who believes it to be 'medicine'; but the term is sometimes also used to describe sham operating procedures or items of equipment, such as 'dummy' stimulators, which while looking impressive, do nothing when connected to the person concerned. About one third of all sufferers given a placebo report a marked relief of pain. The effects of an analgesic drug and a placebo can be compared under conditions in which neither experimenter nor subject knows which substance is administered to any given individual. Such trials are therefore called *double blind*. The magnitude of the placebo effect under double blind conditions is illustrated in Figure 5.36, which summarizes results of experiments to determine the effect of an analgesic on pain thresholds to electrical stimuli. All the subjects were told that they would receive an analgesic drug. Half of the subjects did receive an analgesic, dihydrocodeine, but the remainder were unknowingly given a placebo, a glucose pill identical in appearance to the analgesic.

□ How were the pain thresholds affected by the analgesic and the placebo?

■ Both the analgesic and the placebo raised the mean pain thresholds in the two hours following administration, but the analgesic had a greater effect.

Why should a placebo have any effect on pain? Suggestion appears to play a major role, especially in relation to the anticipated level of response. In other words, placebos tend to be more effective when the recipients believe that they are receiving a powerful drug or treatment. A remarkable feature of the placebo effect is that the placebo is generally about 30–50% as effective as the drug it is tested against. The mode of delivery is also important: large capsules are more effective than small ones, and placebos have a stronger effect when injected than when taken by mouth. The placebo effect is strong in relieving acute pains where there is a high level of emotional stress or anxiety.

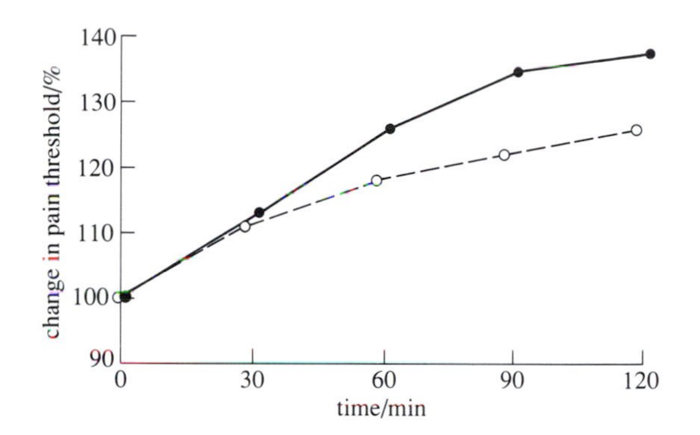

Figure 5.36 The placebo effect. The graph shows the mean results from 300 subjects of the effects of an active analgesic drug, dihydrocodeine (black dots, solid lines) and a placebo (open circles, dashed lines) on the pain threshold to an electrical stimulus. The drugs were administered at time 0, and changes in pain threshold are expressed as a percentage of the value at time 0, which is 100%.

5.9.2 Surgical treatment of pain

When pain cannot be relieved by drug therapy, it may be treated by 'controlled lesioning' of the neural pathways assumed to be involved in transmitting nociceptive information. The pathways may be destroyed chemically, electrically or by cutting. Temporary lesions may be created with local anaesthetics or deep cooling. Figure 5.37 provides a summary of the many sites in the nervous system that have been lesioned in order to alleviate chronic pain. Operations have been carried out at many levels, from the peripheral nerve (1), dorsal roots (rhizotomy—4), spinothalamic tracts (cordotomy—5), the thalamus (8) to destroying parts of the forebrain (9 and lobotomy—10).

Surgical procedures do relieve pain, but the relief is generally temporary and the pain may return after several months or years. Even if the treatment is successful, the person can be left with numbness or abnormal sensations, which can prove as unpleasant as the original pain. Although surgical treatment has its place, for example when drugs have proved ineffective, or in the treatment of severe pain in people with terminal cancer, emphasis is now shifting towards non-destructive procedures, some of which are described in the following sections.

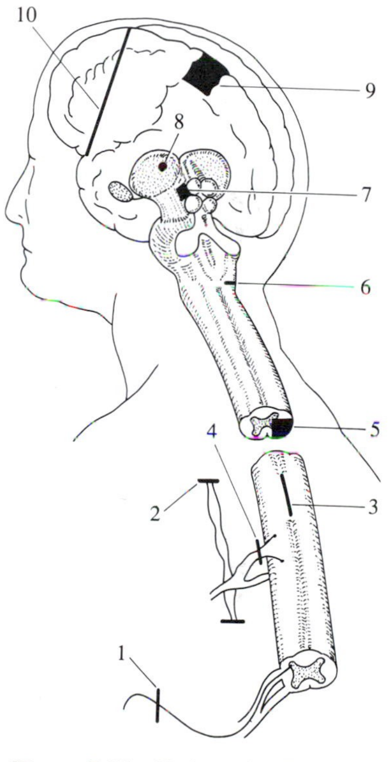

Figure 5.37 Various sites in the nervous system of operations designed to alleviate pain:

1 Peripheral nerve.
2 Sympathetic chain.
3 Commissural fibres crossing the midline.
4 Dorsal roots.
5 Spinothalamic tracts in the spinal cord.
6 Tracts in the medulla.
7 Tracts in the midbrain.
8 Thalamus.
9 Sensory cortex.
10 Frontal lobes.

5.9.3 Psychological treatment of pain

Psychotherapy refers to treatments that are based principally on verbal interaction with the sufferer, rather than relying on drugs or surgical intervention. Supportive psychotherapy employs a sympathetic approach in which the therapist will listen to the sufferer and attempt to provide reassurance. The aim is to reduce anxiety by helping the person to understand the nature of any related medical problems. Therapy can also be extended to involve the sufferer's family.

Another form of therapy employs the principles of operant conditioning (Book 1, Section 6.3.2) to modify certain pain-related behaviour patterns through selective reinforcement. This approach makes no specific attempt to reduce pain, but concentrates on eliminating pain-related behaviour patterns.

□ What might be meant by pain-related behaviour patterns?

■ They include complaining vocally about pain, repeatedly requesting analgesics and a tendency to avoid any form of physical activity including work.

□ How could operant conditioning be used in these circumstances?

■ Complaints of pain and requests for analgesics would be ignored (and thus not reinforced) whereas any attempt at physical activity would be encouraged (by positive reinforcement).

This approach has been used for some 'chronic pain' sufferers who are regarded as having 'given in' to their condition and who are thought to be using it as an excuse not to resume normal employment, or to gain sympathy from others.

Hypnosis is another method of pain control that may be classed as psychological treatment. Hypnosis produces a trance-like state in which the subject focuses attention on the hypnotist, while apparently 'shutting out' other stimuli. In this state, the subject shows a degree of susceptibility to suggestions made by the hypnotist. Individuals differ in their susceptibility to hypnotic suggestion. About one third of the human population can be deeply hypnotized, and some two thirds of this group (approximately 20% of the general population) can experience effective pain relief with hypnosis.

Hypnotic analgesia involves the hypnotist telling subjects that they will not feel any pain during the period of the hypnosis. Figure 5.38 shows how pain tolerance levels in the tourniquet test (see Section 5.5.5) can be affected by hypnotic suggestion and contrasts the responses in individuals who are either strongly or weakly responsive to hypnotic suggestion. The effects of hypnosis are also compared with those of a placebo (saline injection), which subjects were told was an analgesic drug. In highly susceptible subjects ('high'), hypnosis increased the time for which subjects could endure the tourniquet conditions, but little change was evident in the tolerance levels of subjects not susceptible to hypnosis ('low').

□ Is it possible that the analgesia produced by hypnosis is a form of placebo effect?

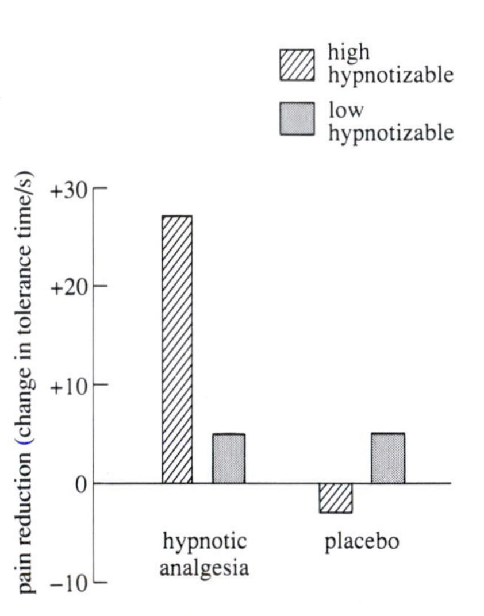

Figure 5.38 Hypnotic analgesia in subjects, 12 of whom had a high susceptibility to hypnotic suggestion ('high hypnotizable'), and 12 others who did not respond to hypnosis ('low hypnotizable'). The experiment involved placing a pressure cuff on the arm to arrest blood flow, and the subjects then exercised the blood-deprived hand for as long as possible. Pain tolerance was reached when the subjects found the pain so unbearable that they were unable to continue. The histogram shows the changes in pain tolerance produced by hypnosis and a placebo (saline injection).

■ For subjects who are not very susceptible to hypnotic suggestion ('low'), this could be true, as Figure 5.38 shows that hypnosis and the placebo (saline injection) raised pain tolerance by a small but equivalent amount. However, in highly susceptible subjects, hypnosis greatly increased pain tolerance levels, while placebos had little effect, and if anything slightly decreased pain tolerance.

It has also been reported that naloxone (an opiate receptor blocker) does not diminish the effects of hypnotic analgesia, suggesting that this type of analgesia is not mediated through the action of opioids.

Susceptible individuals under hypnotic analgesia report that they feel no pain during the application of noxious stimuli that cause intense pain when the subject is not hypnotized. However, during hypnotic analgesia, the associated autonomic effects typical of pain (increased heart rate and blood pressure) may still be present. This raises the possibility that during hypnotic analgesia, the nociceptive inputs are processed normally in some parts of the brain, but are somehow prevented from producing sensations of pain.

Further insight into hypnotic analgesia has been gained by Ernest Hilgard of Stanford University, California, who has exploited a phenomenon known as the 'hidden observer'. Some people who are highly susceptible to hypnotic suggestion are able to respond under hypnosis in a non-verbal manner by 'automatic writing', in which the subjects are somehow able to write answers to specific questions without actually being aware of what they are doing. The 'hidden observer' refers to a 'dissociated' part of the subject's mind that appears to control activities like automatic writing. In his experiments on pain produced by immersing one of the subject's hands in ice-cold water (the cold pressor test—Section 5.5.5), Hilgard instructs subjects to give a verbal estimate of the pain intensity, by using a numerical scale ranging from 0 (no pain) to 10 (pain is intolerable). However, subjects are additionally told that there is a part of their mind of which they are not aware—the 'hidden observer'—that can communicate with the hypnotist, through the unstimulated hand, either by 'automatic writing' or by pressing the keys of an unseen recording device.

Figure 5.39 shows the results of an experiment on eight subjects who immersed one hand in ice-cold water. In the unhypnotized state, the pain became intense after about 25 seconds. (Note that Hilgard permits subjects to report pain levels that exceed the original maximum level—hence some scores are greater than 10.) During hypnotic analgesia, the verbal reports indicate that low pain intensities were felt during cold water immersion, but at the same time, the 'hidden observer' reports pains approaching the intensities felt in the unhypnotized state. Somehow, hypnosis seems to be blocking access of the nociceptive signals to higher cognitive areas of the brain, although the signals nevertheless appear to reach other areas of the brain, capable of expression through 'automatic writing' or key pressing.

These experiments provide some evidence that there is dual processing of nociceptive inputs, which would explain why autonomic components of pain may persist even when the subject says he or she feels no pain. Hypnotic analgesia appears to occur at high cognitive levels, and does not seem to be mediated by endogenous opioids as naloxone does not reduce the level of the effect.

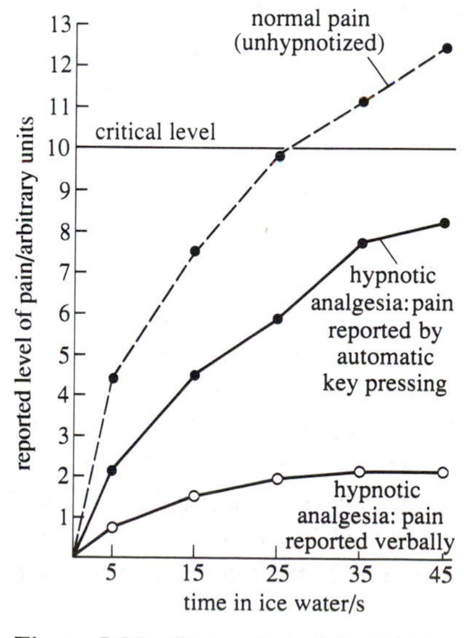

Figure 5.39 Hypnosis and the 'hidden' observer. See text for explanation.

5.9.4 The impact of gate control theory on pain treatment

The gate control theory has undoubtedly had a major influence on pain research, but has it had any impact on the treatment of pain?

□ What prediction does gate control theory make that might be applied clinically in pain control?

■ Pain may be relieved by 'closing the gate', and this may be achieved by activation of (a) large-diameter (Aβ) afferent fibres, or (b) the endogenous analgesic system. Both of these approaches have been used in the treatment of pain.

Transcutaneous Electrical Nerve Stimulation (TENS)

Sensory fibres in superficial peripheral nerves, such as the ulnar or median nerves in the arm, may be activated by electric shocks applied via electrodes placed on the overlying skin. (The stimuli thus pass through the skin—transcutaneous.) Since the large mechanosensitive afferent fibres (Aβ) have low thresholds to electrical stimuli, they can be activated by stimuli insufficient to excite Aδ and C fibres (see Section 5.2.3), and so any sensations evoked are not painful. This technique has been used since the late 1960s and it is found that stimulation periods of 15–30 minutes can abolish chronic pains for many hours. The apparatus is simple, comprising small electrodes which are connected to a battery-powered stimulator. The frequency and duration of the stimuli are usually pre-set, but the sufferer can vary the intensity of the stimulus pulses until the desired effect is achieved. This technique is called Transcutaneous Electrical Nerve Stimulation (TENS).

TENS devices have proved successful in treating a broad range of pain conditions, many of which may not have responded to other forms of treatment. Typically, TENS will achieve a good degree of relief of acute pains in about 60% of sufferers, although the success rate with chronic pains is only around 30%. This procedure is significantly more effective than placebos. In some, but not all instances, the pain relief is reduced by naloxone, suggesting that the effect is mediated at least in part by opioids.

Stimulation of the CNS

The alternative to TENS is to stimulate parts of the CNS to activate the endogenous analgesic system. In view of the inaccessibility of these sites, the stimulation is carried out using permanently implanted electrodes, often in the PAGM, or placed on spinal cord tracts.

Some outstanding successes have been claimed for such methods, although they are surgically more demanding and more prone to develop technical faults. Sufferers are usually carefully selected before receiving permanently implanted electrodes and more than 60% experience satisfactory degrees of pain relief. As with TENS, the sufferer controls the intensity of the stimulus, which is transmitted by radio waves to a subcutaneous receiver. The analgesia produced by brain stimulation appears to be opioid-mediated, as it is associated with increased levels of opioid peptides in the cerebrospinal fluid, and the effects are reduced by naloxone. The method suffers a disadvantage in that repeated stimulation may produce tolerance, both to the electrical stimulation and to narcotic analgesics.

5.9.5 Alternative medicine

Many sufferers and some clinicians who are dissatisfied with traditional methods of pain control may resort to less conventional treatments that may be loosely classified as 'alternative medicine'. It is alternative in the sense that these methods are not generally taught in many Western medical schools. Many forms of alternative medicine are practised throughout the world, including homeopathy, osteopathy, chiropracty, acupuncture and forms of faith healing, each with their own principles of action. Because their effects have not often been verified objectively, they are regarded by some as being 'unscientific'. However, in 1990 T. W. Meade and his colleagues published a report of a controlled clinical trial which showed that chiropracty was more effective in treating chronic back pains than 'conventional' methods. (Chiropracty involves manipulation of the bones of the vertebral column.) Not only did chiropracty provide a greater reduction in the sufferers' symptoms, but the differences persisted for at least three years following the treatment. It seems unlikely that all such differences were due to a placebo effect, and further objective comparisons of traditional and alternative methods are clearly necessary.

Acupuncture is one form of alternative medicine which has been subjected to intensive investigation. Acupuncture is an ancient form of medicine originating in China, which came to the attention of Western medicine for its ability to relieve pain. The technique involves placing long needles through the skin, to a depth of up to several centimetres at certain specified sites, called acupuncture points.

Acupuncture undoubtedly raises the pain threshold to a variety of noxious stimuli. The analgesic effects of acupuncture are blocked when local anaesthetics are applied at the acupuncture point, suggesting that acupuncture is mediated by nerve transmission. Attempts have been made to compare acupuncture with hypnosis and placebo effects. When electrical stimulation of teeth is used as the experimental pain, placebos (saline injections) have little effect on the pain threshold, but the threshold is raised by both acupuncture and hypnosis. The effect of acupuncture, but not hypnosis, can be reversed by naloxone, suggesting that acupuncture is mediated by opioids. This is consistent with reports that levels of opioids in the cerebrospinal fluid are elevated during and after acupuncture.

Manual stimulation of acupuncture needles ('needling') may be augmented by electrical stimulation of the needle (electroacupuncture) with low-frequency, high-intensity electric currents. Although the effects of this stimulus regime are reduced by naloxone, it is of interest that the analgesia produced by TENS, which employs high-frequency, low-intensity currents, is much less susceptible to naloxone.

The traditional acupuncture points do not appear to be very precise, as analgesia may be produced by stimulation of other sites. This is seen in Figure 5.40 (*overleaf*), which shows how the pain thresholds to radiant heat stimulation are affected by various forms of acupuncture procedures. Needling of recognized acupuncture points and areas adjacent to the application of the painful stimulus elevated the pain thresholds, in that longer exposure to the heat was required to produce pain. However, needling distant sites had little effect on the pain thresholds, which remained close to the normal individual variations (shaded areas). Overall, it appears that the intensity of stimulation, rather than its precise location, is the crucial factor. Melzack has concluded that 'acupuncture is one of the many ways to produce analgesia by an intense sensory input, which may be labelled generally as hyperstimulation analgesia'. In other words, acupuncture may be regarded as a form of counter-irritation in which one form of intense stimulus can mask other sensations.

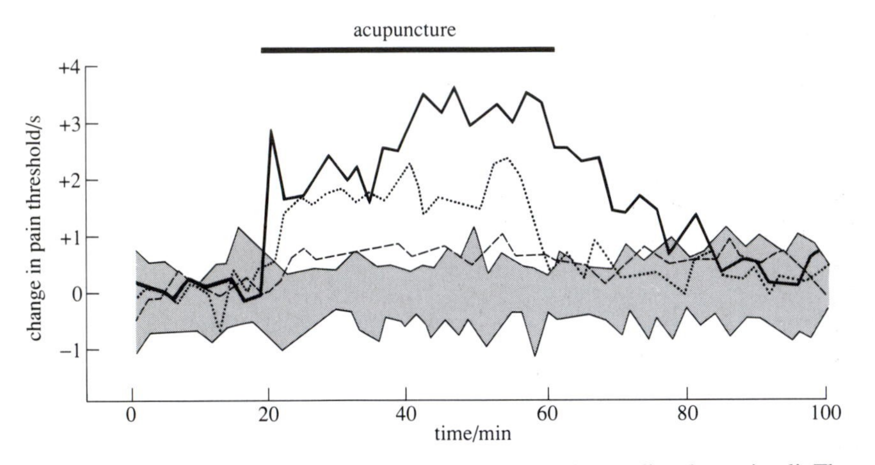

Figure 5.40 The effect of acupuncture on pain thresholds to radiant heat stimuli. The pain thresholds were measured as the time taken for a radiant heat stimulus to produce pain. The baseline values are taken as zero on the Y-axis. The normal individual variations are shown by the grey area. These values were compared with the effects of applying acupuncture needles at various sites: (continuous line, traditional acupuncture points; dotted line, adjacent sites; dashed line, distant sites.) In these experiments, acupuncture in traditional and adjoining sites increased the pain thresholds.

5.9.6 Strategies in pain management

In the previous sections we have described several different ways for controlling pain. With such a wide range of treatments to choose from, how might the clinician select the most appropriate method of pain control for any particular sufferer?

First, the cause of the pain must be identified and treated, if possible. For example, there is not much point in simply treating the painful symptoms of say, appendicitis, without also removing the inflamed or ruptured appendix. The symptoms of pain cannot be ignored, however, and they have to be dealt with as part of the overall treatment plan. The methods chosen to treat the pain will be influenced by its severity and duration; individual factors such as the prognosis and the person's age and general state of health may also influence the choice of treatment.

Analgesic drugs are the mainstay of treatment for acute pains. Pains of mild to moderate severity, for example toothache, may be controlled by mild analgesics like aspirin. However, narcotic analgesics may be required to relieve more severe pains, associated with injuries or after major surgery.

The management of chronic pains presents further difficulties. The fact that the pain is 'chronic' suggests that it has already failed to respond to the initial treatments. Alternatively, the pain may be adequately controlled by analgesics, but if the underlying cause cannot be removed, additional problems may be encountered due to the side effects of prolonged drug therapy.

In such cases of 'intractable' or persistent pain, alternative modes of treatment have to be considered, and the choice of therapy may be determined more by the sufferer's age and life expectancy than by the severity of the pain.

In people with pain associated with the terminal stages of cancer, there is limited life expectancy, and already a degree of incapacity due to the disease. Here, the priority is to ensure that sufferers are able to live out their remaining time in a dignified, and relatively pain-free state. Surgical destruction of parts of the CNS may be an appropriate and effective treatment, the inevitable sensory and possible motor impairment being an acceptable 'trade-off' for pain relief. Largely because of fears of sufferers developing tolerance or dependence (addiction), some physicians have been sparing in their use of narcotics, such as morphine or heroin. However, these are not found to be problems with terminally ill people, and in hospices which specialize in the care of the dying, narcotics, such as the famous Brompton mixture, are the basis of a successful regime which also includes intensive and supportive nursing. The Brompton mixture is a potent 'cocktail' of morphine, cocaine and alcohol, with the dose of morphine being adjusted to a level that removes the pain, and keeps it away. Each dose is taken before the pain returns. It has been found that cancer sufferers can be maintained in a pain-free state for months, with no tendency to develop tolerance or dependence.

However, methods of pain control that are suitable for terminally-ill people may not be appropriate for use with younger people who have greater life expectancy and in whom some physical incapacity may be less acceptable. In the past, surgical intervention was perhaps regarded as a last resort, to be used only when other methods had proved ineffective. Since the advent of gate control theory, however, new lines of treatment, which are less destructive, have been developed. The technique of TENS has proved effective for relieving both acute and chronic pain, and is being used with increasing frequency. The electrodes can be set up quickly, and the method can be used for acute (short-term) care. It is also suitable for self-administration by sufferers, who can become more self-sufficient as a result. Implanted electrodes offer another alternative to surgical intervention, and with suitable screening can be used for people who show a good response to electrical brain stimulation.

But perhaps the greatest advance has been the creation of specialist pain clinics, in which chronic pain is treated as a condition in its own right, rather than as an annoying symptom of some diseases. This involves a multidisciplinary approach where experts in different specialities, such as neurologists, neurosurgeons, anaesthetists and psychiatrists, can collaborate to establish the optimum treatment programme for each individual.

Summary of Chapter 5

Although pain is an unpleasant experience, acute pains have undoubted survival value, as people who are unable to feel pain tend not to survive long into adulthood. The simple idea of a specific 'pathway for pain' is untenable, because the amount of pain experienced can vary widely depending on a variety of factors, and this cannot readily be explained in terms of traditional pain theories. Also, noxious stimuli do not always produce pain and surgical operations on nerve pathways do not always abolish pain. Often, pain can arise for no obvious, or at least clinically identifiable, reason.

Reactions to noxious stimuli can be modified during an animal's development and through its experiences. These factors may explain some of the observed differences in the way different individuals may respond to pain. Pain may affect personality and vice versa. Acute pain is associated with increasing anxiety, whereas depression is a more prominent feature of chronic pain.

It is difficult to measure pain, especially pains of a chronic nature. Pain can be expressed in terms of the stimuli that produce it (thresholds), or by the involuntary reactions it provokes. However, these do not accurately mimic naturally occurring pains. Subjective measures, such as the visual analogue scale and McGill pain questionnaire, have proved extremely effective ways of assessing pain and treatments for pain.

Gate control theory provides a plausible physiological explanation for ways in which signals arising from stimulation of nociceptors can be suppressed at the first synapse in the spinal cord. The 'gate' may be 'closed' by activity in large-diameter (mechanosensitive) fibres or by activation of descending pathways from areas in the brain stem, such as the periaqueductal grey matter (PAGM) or the raphe nuclei. This gating mechanism seems to involve activity in small enkephalin-releasing interneurons in the substantia gelatinosa of the dorsal horn. The identification of opiate receptors in the brain stem and dorsal horn and the discovery of endogenous opioid peptides has provided a rational basis for the analgesic action of narcotic drugs such as morphine. Although the precise mechanisms are uncertain, there is evidence for the existence of an endogenous analgesia system, which is capable of 'shutting off' pain perception in times of extreme stress.

This increased understanding of the neural processes involved in pain perception has been applied clinically to great effect in pain treatment. Rather than relying upon narcotic drugs or surgical procedures, clinicians can now harness methods such as Transcutaneous Electrical Nerve Stimulation (TENS), in which mild electric currents are applied to peripheral nerves to produce effective pain relief. The advent of specialist pain clinics has led to a more integrated approach in the use of a combination of surgical, pharmacological and psychological approaches to pain therapy. There is also considerable interest in the use of 'alternative medicine' for the treatment of pain, and this is likely to be an area in which considerable advances can be expected in the next decade.

Objectives for Chapter 5

Now that you have completed this chapter, you should be able to:

5.1 Define and use, or recognize definitions and applications of each of the terms printed in **bold** in the text.

5.2 Explain the receptor mechanisms involved in the detection and coding of sensory events occurring in the skin. (*Question 5.1*)

5.3 Distinguish between the types of nerve fibres involved in touch and pain. (*Question 5.1*)

5.4 Describe the main differences in the organization of neural pathways involved in the transmission to the brain of specific or non-specific sensory information. (*Question 5.2*)

5.5 Give examples to show how the structure of the CNS relates to an animal's behaviour. (*Question 5.3*)

5.6 Describe how information from one sensory modality (e.g. touch) can affect perception of other stimuli. (*Question 5.4*)

5.7 Identify and distinguish between the various components of pain and the different types of pain. (*Question 5.5*)

5.8 Explain the possible survival values of pain. (*Question 5.6*)

5.9 Describe approaches to the measurement of pain, indicating their limitations and potential value. (*Question 5.7*)

5.10 Indicate how developmental and psychological factors may affect pain perception and reactions to it. (*Question 5.8*)

5.11 Explain how activity in nociceptive pathways can be modified by a 'gate' in the dorsal horn of the spinal cord. (*Question 5.9*)

5.12 Describe how an endogenous analgesic system in the CNS might operate, indicating the role of opioid peptides and opiate receptors. (*Question 5.10*)

5.13 Explain the principles underlying the control of pain by pharmacological, surgical and psychological methods. (*Question 5.11*)

5.14 Identify factors that might influence a decision about possible choices of treatment for pain. (*Question 5.12*)

Questions for Chapter 5

Question 5.1 (*Objectives 5.2 and 5.3*)

Which of the elements (a)–(i) are necessary for the detection and transmission of information about the duration of a light mechanical stimulus applied to a region of skin for at least 10 seconds?

(a) Mechano-nociceptors

(b) Rapidly adapting mechanoreceptors

(c) Slowly adapting mechanoreceptors

(d) Receptor potential

(e) Group Aβ nerve fibres

(f) Group Aδ nerve fibres

(g) Ventral root nerve

(h) Dorsal root nerve

(i) Spinoreticular tracts

Question 5.2 (*Objective 5.4*)

The Brown–Séquard syndrome results from a specific spinal cord injury in which all the tracts on one side of the cord are cut through, leaving those on the other side intact. Such a lesion in the thoracic part of the cord will affect sensations from the legs, although the effects will be different for the right and left legs. From your knowledge of spinal cord pathways, predict the effects that might result from a lesion that cuts through the *right* side of the spinal cord in the thoracic region.

Question 5.3 (*Objective 5.5*)

Figure 5.15 showed diagrams of the somatosensory cortices in some animals, each of which uses touch in a different way. Which one(s) does the ‘homunculus’ most resemble?

Question 5.4 (*Objective 5.6*)

A parent is observed to hold a child gently by the wrist while chiding it for behaving in a dangerous manner. How might this affect the child’s ability to understand what is being said?

Question 5.5 (*Objective 5.7*)

A person who sprained his ankle two days ago now describes the pain as ‘chronic’.

(a) Is ‘chronic’ an appropriate word to use to describe this pain?

(b) Classify this pain as: ‘superficial’, ‘deep’ or ‘visceral’.

Question 5.6 (*Objective 5.8*)

Explain whether there is any possible survival value for each of the following:

(a) intestinal pain after eating an unknown type of berry;

(b) the pain associated with a broken arm;

(c) pain due to advanced stages of cancer.

Question 5.7 (*Objective 5.9*)
Refer back to Figure 5.26 (p. 180), and answer these questions:

(a) There is no record of the pain score at 1400 hours. From the other data, estimate what you think it might have been, had it been recorded.

(b) For how long is this analgesic drug effective?

Question 5.8 (*Objective 5.10*)
The prevalence of pain experienced by soldiers wounded in battle is less than for civilians who sustain injuries of comparable severity and extent. Can you account for this in terms of psychological factors?

Question 5.9 (*Objective 5.11*)
Why might rubbing an injured part of the body relieve pain?

Question 5.10 (*Objective 5.12*)
Evidence was presented for the existence of an endogenous analgesic system by which the brain can exert descending inhibition on transmission of nociceptive information at synapses in the spinal cord. Which of the elements in this system listed below contain opioid peptides or opiate receptors?

(a) Periaqueductal grey matter (PAGM)

(b) Raphe nucleus (RN)

(c) Dorsolateral funiculus (DLF)

(d) Substantia gelatinosa (SG)

(e) C fibres

Question 5.11 (*Objective 5.13*)
Many different drugs are used in the management of pain. For each of those listed (a–d), select *one* of the items (i–v) that provides the most appropriate description of its mechanism or site of action.

(a)	Aspirin	(i)	blocks impulse propagation in nerve fibres
(b)	Morphine	(ii)	paralyses muscles
(c)	Lignocaine	(iii)	blocks sensitizing effects of prostaglandins
(d)	Carbamazepine	(iv)	affects synaptic transmission in the CNS
		(v)	acts on opiate receptors

Question 5.12 (*Objective 5.14*)
Two people are referred to a pain clinic. One of them, a 35-year-old lawyer who lost a leg in a road traffic accident, suffers phantom limb pain. The other is a 70-year-old man suffering pain due to cancer which has spread throughout his body and is deemed inoperable. How might the approaches to pain treatment differ in these two cases?

References

Bond, M. R. and Pearson, I. B. (1969) Psychological aspects of pain in women with advanced carcinoma of the cervix, *Journal of Psychosomatic Research*, **13,** pp. 13–19.

Mayer, D. J., Wolfle, T. L., Akil, H., Carder, B. and Liebeskind, J. C. (1971) Analgesia from electrical stimulation in the brain stem of the rat, *Science*, **174,** pp. 1351–1354.

Melzack, R. and Wall, P.D. (1965) Pain mechanisms: a new theory, *Science*, **150,** pp. 971–979.

Further reading

Melzack, R. and Wall, P. D. (1982) *The Challenge of Pain,* Penguin. (Reprinted 1996)

Wall, P. D. and Melzack, R. (1994) *Textbook of Pain*, 3rd edn, Churchill Livingstone.

EPILOGUE

In this book, you have read a lot of detailed material about the senses, especially of humans. The purpose of this brief epilogue is to extract from the detail a number of general themes that have run through the book.

The principal aim of the book has been to integrate neurobiology and behaviour into an account of sensory systems. Neurophysiology has featured prominently in all the chapters in this book, showing how important a good knowledge of the properties of receptors and of neurons is to understanding sensory systems and communication. Neuroanatomy has also been prominent in all chapters, because it is fundamental to understanding how sensory information is passed via the nervous system to the brain. In addition, psychology has provided critical parts of the overall story, particularly in relation to those aspects of hearing, vision and pain where the focus of attention switches from the *sensation* provided by stimuli in the various sense organs to the *perception* of those stimuli in the central nervous system. Apart from Chapter 2, the book has said rather little about behaviour, but you should appreciate the importance to animals of information about their external and internal environment that is provided by their senses. Much of the behaviour of animals is guided by their senses; moreover, animals are not passive receivers of sensory stimuli but, in many contexts, actively seek sensory information. The sense organs of animals are very selective; they are, to varying degrees, responsive to a narrow range of stimuli. Sense organs are an animal's 'windows onto the world', providing only a partial 'view' of their environment. In this book, you have read how the ears of frogs are tuned to very specific frequencies of sound, how the human ear is more responsive to certain sound frequencies than others. how the human eye is organized so as to be particularly sensitive to certain kinds of visual stimuli such as edges, and how human sensitivity to touch and pain varies from one part of the body to another.

A major factor in this selectivity, seen in all kinds of senses, is the property of those cells that detect sensory stimuli, the receptors. This book has shown how the basic properties of nerve cells are modified in receptor cells such that the receptor cells are especially responsive to specific kinds of stimulation. A single mechanoreceptor in a frog's ear, for example, may be tuned to a very specific frequency of sound. This specificity of response to particular kinds of input is seen in many other neurons in sensory systems; a cell in the visual cortex may respond only to edges moving in one specific direction. In each of the senses considered in this book, any individual receptor gathers a very limited amount of information. You know as much as you do about your environment because human sense organs contain very large numbers of receptors, each with slightly different properties. Despite the large number of sensory receptors in human sense organs, there is still a great deal going on in the environment to which humans are totally insensitive, such as patterns of ultraviolet light. Other species, however, have evolved sense organs that do detect sensory stimuli beyond the human range and so live in very different 'sensory worlds' to ourselves. Dogs and cats may well be very puzzled by the failure of people to respond appropriately to their olfactory signals.

All receptors, whether they are rods in the eye, hair cells in the ear or nociceptors in a limb have to provide the nervous system with information such that four basic attributes of any stimulus can be determined by the nervous system. These can be summarized as: what is it? (the quality of a stimulus), how large or strong is it? (its intensity), how long does is it last? (its duration), and where is it? (its location). From the information provided in this book, you should be able to compare and contrast the way that different kinds of sensory system provide answers to these four questions.

The information gathered by the various sense organs must be passed to, and processed by, the brain. This book has described this process in some detail. In particular, it has emphasized how the topographic arrangement of sensory information—the spatial relationship between its components—is preserved in sensory pathways, such that the brain contains a representation of the spatial arrangement of the outside world. This is shown especially clearly in the context of vision, in which there is mapping of information from each retina to the visual cortex. In addition, information from the two eyes about equivalent points in space must be compared to give stereoscopic vision. It is important too, in your perception of your body; there is little value in being in pain if you do not know the location of the pain.

In Chapter 1, a distinction was made between sensation and perception. A simple visual illusion, of a spotted dog, was used to illustrate the essential difference between these two processes. Much of Chapter 4 was concerned with how the nervous system processes the information gathered at the retina, to provide people and other animals with a detailed perception of their world. In Chapter 3, you read something of how, both in the ear and in the brain, sounds are processed so that humans recognize distinct speech sounds, called phonemes. Chapter 5 discussed a number of factors, such as mood, personality and experience, that influence the perception of pain. Sensory information may be filtered, or processed, both at the level of the sense organ and, more centrally, in the nervous system. In the case of frogs, recognition of conspecific calls is achieved wholly by peripheral filtering, that is by the selective responsiveness of a frog's ear. For humans, in contrast, recognition of familiar stimuli such as a face or the sound of one's name involves central filtering and brings into play other processes, particularly memory.

Finally, it is important to point out that the coverage of the senses provided by this book is not complete. There are a number of senses, possessed by non-human animals, that have not been mentioned. Notable among these is the ability of some birds, such as pigeons, to use variations in the Earth's magnetic field to navigate accurately over very long distances. There are also aspects of human senses and communication that have not been discussed. For example, there is abundant evidence that people, like most animals, are sensitive to the daily alternation of light and dark and that this has a profound effect on many aspects of behaviour and physiology. The book has devoted considerable space to human speech but has said very little about non-verbal communication among people. The responses of one person to another are largely determined by subtle visual cues, notably postures and facial expressions, that indicate whether someone is friendly or hostile, relaxed or tense, etc. Recall from Chapter 5 how a simple touch to the hand can alter people's perception of a librarian. There is also some evidence that humans communicate chemically, by means of pheromones, a form of communication of which they may not be consciously aware. Pheromones may be involved, for example, in the synchronization of menstrual cycles observed among women who live in close proximity to one another, and, more controversially, in sexual attraction between individuals. The subject of sensory perception and communication is a huge one and this book has been able to give you only a glimpse of its many ramifications and complexities.

GENERAL FURTHER READING

Dusenbery. D. B. (1995) *Sensory Ecology: How organisms acquire and respond to information*, W. H. Freeman and Co. Ltd. This book looks at a wide diversity of senses, in a wide diversity of animals, and considers the ways in which the sensory systems of animals have been shaped, by natural selection, to equip animals to survive in their natural habitats.

Goldstein, E. B. (1989) *Sensation and Perception,* 3rd edn, Brooks Cole Publishing Co., California. This book is mostly concerned with vision, but also contains material on other senses.

Halliday, T. R. and Slater, P. J. B. (eds) (1983) *Animal Behaviour,* Vol. 2, *Communication,* Blackwell Scientific Publications, Oxford. Although primarily concerned with behavioural aspects of communication, this book includes a lot of material on the physiology of a variety of senses.

McFarland, D. (ed.) (1982) *The Oxford Companion to Animal Behaviour,* Oxford University Press, Oxford. This book contains useful, very concise entries on several topics covered in this book, including: chemical senses, colour vision, communication, echolocation, hearing, language, pain perception and vision.

ANSWERS TO QUESTIONS

Chapter 1

Question 1.1

The simple, incomplete answer is 50 dB, i.e: the threshold at its best frequency. Strictly speaking, however, its threshold varies with the frequency of sound to which it is exposed. Thus, for example, the intensity threshold is about 70 dB at both 400 and 1 000 Hz.

Question 1.2

(a) Single receptors typically have a response threshold, so they will only respond when the intensity of a stimulus equals or exceeds that threshold.

(b) In an array, receptors typically have different intensity thresholds so the more intense the stimulus, the more receptors in the array will respond.

Chapter 2

Question 2.1

(a) Their ears contain receptors, each of which is tuned to a specific sound frequency.

(b) There are many receptors tuned to frequencies that are present in the calls of conspecific males, but few or none tuned to frequencies in the calls of males of other species.

Question 2.2

Because, first, the amphibian papilla contains neurons linked to receptors that are tuned to low frequencies, whereas the basilar papilla contains neurons linked to receptors that are tuned to high frequencies and, second, low frequencies are less attenuated with distance than high frequencies.

Question 2.3

(a) Individual males will call relatively rarely, and will tend to produce calls consisting only of 'whines'. (b) Individual males will call frequently and will produce calls containing several 'chucks'.

Question 2.4

Because of the Doppler effect, an insect that is approaching the bat will produce an echo with a higher frequency than one that is flying away from it.

Chapter 3

Question 3.1

No. The human ear has a frequency range from about 20 Hz to more than 16 kHz but within this range the greatest sensitivity is between 1.5 and 3 kHz.

Question 3.2

Sound that reaches the tympanum is converted into pressure waves in the fluid of the cochlea. The hair cells are positioned between two membranes and the differential movement of the membranes in response to the pressure changes in the fluid produce a force on the hair cells that depolarizes them.

Question 3.3

Some regions of the auditory cortex are mapped in such a way that cells that receive frequency information are arranged in lines of increasing frequency across the surface of the cortex. This is called tonotopic organization and it is found in a number of auditory areas of the cortex.

Question 3.4

Formant frequencies are the spectral peaks that arise from resonances in the vocal tracts. There is no direct relationship between a *specific* vocal tract cavity and a *specific* formant.

Question 3.5

(a) [sh] is produced without the vocal cords, so it is voiceless.

(b) [z] is produced by vibrating the vocal cords, so it is a voiced sound. If you place your fingers on your 'Adam's apple' while speaking, you will feel the vibrations.

(c) [s] is produced by a similar configuration of the vocal tract to [z], but the vocal cords are not involved, so it is voiceless.

Question 3.6

A phoneme is a discrete segment of speech that very roughly corresponds with the letters used in writing. It is the basic unit of speech, though it is important to recognize that it is a useful description that is generally applicable but it does not imply that all speech and recognition processes are based on the phoneme as the fundamental unit. A phoneme is defined as the minimal sound change that will alter meaning. Thus 'cat' contains three phonemes since you could form 'bat', 'bit' or 'car' by single changes of the first, second or third letter. Although in this example each phoneme is a single letter, this is not always so.

Chapter 4

Question 4.1

The lens brings the images of objects at different distances into focus upon the retina. In human beings and other land-dwellers, most of the refraction takes place at the cornea, where light passes from air to (in effect) water. An eye without a lens would thus work in principle, but it would only be able to focus on objects at one specific distance. The addition of the lens adds greater flexibility by allowing the eye to focus over a range of distances, though (unless the light is very bright) only one narrow range of distances is in focus at any one time.

Question 4.2

Rhodopsin is bleached by light. This bleaching forms the first step in the chain of transduction, leading eventually to hyperpolarization of the retinal receptor. Bleached rhodopsin is regenerated automatically so that the cycle can be repeated.

Question 4.3

Luminance discontinuities in the retinal image will frequently signify object boundaries in the world. However, the relationship is not simple. Some object boundaries will not produce luminance discontinuities (e.g.. the hidden edges of a cube) and not all luminance discontinuities are caused by object boundaries (e.g. changes in surface pigmentation, shadows).

Question 4.4

Mach bands provide indirect evidence that the responses of retinal ganglion cells are important in the perception of lightness. The perceived light and dark stripes on either side of a luminance boundary are very similar to the spatial pattern of response across retinal ganglion cells which is produced by the boundary. However, this pattern of response must be interpreted by higher processes because not all aspects of lightness perception can be explained by retinal processes (e.g. the Mach card).

Question 4.5

Natural illumination varies over a huge range, from dark nights to brilliant sunlight. A system which can function effectively over this range, while still being sensitive to relatively small changes in luminance, is just not physically possible. Dark adaptation provides a solution to this problem by allowing the visual system to be adequately sensitive over a relatively narrow range of illuminations, but allowing the range to change according to the prevailing conditions.

Question 4.6

When cells in the LGN are directly stimulated electrically, complex cells in the cortex tend to respond before simple cells. If simple cells provided the input to complex cells, then complex cells should tend to respond after simple cells.

Question 4.7

Yes, but you should not think of a separate hierarchy leading to a distinct 'detector' for each different face. Probably, faces are encoded by the pattern of response across a limited set of neurons, each specific for a particular facial feature (e.g.

length of nose, separation of eyes). This set of neurons presumably receives its input from cells sensitive to more general features (e.g. lines, simple shapes) and, in that sense, the processing is hierarchical because it builds up from general to more specific features.

Question 4.8

In general, any simple physical similarity will tend to produce grouping. So, for example, elements will tend to be grouped together if they are close together, have the same colour, are at the same orientation, are the same shape, or are moving in the same direction. The underlying principle appears to be that, if elements share a common physical attribute, then they may well have the same physical cause and thus belong together.

Question 4.9

Computational models provide a rigorous test of hypotheses: they show whether or not a particular hypothesis can really account for neurophysiological or psychological findings. Computational models also provide a way to put small theories together to see whether the overall system really does function properly.

Chapter 5

Question 5.1

The correct items are (c), (d), (e), (f) and (h).

The others are incorrect because:

(a) Mechanonociceptors have high thresholds and will not respond to a light mechanical stimulus.

(b) Rapidly adapting mechanoreceptors will respond to the stimulus, but the action potential discharge will cease (adaptation) while the stimulus is being applied. Therefore these will be unable to signal the duration of a maintained stimulus.

(g) The axons in ventral spinal nerve roots are efferent fibres from motor neurons in the ventral horn of the spinal cord, which do not form part of the sensory pathway.

(i) Spinoreticular tracts form a non-specific sensory pathway, where there is considerable convergence of inputs from different modalities. There may be some input from mechanoreceptors, but this information will be mixed up with inputs from other receptors, with loss of resolution. Action potentials initiated in slowly adapting mechanoreceptors will be conveyed mainly along axons that ascend to the brain in the dorsal columns.

Question 5.2

A key point here is where the tracts cross the midline of the spinal cord. The lesion will cut through all tracts on the right side of the spinal cord. The right dorsal columns receive inputs from the right leg, where there will be loss of fine touch, vibration sense and proprioception, but pain and temperature sense in the right leg will be unaffected. The right spinothalamic tracts contain the axons of neurons in the left dorsal horn, and so there will be loss of pain and temperature sense in the left leg. Touch sense in the left leg will be normal. Award yourself a bonus if you also mentioned that there will be paralysis of the muscles in the right leg, due to damage to motor nerves on the right of the spinal cord

Question 5.3

In the human sensory cortex, the largest areas are devoted to the hands and lips; humans are thus hand and lip 'feelers'. None of the other animals show a similar pattern, but these areas are comparable to (but not as extensive as) the forepaw areas in the raccoon and mouth regions in the sheep.

Question 5.4

Holding reinforces the social bond and helps to maintain it despite the punishing effect of the scolding. The touch may also increase the level of β-endorphin and thus reduce the level of anxiety (stress) allowing the child to concentrate on what it is that is being said.

Question 5.5

(a) No. This pain is an acute one. The term 'chronic' is normally applied only when the pain has lasted for six months or more.

(b) An ankle sprain will damage deep structures such the ligaments of the joint. 'Superficial' applies to the skin , while 'visceral' relates to internal organs, such as intestines, kidneys, etc.

Question 5.6

(a) This has definite survival value and is comparable to the effect evident in the 'bright, noisy water' experiment mentioned in Book 1. It is possible that this type of berry is poisonous and, as a result of this painful experience, the individual will be unlikely to eat this type of berry again.

(b) Here the pain will force the individual to rest the injured arm, thereby aiding the recovery process.

(c) The pain will indicate that something is wrong, but there is no obvious survival value, as the cause cannot be removed and it is difficult, if not impossible, for the sufferer to alter their behaviour to lessen the pain. This type of pain may be associated with depression.

Question 5.7

(a) It is always risky extrapolating data, but the pain score at 1400 hours is unlikely to be very different from the scores at 1000 hours and 1800 hours when the analgesic is administered. Thus the pain score at 1400 hours would probably be around 8 (7–9).

(b) One hour after the analgesic has been given, the pain score has reduced to a low value (the score is around 2). The pain score remains low for the next hour, but begins to increase during the hour following that. After 4 hours, the pain score has reached a high value. The analgesic is thus effective for 2–3 hours.

Question 5.8

Although the extent of the injuries sustained by the soldiers and civilians may be comparable, the context of the situation is different. The wounded soldiers have been removed from the combat zone to safety, and may even be sent home. However, the consequences of injury are much less favourable for the civilian, who will perhaps be unable to work, and will tend to worry about loss of income. This increased anxiety may affect the level of pain. The lower prevalence of pain among the soldiers could well be due to the perceived reinforcement 'value' of the injury.

Question 5.9

Rubbing will stimulate mechanoreceptors. The action potentials in the large diameter (Aβ) fibres will have the effect of 'closing the gate' in the dorsal horn, thus blocking normal excitation of the T cells by activity in the small diameter nociceptive nerves.

Question 5.10

(a) and (d) both contain high levels of opiate receptors and various opioid peptides.

(b) contains some opiate receptors.

(c) contains neither opiate receptors nor opioid peptides.

(e) The terminals of many C fibres possess opiate receptors.

Question 5.11

(a) (iii).

(b) (v) is best, but (iv) is also a possibility.

(c) (i).

(d) (iv). Carbamezepine blocks presynaptic inhibition and is effective in treating trigeminal neuralgia.

Question 5.12

In each case, the aim is to abolish the pain, or at least to reduce it to acceptable levels, while minimizing any side effects. In each case, counselling and supportive therapy will be necessary. The lawyer has a normal life expectancy, and treatment should interfere as little as possible with a professional career, which could continue for another 30 years or so. Conservative methods are preferred, such as local treatment of the stump, and TENS. Permanently implanted electrodes might be considered. However, the use of narcotic analgesics or neurosurgery are probably unsuitable for this case. Conservative treatment may be attempted also for the cancer patient, but this is likely to be inadequate. With the shorter life expectancy and greater degree of debility, narcotics may be necessary. Neurosurgery is another possible treatment in this case.

GLOSSARY

accommodation Changes in the shape of the lens of the eye that enable the image of an object to remain in focus on the retina despite changes in the object's distance. (Section 4.2)

advertisement call A call, produced by males, particularly in toads and frogs, that attracts females and, in many species, elicits aggressive responses by other males. (Section 2.2)

articulators The larger moving parts of the vocal tract—lips, tongue and jaw. (Section 3.4)

background firing rate The spontaneous rate of generation of action potentials in a neuron, in the absence of stimulation, e.g. the activity observed in retinal ganglion cells when their receptive fields are uniformly illuminated. (Section 4.3.1)

blobs Groups of cells in the visual cortex that take up the stain cytochrome oxidase. (Section 4.7.3)

brightness The perceptual correlate of the degree of physical illumination of a scene. (Section 4.3.3)

co-articulation The smooth transition between successive phonemes is the result of co-articulation. The movement of the major articulators that produces a particular phoneme is modified by the preceding and following phonemes. (Section 3.4.4)

cochlea A cone-shaped coiled tube forming part of the inner ear. It contains the basilar membrane and the hair cells. (Section 3.2.1)

cold pressor test A test of pain sensitivity in which a subject's hand is immersed in ice-cold water for as long possible. The duration of exposure can be used to assess pain tolerance. (Section 5.5.5)

colour constancy Capacity of the visual system to perceive colours consistently in spite of different lighting conditions. (Section 4.4.4)

colour opponency Property of cells (e.g. in the retina or lateral geniculate nucleus (LGN)) such that they are excited when their receptive fields are illuminated by light of one wavelength and are inhibited when their receptive fields are illuminated by light of another wavelength. (Section 4.4.2)

complementary colours Those colours (wavelengths) of light which, when added in equal proportions, produce white (uncoloured) light (e.g. red and green, blue and yellow). (Section 4.4.2)

complex cell Type of cell in the visual cortex with a large receptive field that is not divided into discrete antagonistic sub-regions. (Section 4.3.4)

dark adaptation Process by which the visual system progressively alters its sensitivity to light according to the prevailing lighting conditions by means of regeneration of 'bleached' pigments in receptors. (Section 4.3.3)

dark current Neural activity in retinal receptors that occurs in the dark. It is produced by sodium entering the receptors to maintain them in a depolarized state. (Section 4.3.1)

delayed inhibition Mechanism (e.g. in retinal motion detectors) whereby inhibitory input from a cell is delayed. (Section 4.5.1)

depth cue A specific feature of the two-dimensional image of an object that enables the visual system to interpret the three-dimensional structure of that object. (Section 4.8.2)

dominant frequency That frequency in a sound that contains the greatest proportion of the energy in that sound. (Section 2.2.1)

double opponent cell A type of cell in the visual system (e.g. in the lateral geniculate nucleus (LGN) or visual cortex) with a receptive field divided into two discrete sub-regions. Stimulation of one sub-region of the receptive field with light of one wavelength produces an excitatory response whereas stimulation with another wavelength produces an inhibitory response.The other sub-region has complementary properties, with the second wavelength producing an excitatory response and the first wavelength producing an inhibitory response. (Section 4.4.3)

ear ossicles The three bones, the malleus, incus and stapes, that couple the tympanum to the cochlea. (Section 3.2.4)

end-stopped cell Alternative name for hypercomplex cell.(Section 4.3.4)

feature detection A theory which proposes that the first stage of visual perception involves the detection of such image features as lines and edges. In an extreme form, individual simple cells in the visual cortex are thought to act as feature detectors. (Section 4.3.4)

flow pattern Smooth change in retinal images that is a consequence of movements by an observer. (Section 4.5)

formant A natural resonance in the vocal tract. (Section 3.4.1)

fricative A sound produced when air is forced through a constriction in the vocal tract without any vibration of the vocal cords. Examples of fricative consonants are [f] and [s]. (Section 3.4.1)

gate control theory A mechanism proposed by Melzack and Wall to account for variations in pain experience. The basic idea is that action potentials in the neurons of small afferent nerves are subjected to filtering at central nervous system (CNS) synapses by inputs in other afferent neurons and by pathways descending from the brain. (Section 5.7)

grouping process Synthetic process in visual perception by which large-scale descriptions of images are built up by combining several smaller-scale descriptions. (Section 4.8)

hair cells Auditory receptor cells in the cochlea with synapse connections to neurons in the cochlear nerve and hairs linked to the tectorial membrane. (Section 3.2.5)

hierarchical wiring A theory of visual perception in which simple cells in the visual cortex send inputs to complex cells, which send inputs to hypercomplex cells. (Section 4.3.4)

hyperalgesia Increased sensitivity of injured or inflamed tissues such that pain may be produced by mild stimulation. (Section 5.2.2)

hypercolumn A set of columns of cells in the visual cortex, all of which have receptive fields in the same position on the retina but which have different selectivities for stimulus orientation, motion direction, size, and so forth. (Section 4.3.4)

hypercomplex cell (end-stopped cell) A type of cell in the visual cortex with a large receptive field that is not divided into discrete antagonistic sub-regions and which responds best to appropriately-oriented luminance discontinuities which end within the cell's receptive field. (Section 4.3.4)

illumination Measure of the amount of light falling upon a surface. (Section 4.3.3)

integrative process Synthetic process in visual perception by which descriptions of images are built up by combining information from different sources (e.g. different image positions, or different visual sub-modalities). (Section 4.8)

interblobs Unstained areas of the visual cortex between groups of cells (blobs) which take up the stain cytochrome oxidase. (Section 4.7.3)

lateral suppression The term used instead of 'lateral inhibition' (Book 2, Section 9.5.3) to refer to auditory processing; the process by which a sensory receptor inhibits the response of neighbouring receptors. (Section 3.5.3)

lightness constancy The capacity of the visual system to interpret surface reflectance consistently despite differences in lighting conditions. (Section 4.3.3)

lightness contrast A process in the visual system whereby differences in the reflectance of adjacent surfaces are perceptually accentuated. (Section 4.3.3)

lightness The perceptual correlate of surface reflectance. (Section 4.3.3)

luminance A measure of the amount of light projected by a surface. (Section 4.3.2)

luminance discontinuity A spatially-abrupt change in luminance. (Section 4.3.2)

luminance profile The spatial pattern of luminance along a straight line across a surface. (Section 4.3.2)

McGill Pain Questionnaire (MPQ) A table devised at McGill University in Canada, containing sets of words describing various aspects and levels of pain, and used to quantify pain severity. Pain sufferers select words that best describe the quality and severity of their present pain, and from these responses, an overall assessment can be made of the current pain state. (Section 5.5.4)

monochromatic light Light consisting of a single wavelength, which is therefore of a particular colour. (Section 4.4.1)

motion after-effect Perceptual illusion that a stationary object is moving, as a consequence of prolonged exposure to a stimulus moving in the opposite direction. (Section 4.5.1)

neural image Pattern of neural activity over a number of retinal receptors that corresponds to the pattern of light in an image. (Section 4.3.1)

noise Random energy that tends to obscure or to interfere with signals. (Section 1.2.3)

off-centre cell A retinal ganglion cell with a receptive field consisting of a central off-region surrounded by an annular on-region. (Section 4.3.2)

off-region Part of the receptive field of a cell in the visual pathway (e.g. a retinal ganglion cell) in which an increase in light causes a decrease in the cell's response and a decrease in light causes an increase in the cell's response. (Section 4.3.2)

on-centre cell A retinal ganglion cell with a receptive field consisting of a central on-region surrounded by an annular off-region. (Section 4.3.2)

on-region Part of the receptive field of a cell in the visual pathway (e.g. a retinal ganglion cell) in which an increase in light causes an increase in the cell's response and a decrease in light causes a decrease in the cell's response. (Section 4.3.2)

opiate A term originally used to describe drugs extracted from opium, including morphine and chemical derivatives of morphine. (Section 5.8.3)

opiate receptors Cell membrane proteins that bind selectively to opiates or opioids (Section 5.8.3)

opioid A term used to describe all drugs, either natural or synthetic, with morphine-like actions. (Section 5.8.3)

orientation column A column of cells in the visual cortex, all with receptive fields of the same orientation. (Section 4.3.4)

pain perception threshold The lowest stimulus intensity required for a given stimulus to produce a sensation that is distinctly, but just, painful. (Section 5.5.2)

pain tolerance The limit of pain endurance, usually expressed in terms of the maximum stimulus intensity which a subject can withstand. (Section 5.5.2)

pale stripes Strips of cells in the visual cortex that do not take up the stain cytochrome oxidase. (Section 4.7.3)

phoneme The smallest functional unit recognizable in speech. (Section 3.4.2)

referred pain A phenomenon in which painful sensations are felt as if they originated from an area of the body surface, but which in reality result from disease or injury of an internal organ. (Section 5.2.5)

reflectance The proportion or percentage of the light falling upon a surface which is reflected by that surface. (Section 4.3.3)

retinotopic map A region of the visual system (e.g. in the visual cortex) in which the spatial arrangement of the receptive fields of visual cells preserves the gross spatial arrangement of the retina. (Section 4.3.4)

sensitization A state in which there is a lowered threshold and increased responsiveness to stimulation; in the context of this book, an increased responsiveness of sensory receptors, particularly nociceptors, to stimulation. Sensitization in the context of learning was discussed in Book 1, Section 6.3.1. (Section 5.2.2)

sensory adaptation A property of sensory receptors which results in a decrease in action potential firing frequency in an afferent neuron when a constant stimulus is applied to its receptive field. Two types of adaptation are distinguished: in rapidly adapting receptors, firing ceases soon after the onset of a maintained stimulus; in slowly adapting receptors, firing continues throughout the period of stimulation. (Section 5.2.1)

sensory coding The properties of sense organs that enable them to signal variations in sensory input. (Section 1.2.3)

sensory detection threshold The lowest stimulus intensity at which a sensation of any sort can be reliably detected in response to that stimulus. (Section 5.5.2)

short-term adaptation The decline in response of a receptor as stimulation persists. (Section 3.5.3)

simple cell Type of cell in the visual cortex with an elongated receptive field divided into discrete antagonistic sub-regions. (Section 4.3.4)

stereopsis Process in the visual system whereby the three-dimensional structure of an object is recovered from retinal disparity. (Section 4.6)

stimulation-produced analgesia (SPA) Analgesia produced by electrical stimulation of specific regions of the brain, such as the periaqueductal grey matter and raphe nuclei. (Section 5.8.1)

stop consonants Consonants produced by the sudden release of pressure built up in the vocal tract. Examples of these are [p] and [b]. (Section 3.4.1)

sustained cell Type of cell in the visual pathway (usually a retinal ganglion cell) with a response that persists throughout the presentation of a stimulus. (Section 4.5.1)

tail flick test A test involving the reflex withdrawal of a rat's tail in response to a noxious stimulus. Changes in the latency of the response time can be used to assess the effectiveness of analgesic drugs and procedures. (Section 5.5.5)

thick stripes Wide strips of cells in the visual cortex that take up the stain cytochrome oxidase. (Section 4.7.3)

thin stripes Narrow strips of cells in the visual cortex that take up the stain cytochrome oxidase. (Section 4.7.3)

tonotopic organization Anatomical arrangement of auditory nerve cells in which the cells are sited in order of increasing frequency response along one dimension. (Section 3.3)

tourniquet test A test of pain tolerance in which pain is produced when the arterial blood supply to a limb is cut off by a tight band placed round the limb. (Section 5.5.5)

transient cell Type of cell in the visual pathway (usually a retinal ganglion cell) that responds briefly only at the onset or offset of a stimulus. (Section 4.5.1)

trichromatic vision Colour vision based upon three types of cone, each sensitive to a different band of light wavelengths. (Section 4.4.2)

tympanum A thin membrane of tissue between the outer and middle ear. Also called the eardrum. (Section 3.2.2)

visual analogue scale (VAS) A means of converting the magnitude of a sensation, often pain, to a linear dimension. A VAS is usually a line 10 cm long: one end represents, for example, no pain and the other end the most severe pain imaginable. The pain felt is expressed as being at some point between these extremes. (Section 5.5.4)

vocal tract The peripheral structures that generate speech—vocal cords, pharynx, soft and hard palate, tongue, teeth, jaw, lips and nasal passages. (Section 3.4)

voiced sound A sound produced by vibration of the vocal cords. (Section 3.4)

voiceless sound A sound produced without vocal cord vibration. (Section 3.4)

X cell A type of retinal ganglion cell which sums the responses to spots of light falling within its receptive field in a predictable way. (Section 4.7.1)

Y cell A type of retinal ganglion cell which sums the responses to spots of light falling within its receptive field in an unpredictable way. (Section 4.7.1)

ACKNOWLEDGEMENTS

Grateful acknowledgement is made to the following sources for permission to reproduce material in this book:

FIGURES

Figures 1.2, 2.5, 2.6a: adapted from Capranica, R. R. (1977) in Taylor, D. H. and Guttmann, S. I. (eds), *The Reproductive Biology of Amphibians*, Plenum Publishing Corporation; *Figures 1.3, 1.4:* McFarland, D. (1985) *Animal Behaviour*, Longman Group UK; *Figure 2.4:* Gerhardt H. C. and Rheinlaender J. (1980) *Naturwissenschaften*, **67**, pp. 362–363, Springer–Verlag; *Figure 2.6b:* Capranica, R. R. and Moffat, A. J. M. (1975) *Journal of Comparative Physiological Psychology*, Springer–Verlag, Heidelberg; *Figures 2.9, 2.10:* Littlejohn, M. J. (1977) in Taylor D. H. and Guttmann, S. I. (eds), *The Reproductive Biology of Amphibians*, Plenum Publishing Corporation; *Figure 2.11:* Gerhardt, H. C. (1978) *Science*, **199**, pp. 992–994, © American Association for the Advancement of Science; *Figure 2.12:* Halliday, T. (1980) *Sexual Strategy*, Oxford University Press; *Figure 2.13:* Rand, A. S. and Ryan, M. J. (1981) *Zeitschrift für Tierpsychologie*, **57**, pp. 209–214, Paul Parey, Berlin; *Figure 2.14:* adapted from Møller, A. P. (1987) *Animal Behaviour*, **35**, pp. 1637–1644, Baillière Tindall; *Figure 2.15:* Halliday, T. (1983) in Halliday, T. R. and Slater, P. J. B. (eds), *Animal Behaviour, Vol. 2, Communication*, Blackwell Scientific Publications Ltd; *Figure 3.1:* adapted from Duncan, G. (1990) *Physics in the Life Sciences*, Blackwell Scientific Publications Ltd; *Figure 3.2:* From *Tissues and Organs: A Text-Atlas of Scanning Electron Microscopy*, by Richard G. Kessel and Randy H. Kardon. Copyright © 1979 by W. H. Freeman and Company. Reprinted by permission; *Figures 3.3, 3.4, 3.5a, 3.5b, 3.5c,:* adapted from Jennett, S. (1989) *Human Physiology*, Churchill Livingstone; *Figure 3.5d:* adapted from Berne, R. M. and Levy, M. N. (1990) *Principles of Physiology*, Mosby–Year Book, Inc.; *Figure 3.6:* Pickles, J. O. (1988), *An Introduction to the Physiology of Hearing*, Academic Press, by courtesy of the author; *Figure 3.8:* Pickles, J. O., Osborne, M. P. and Comis, S. D. (1987) *Hearing Research*, **25**, pp. 173–183, Elsevier Science Publishers, by courtesy of J. O. Pickles; *Figure 3.9:* From *The Human Brain Coloring Book* by Marian C. Diamond, Arnold B. Scheibel and Lawrence M. Elson. Copyright © 1985 by Coloring Concepts, Inc. Reprinted by permission of HarperCollins Publishers; *Figure 3.10:* Carpenter, R. H. S. (1990) *Neurophysiology*, 2nd edn, Hodder and Stoughton; *Figure 3.11a:* adapted from Woolsey, C. N. (1960) *Neural Mechanisms of the Auditory and Vestibular Systems*, courtesy of Charles C. Thomas, Publisher, Springfield, Illinois; *Figures 3.12a, 3.12c, 3.12d:* Starr, C. and Taggart, R. (1987) *Biology, The Unity and Diversity of Life*, 4th edn, © 1987, 1984 and 1981 by Wadsworth Inc.; *Figure 3.12b:* From *Principles of Anatomy and Physiology* 6 edited by Gerard Tortora and Nicholas P. Anagnostakos. Copyright © 1990 by Biological Sciences Textbooks, Inc., A & P Textbooks, Inc. and Elia-Sparta, Inc. Reprinted by permission of HarperCollins Publishers; *Figure 3.13:* adapted from Levinson, S. E. and Liberman, M. Y. Copyright © April 1981, by *Scientific American, Inc.* All rights reserved; *Figure 3.15:* Anderson, J. R. (1969) *Journal of Verbal Learning and Verbal Behaviour,* **8**, pp. 240–247, © 1969 Academic Press; *Figure 4.7:* adapted from Boycott, B. and Dowling, B. (1969) *Philosophical Transactions of the Royal Society of Biology*, **225**, pp. 109–184, Royal Society of Biology; *Figure 4.8:* Frisby, J. P. (1980) *Seeing, Illusion, Brain and Mind,* Oxford University Press; *Figure 4.9:* Lentz, T. L. (1971) *Cell Fine Structure,* W. B. Saunders & Co. Inc.; *Figure 4.30:* Hubel, D.H., Wiesel, T. N. and Stryker, M. P. (1978) *Journal of Comparative Neurology*, **177**, pp. 361–380, copyright © 1978 Wiley–Liss, Massachusetts. Reprinted by permission of Wiley–Liss, a division of John Wiley and Sons Inc.; *Figures 5.1, 5.3, 5.6:* adapted from Schmidt, R. F. (1978), *Fundamentals of Sensory Physiology*, Springer-Verlag, Heidelberg; *Figure 5.5:* adapted from Knibestol, M. (1973) *Journal of Physiology*, **232**, pp. 427–452, © The Physiological Society; *Figure 5.7:* Hensel, H. and Kenshalo, D. R. (1969) *Journal of Physiology*, **204**, pp. 99–112, © The Physiological Society; *Figure 5.8:* Perl, E. R. (1985) *Journal of Physiology*, **197**, pp. 593–615, © The Physiological Society; *Figure 5.9:* adapted from Zimmermann, M. (1980) in Anderson, D. J. (ed.) *Physiology—Past, Present and Future*, Pergamon Press; *Figure 5.10:* Lyon, B. (1979) *Journal of Physiology*, **287**, pp. 493–507, © The Physiological Society; *Figure 5.15:* adapted from Walker, W. I. (1973) *Brain, Behaviour and Evolution* **1**, pp. 253–336, S. Karger AG, Basel; *Figure* 5.16: the British Broadcasting Corporation; *Figure 5.17*: Illustrations by Robert Gillmor in Broom, D. M. (1981) Biology of Behaviour, 1st edn, Cambridge University Press © Robert Gillmor 1981; *Figure 5.18*: Morris, D. (1977) Manwatching, Jonathan Cape; *Figure 5.24*: adapted from Torebjork, H. E., LaMotte, R. H. and Robinson, C. J. (1984) *Journal of Neurophysiology,* **51**, pp. 325–339, The American Physiological Society; *Figures 5.25, 5.26:* adapted from Bond, M. R. (1984) *Pain—Its Nature, Analysis and Treatment*, Churchill Livingstone; *Figure 5.27:* Reading, A. E. (1989) in Wall, P. D. and Melzack, R. (eds), *Textbook of Pain*, Churchill Livingstone; *Figure 5.31:* adapted from Carstens, E., Bihl, H., Irvine, D. R. F. and Zimmermann, M. (1981) *Journal of Neurophysiology,* **45**, pp. 1029–1042, The American Physiological Society; *Figure 5.32:* Gebhardt, G. F., Sandkuhler J., Thalhammer J. G. and Zimmermann M. (1984) *Journal of Neurophysiology*, **51**, pp. 75–89, The American Physiological Society; *Figure 5.33:* Fields, H. L. and Basbaum, A. I. (1989) in Wall, P.D. and Melzack R. (eds), *Textbook of Pain,* Churchill Livingstone; *Figure 5.35:* adapted from Cousins, M. and Phillips, G. D. (1986) *Acute Pain Management*, Churchill Livingstone; *Figure 5.37:* Ganong, W. F. (1991) *Medical Physiology*, 15th edn, Appleton and Lange; *Figures 5.38, 5.39:* Hilgard, E. R. (1978) in Sternbach, R. A. (ed.), *The Psychology of Pain*, Raven Press Ltd; *Figure 5.40:* Croze, S., Antonietti, C. and Duclaux, R. (1976) *Brain Research,* **104**, pp. 335–340, Elsevier Science Publishers.

INDEX

Note Entries in **bold** are key terms. Indexed information on pages indicated by *italics* is carried mainly or wholly in a figure or table.

absolute refractory period, 150
accommodation, **81**–2
acoustic communication *see* auditory communication
acoustic cue (speech perception), 70–1
Acris crepitans (cricket frog), auditory neurons, 7–8, 22, *23*
action potential conduction velocity, 157–9
acupuncture, 203–4
adaptation (sensory), 152–3
  short-term, 67, 69
addiction *see* dependence (narcotics)
adequate stimulus, 148
advertisement call (frogs), **18**
aerial perspective, 139
afferent nerve, specificity of, 157–60
after-image, 117, *Pl. 3*
aggressive call (frogs), 18
  see also auditory communication, frogs
allogrooming, 167–8
alternative medicine, 203–4
altruism, reciprocal, 167–8
amacrine cell, *83*, *84*, 122–3
amphibian papilla, 22, 25–6
analgesic drug, 156–7, 196–7, 204–5
  monitoring effects, 179, *180*, 182
'animalculus' *see* brain maps
animals, experiments on, 2–3, 182
anticipatory co-articulation, 64
anticonvulsants, in pain treatment, 198
antidepressants, in pain treatment, 185, 198
anxiety, related to pain, 185
Apis mellifera (honey-bee), communication, 33–4
apparent motion, 123
articulator, **56**
artificial intelligence, 72–3
aspirin, 156–7, 196
auditory canal, external, 44–6
auditory communication
  frogs, 14–15, 17–31
  humans *see* hearing, human; speech, human
auditory cortex, 45, 53–4, *55*
auditory neuron, tuning curve, 6–8, 22, *23*
auditory pathway, 53–4, *55*
auditory system
  frogs, 21–6, 51
  humans, 42–53
  speech processing, 66–7
autoradiography, 107, *108*

background firing rate (retinal ganglion cells), **86**
'badge' of status, 32–3
balance, 120–1
barbiturates, 198
basilar membrane, 48, *49*
basilar papilla, 22, 25–6
bat
  echolocation, 35–7
  predator on tungara frog, 14, 29–31
bee *see* honey-bee
binaural processing, 45
binocular vision, 127–31
bipolar cell, *83*, *84*, 85, 94–6, *98*, 116–17, 122
birds
  plumage variation, 32–3
  song dialect, 34–5
  status signals in, 32–3
bit (of information), 13
blind spot, *81*, 85
blob, **134**–5, *136*
Bombyx mori (silkmoth), olfactory sense, 10
Bond, M. R., 184–5
bradykinin, 156, 196
brain, 212
  analgesia produced by stimulating, 188–91
  auditory pathways, 53–4, *55*
  binaural processing in, 45
  Brodmann's areas, 160
  opiate receptors, 191, 194
  see also auditory cortex; primary visual cortex
brain map, 165–7
brightness, **101**
  coding, 99–100
Brodmann's areas, 160
Brompton mixture, 205

Bufo calamita (natterjack toad), call, 29

cancer, pain related to, 184–5, 205
carbamazepine, 198
cat
  auditory cortex, 53, *55*
  response of retinal ganglion cells, 86
Cavia porcellus (guinea-pig), ear, *49*, 51
Cercopithecus aethiops (vervet monkey), allogrooming, 167–8
channel (of communication), 14, *15*
chemoreceptor, 5, 9–10
chimpanzee, greeting ceremonies, *167*
chiropracty, 203
ciliary muscle, 81
co-articulation, **64**, 65
cochlea, **44**, 48–54
cochlear nerve, 53, *54*
cochlear nuclei, 53–4, *55*
coevolution, 37
cold pressor test, **182**, 201
Colostethus nubicola (frog), response to advertisement call, *21*
colour constancy, **118**–19
colour opponency, **116**–18
colour vision, 8, 113–19, 135
  cortical processing, 117–19
  retinal coding, 114–17
columella, 22
communication, 13–15, 31–7, 212–13
  see also auditory communication
complementary colour, **117**
complex cell (cortex), **105**
computer model, object recognition, 141–2
conditioning, 183, 200
conduction velocity (action potential), 157–9
cone (retinal receptor), *83*, *84*, 100, 114–17, 119
consonant, 59
coqui frog, frequency sensitivity, 23–4
cornea, 81
Couch's spadefoot toad, tuning curve for auditory neuron, 22, *23*
Craik–Cornsweet–O'Brien illusion, 97–9
cranial nerves, *54*
cricket frog, tuning curve for auditory neuron, 7–8, 22, *23*
cytochrome oxidase staining, 134

dark adaptation, **100**
dark current, **84**
decibel, 7, 46
delayed inhibition, **121**–3
dependence (narcotics), 197, 205
depression, related to pain, 185
depth cue, **138**–40
Descartes, René, 173
diazepines, 198
direct perception, 11
direction coding, 123–4
disparity, 127, 130–1
distal object, 11, 77–8, 112–13
dominant frequency (of call), **19**
Doppler effect, 36
dorsal column nucleus, 162
dorsal column (DC) pathway, 162–3
dorsal horn
  neuron, 160–4, 186–8, 190, 194–5
  see also substantia gelatinosa
dorsolateral funiculus (DLF), 190–1, *193*
double blind trial, 198
double opponent cell, **117**–18, *Pl. 4*
drugs
  used in pain treatment, 185, 196–9
  see also analgesic drug
dynorphins, 192, 194

ear
  frog, 22–5
  human, 42, 44–53
eardrum *see* tympanum
ear ossicle, **47**–8
'eavesdropper', 14–15, 29–31
echolocation, bats, 35–7
edible frog, call, *19*
electroacupuncture, 203
electrode implants (pain relief), 202, 205
Eleutherodactylus coqui (coqui frog), frequency sensitivity, 23–4
end-stopped cell (cortex), **105**
endogenous analgesic system, 188–95
endolymph, *49*
endorphin, 168–9, 192
enkephalin, 192, 194
enteroceptor, 9, 161
experiments on animals, 2–3, 182
exteroceptor, 9
extroversion, related to pain perception, 184–5
eye, 80–2

face recognition, 112–13
feature detection, **105**–10

'fight or flight' response, 195
fish, cone receptor, 114
flow pattern (retinal images), **120**–1, 124–6
formant, **58**
fovea, *81*, 100, 115, 133
free nerve ending, 148
  see also nociceptor
frequency coding, 149–50
frequency spectrum, 43
  of call, 19–21, 23
fricative, **59**, 61–2
Frisch, K. von, 33
frog
  auditory communication, 14–15, 17–31
  auditory neuron, tuning curve, 6–8, 22, *23*
  auditory system, 21–6, 51
  see also coqui frog, cricket frog, leopard frog, tungara frog

gate control theory, **186**–7, *188*, 194–5, 202–3
Gerhardt, H. C., 21, 28
Gestalt psychology, 137
Gibson, J. J., 11
gray treefrog, call, *19*, 27–9
great tit, status signals, 32
greater horseshoe bat, echolocation, 36–7
green treefrog, auditory communication, 25–6, 29
grooming, 167–8
grouping process, **136**–8
guinea-pig, ear, *49*, 51

hair cell (auditory), **48**–53
hair-follicle receptor, *148*, 152–3
Harris sparrow, status signals, 32
hearing, human, 41–53
  damage, 46, *52*, 53
  impairment, 67
  range, 42
  see also auditory pathway
helicotrema, *49*
hertz, defined, 7
'hidden observer' effect, 201
hierarchical wiring, **105**, 110–12
Hilgard, E., 201
histamine, 156
honey-bee, dance language, 33–4
horizontal cell (retina), 83–5, 94–6, 122
house sparrow, throat patch as status signal, 32–3
housefly, olfactory receptors, 9
HRP (horseradish peroxidase) labelling (neurons), 161
Hubel, D., 104–6
hugging, 169
Hughes, J., 192
Hyla chrysoscelis (treefrog), call, 27–9
Hyla versicolor (gray treefrog), call, *19*, 27–9
hyperalgesia, **156**
hypercolumn, *106*, **107**, 109, 129, 134
hypercomplex cell (cortex), **105**
hypnotic analgesia, 200–201

illumination, **100**, *101*
image
  description
    binocular vision/stereopsis, 127–31, 135
    grouping process, 136–8
    integrative process, 136, 138–40
    luminance, 82–113
    motion, 120–6, 135
    spectral composition, 113–19, 135
  formation, 79–82
  interpretation, 140–2
  processing, 77
  segmentation, 113, 120
incus, *44*, 47
induced motion, 125
inflammation, 156–7
information, 13
integrative process, **136**, 138–40
interblob, **134**, 135, *136*
interneuron (retina), 85
inter-pulse interval of call, 20
iris, *81*

Johannson, G., 120

Kelvin, Lord, 176
Kosterlitz, H., 192

laminae *see* Rexed's laminae
language, 60
  understanding, 71–3
larynx, 56–8
  observation, 63
lateral geniculate nucleus, *78*, 103–4, 110–11, *128*, 129, 133–6, 165
  double opponent cell, 117–18
lateral inhibition, 67, 87–8
lateral suppression, **67**
Latin names for species, 3
learning, 183

lens (eye), 81
leopard frog, *21*
Lewis, Sir Thomas, 159
LGN *see* lateral geniculate nucleus
light receptor *see* photoreceptor
lightness, **101**
  coding, 99–103
lightness constancy, **101**, 118–19
lightness contrast, **102**
lignocaine, 197
Limnodynastes dumerili (frog), calling site, *26*, *27*
Limnodynastes peroni (frog), calling site, *26*, *27*
Limnodynastes tasmaniensis (frog), calling site, *26*, *27*
Lindauer, M., 34
linear perspective, 138
Litoria ewingi (frog), calling site, *26*, 27
Litoria raniformis (frog), calling site, *26*, *27*
Litoria verreauxi (frog), calling site, *26*, *27*
Livingstone, D., 192–3
local anaesthetics, 196–8
luminance discontinuity, **89**–93, *94*, 97–9, 102–3
luminance of image, 82–113, **89**
luminance profile, **89**

McGill Pain Questionnaire, **180**, *181*
Mach bands, 97–9
Mach card, 103, 140
magnocellular layers (LGN), 133–5, *136*
malleus *44*, 47
Marr, D., 139
mating call *see* advertisement call
Mayer, D., 188–9
Meade, T. W., 203
mechano-nociceptor, 155
mechanoreceptor, 5, 161
  classification, 152
  sensory coding, 148–54
medial geniculate nucleus, 165
Meissner's corpuscle, *148*, 152–3
Melzack, R., 180, 183, 186–7, 204
Merkel's disc, 152–3
message (communication), 14, *15*
microelectrode use in study of action potential transmission in conscious humans, 159
middle temporal cortex, *78*, 112, 125
Miopithecus talapoin (talapoin monkey), endorphins in grooming, 168
Møller, A. P., 32–3
monkey
  middle temporal cortex cell, 112, 125
  V4 cell, 119
  see also talapoin monkey; vervet monkey
monoamine oxidase inhibitor, 198
monochromatic light, **114**
monocular dominance stripe, 129–30
morphine, 191–2, 195–7, 205
moth
  antennae, 10
  predator avoidance, 37
motion after-effect, **124**
Musca domestica (housefly), olfactory receptors, 9

naloxone, 201, 203
narcotic analgesic, 195–7, 204–5
natterjack toad, call, 29
nerve fibre
  classification, 157–60
  projection pathways, *160*, 161–3
neural image, **84**–5
neural network, 142
neuroticism, related to pain perception, 184–5
nociceptor, 154–7, 161, 173
node of Ranvier, 149
noise, **9**, 26–7
noradrenalin, 196–7

occlusion, 139
off-centre cell, **87**, *88*, 94, 96, *98*
off-region (receptive field), **87**
olfaction *see* smell, sense of
on-centre cell, **87**, *88*, 94–7
on-region (receptive field), **86**–7
operant conditioning, 200
operculum, 22
opiate receptor, **191**, 194
opiate, **191**
opioid, 168, **192**, 194, 202
optic chiasma, *78*, 128
optic nerve, *81*, *84*, 85
  see also optic chiasma
optic radiation, *78*, *128*
orientation column, **106**, 129, 134
oval window, 47, *49*

Pacinian corpuscle, 148–9, 152–3
pain 147, 158–9
  acute and chronic, 170–1, 173
  components, 170, *171*
  endogenous analgesic system, 188–95
  experimental, 180–2

first and second, 159
insensitivity to, 172–3
measurement, 176–80, *181*
mechanism, 173–4
see also gate control theory
nature of, 169–71
psychology of, 183–5
referred, 163–4
relation to neural activity, 178–9
threshold, 183–4
treatment of, 195–205
by alternative medicine, 203–4
impact of gate control theory, 202–3
management strategies, 204–5
psychological, 200–1
surgical, 175, 199
use of drugs, 196–9, 204–5
types of, 170, *171*
value of, 172–3
pain clinic, 205
pain perception threshold, **177**–8, 183–4, 198
Pain Rating Index (PRI), 180, *181*
pain tolerance, **177**–8, 183–4, 199
pale stripe, **134**
papilla (pl. papillae)
ear, 22, 25–6
taste bud, 10, *11*
paracetamol, 156, 196
Parallel Distributed Processing, 142
Parus major (great tit), status signals, 32
parvocellular layer (LGN), 133–5, *136*
Passer domesticus (house sparrow), throat patch as status signal, 32–3
pathway
ascending, 160–3
auditory, 53–5
dorsal column (DC), 162–3
spinothalamic tract (STT), 162–3, *193*
spinoreticular tract (SRT), 162–3
visual, *78*
Pavlov, I., 183
Pearson, I. B., 184
perception, 10–12, 212
of speech, 66–7, 69–71
periaqueductal grey matter (PAGM), 188, *189*, 190–1, 193–4
perilymph, 48, *49*
periventricular grey matter (PVGM), 188, *189*
personality, related to pain, 184–5
perspective drawing, 138–9
Pert, C., 191
pethidine, 196–7
phantom limb pain, 175
phenothiazine, 198
pheromone, 213
phoneme, **60**–2, 65, 68–9
photopic system, 100
photoreceptor (light receptor), 5
output, 6
see also retinal receptor
Physalaemus pustulosus (tungara frog), 14, 29–31
pinhole camera, 79
pinna, *44*, 45
placebo effect, 198–9
plectrum, 22
plexiform layer (retina), *83*, *84*
polymodal nociceptor, 155–7
Present Pain Index (PPI), 180, *181*
primary visual cortex, *78*, 104–10, 117–19, 123–4, 129–31, 134–6
double opponent cell, 117–18
primate, greeting ceremonies, 167
proprioceptor, 161
prostaglandin, 156–7, 196
proximal stimulus, 11, 77–8, 112–13
psychotherapy, 200
psychotropic drug, 198
pupil, 80–82, 100

raccoon, sensory cortex, 165
Rana esculenta (edible frog), call, *19*
Rana sphenocephala (leopard frog), *21*
Ranidella signifera (frog), calling site, *26*, *27*
raphe nuclei, 189–91, *193*
rat, sensory cortex, 165
reaction times, 159
receptive field
LGN cell, 104, 117
middle temporal cortex cell, 125–6
retinal ganglion cell, 86–9, 94–7, 101–3, *111*, 117–18
sensory neuron, 151, 167
V1 cell, 104–5, 117–18
receptor potential, 5–6, 8, 149–51
receptor, 5–8, 211–12
see also auditory neuron; retinal receptor; skin receptor
reciprocal altruism, 167–8
reciprocal synapse (amacrine cell), 123
recognition
computer modelling, 141–2
of faces, 112–13

referred pain, **163**–4
reflectance, **100**–2
spectral, 114, *Pl. 2*
refraction of light, 80
refractory period, 150
relative refractory period, 150
release call (frog), 18
reticular formation, 162–3
retina, *81*, 83–103, 121–3
structure, 83–5
retinal flow pattern, 124–6
retinal ganglion cells, 83–9, 92–9, 104, *111*, 122, 133
retinal process, 121–2
retinal receptors, 83–97, 100
see also cone; rod
retinotopic map, **104**, 129
Rexed's laminae, 160–1
Rhinolophus ferrumequinum (greater horseshoe bat), echolocation, 36–7
rhodopsin, 84
rod (retinal receptor) *83*, *84*, 100, 115
Rohwer, S., 32
round window, *49*
Ruffini ending, *148*, 152–3
Ryan, M. J., 30

saturated response (neuron), 151
scala tympani 48, *49*
scala vestibuli 48, *49*
Scaphiopus couchi (Couch's spadefoot toad), tuning curve of auditory neuron, 22, *23*
scotopic system, 100
Scott, T. H., 183
sea sickness, 121
semicircular canals, *44*, 48
sensation, related to perception, 10–12, 212
sense organ, 4–10
sensillum (pl. sensilla), 10
sensitization (of nociceptor), **156**–7
sensory adaptation, **152**–3
short-term, 67, 69
sensory coding, **6**–8, 147–8
of auditory stimuli, 51
by action potential frequency, 149–50
by mechanoreceptors, 148–54
of visual stimuli
brightness and lightness, 99–103
colour, 114–17
image motion, 124–6
luminance, 83–5
orientation, 105–10
spatial structure, 85–6, 97–9
sensory detection threshold, **177**
sensory modality, 8
sentences, meaning, 71–3
serotonin (transmitter), 156, 189, 191, *193*, 196
shadow, 102
sheep, sensory cortex, 165
short-term adaptation, **67**, 69
signal-to-noise ratio, 8–9
silkmoth, olfactory sense, 10
simple cell (cortex), **104**
simultaneous colour contrast 118, *Pl. 5*
skin
receptors in, 148–9, 152–4
classification, 152
smell, sense of, 9–10
Snyder, S., 191
somatosensory cortex, 165
sonagram
of frog calls, 19–20
of speech, 62–4
sound
characteristics, *42*, 43
measurement of intensity, 46
units of measurement, 7
sound spectrograph, 63
spatial comparison, 90
species, Latin names for, 3
spectral reflectance, 114, *Pl. 2*
spectrogram *see* sonogram
spectrum, 114, *Pl. 1*
speech, human, 41
perception, 66–7
processing, 67–71
production, 56–65
spinal cord, 160–4
see also dorsal horn
spinoreticular tract (SRT), 162–3
spinothalamic tract (STT), 162–3, *193*
spiral ganglion, *49*, 53
Spoonerism, 65
stapedius muscle
in bat, 36
in human, *44*, *47*, 48
stapes, *44*, 47
stapes footplate, 48
status signals, 32–3
stereognosis, 167
stereopsis, **127**–31, 135

stimulation-produced analgesia, **188**–91
stimulus
  duration, coding, 152–3
  intensity, 8
    coding, 149–52
  location, 8
    coding, 153–4
  orientation, coding, 105–10
  see also proximal stimulus
stimulus modality, 148
stop consonant, **59**, 61
striate cortex *see* primary visual cortex
substance P, 194–6
substantia gelatinosa, 161, 187, *193*, 194
superior olivary nucleus, 45, 53, *54*
suspensory ligament, *81*
sustained cell (retinal ganglion cell), **122**–3
sympathetic blocking drug, 196–7

T (transmission) cell, 186–7
tabes dorsalis, 175
'taboo' zones of body, *168*
tail flick test, **182**
talapoin monkey, endorphins in grooming, 168
taste, sense of, 9–10, *11*
taste bud, 10
tectorial membrane, 48, *49*
television, 102
temperature, effect on frog calls, 27–9
temperature sense, 147
  see also thermoreceptors
temporal patterning of call, 19–20
TENS *see* Transcutaneous Electrical Nerve Stimulation
tensor tympani muscle, *47*, 48
terminology, 2
thalamic nuclei, 163, 165
  see also lateral geniculate nucleus
thalamus, 162
thermoreceptor, 154–6, 161
thick stripe, 134–5, *136*
thin stripe, 134–5, *136*
threshold intensity, auditory neuron, 6–7
threshold response (nerve), 151
tic doloureux see trigeminal neuralgia
tilt after-effect, 108–10
toad *see* Couch's spadefoot toad; natterjack toad
tolerance (narcotics), 197, 205
tongue, human, 10, *11*
tonotopic organization, **53**–4, *55*
topographical representation, 153–4
touch
  sense of, 147
    social aspects, 167–9
    value, 165–9
touch receptor, output, 5–6
tourniquet test, **182**
Trachops cirrhosus (bat), predator on tungara frog, 14, 29, 31
tranquillizers, in pain treatment, 185, 198
Transcutaneous Electrical Nerve Stimulation (TENS), 187, 202–3, 205
transducer, 5
transient cell (retinal ganglion cell), **122**–3, *124*
trichromatic vision, **115**–16
tricyclic antidepressants, 198
trigeminal neuralgia, 175, 198
tungara frog
  call, 29–31
  predation on, 14, 29–31
tuning curve (auditory neuron), 6–8, 22, *23*
tympanum (tympanic membrane), *44*, **45**
  frequency response, 47–8

variation
  in behaviour, 32
  in plumage, 32–3
V1 *see* primary visual cortex
velocity gradient, 125–6
ventrobasal nucleus, 165
vervet monkey, allogrooming, 167–8
vibration receptor, 153
  output, 6
vision, 8, 77–143
  see also binocular vision; colour vision
visual analogue scale (pain measurement), **179,** *180*
visual cortex, *78*
  primary *see* primary visual cortex
visual pathway, *78*
visual system, anatomy, 78
vocal cords
  frogs, 18
  humans, *57*, 58
vocal sac, frogs, 18, *19*
vocal tract, 56
  human, 56–60
    action, 60–2
    observation, 63
voice
  frequency range, 42
  see also speech

voiced sound, 56
voiceless sound, 56
voiceprint *see* sonagram
vowel, production, 58–61

WALKER computer program, 141–2
Wall, P., 186–7
Wiesel, T., 104–6
words, meaning, 71–2

X cell, 133–4

Y cell, 133–4

Zeki, S., 119
Zonotrichia querula (Harris sparrow), status signals, 32